Advances in Intelligent Systems and Computing

Volume 1023

The series "Advances in Intelligent Systems and Computing" contains publications on theory, applications, and design methods of Intelligent Systems and Intelligent Computing. Virtually all disciplines such as engineering, natural sciences, computer and information science, ICT, economics, business, e-commerce, environment, healthcare, life science are covered. The list of topics spans all the areas of modern intelligent systems and computing such as: computational intelligence, soft computing including neural networks, fuzzy systems, evolutionary computing and the fusion of these paradigms, social intelligence, ambient intelligence, computational neuroscience, artificial life, virtual worlds and society, cognitive science and systems, Perception and Vision, DNA and immune based systems, self-organizing and adaptive systems, e-Learning and teaching, human-centered and human-centric computing, recommender systems, intelligent control, robotics and mechatronics including human-machine teaming, knowledge-based paradigms, learning paradigms, machine ethics, intelligent data analysis, knowledge management, intelligent agents, intelligent decision making and support, intelligent network security, trust management, interactive entertainment, Web intelligence and multimedia.

The publications within "Advances in Intelligent Systems and Computing" are primarily proceedings of important conferences, symposia and congresses. They cover significant recent developments in the field, both of a foundational and applicable character. An important characteristic feature of the series is the short publication time and world-wide distribution. This permits a rapid and broad dissemination of research results.

** Indexing: The books of this series are submitted to ISI Proceedings, EI-Compendex, DBLP, SCOPUS, Google Scholar and Springerlink **

More information about this series at http://www.springer.com/series/11156

Munir Merdan · Wilfried Lepuschitz ·
Gottfried Koppensteiner ·
Richard Balogh · David Obdržálek
Editors

Robotics in Education

Current Research and Innovations

 Springer

Editors
Munir Merdan
Practical Robotics Institute Austria
Vienna, Austria

Wilfried Lepuschitz
Practical Robotics Institute Austria
Vienna, Austria

Gottfried Koppensteiner
Practical Robotics Institute Austria
Vienna, Austria

Richard Balogh
Slovak University of Technology
in Bratislava
Bratislava, Slovakia

David Obdržálek
Faculty of Mathematics and Physics
Charles University
Prague, Czech Republic

ISSN 2194-5357   ISSN 2194-5365   (electronic)
Advances in Intelligent Systems and Computing
ISBN 978-3-030-26944-9   ISBN 978-3-030-26945-6   (eBook)
https://doi.org/10.1007/978-3-030-26945-6

This Springer imprint is published by the registered company Springer Nature Switzerland AG
The registered company address is: Gewerbestrasse 11, 6330 Cham, Switzerland

Preface

We are honored to present the proceedings of the 10th International Conference on Robotics in Education (RiE) held in Hall of Sciences, Vienna, Austria, during April 10–12, 2019. The RiE is organized every year with the aim to bring scientists, teachers, organizers of educational activities, policy makers, and industry together to present their research and share their ideas and experiences, but also to discuss the practical challenges which need to be tackled.

The industry is faced with an increasing lack of science, technology, engineering, and mathematics (STEM) graduates every year. It is of vital importance to motivate young learners toward attending technical schools or schools in the STEM fields. Robotics represents a highly innovative domain encompassing physics, mathematics, informatics, and even industrial design as well as social sciences. The fascination for autonomous machines and the variety of fields and topics covered with this domain makes robotics a powerful idea to engage with. In this context, the use of robots in classrooms has evolved from purely presenting technology to education with a strong multidisciplinary perspective. Consequently, educational robotics is used as an innovative approach for increasing the interest of young people for STEM topics. It promotes the development of systems thinking, and young girls and boys can easily connect robots to their personal interests and share their ideas. Besides, due to various application areas, teamwork, creativity and entrepreneurial skills are required for the design, programming, and innovative exploitation of robots and robotic services. As a consequence, it is regarded as very beneficial if engineering schools and university program studies include the teaching of both theoretical and practical knowledge on robotics. Since the use of robots in the classroom can motivate students to possible STEM career paths, curricula need to be developed and new didactic approaches for an innovative education need to be designed for improving the STEM skills among young people. In this context, an exploration of the multidisciplinary potential of robotics toward an innovative learning approach is advised for fostering the pupils' and students' creativity leading to collaborative entrepreneurial, industrial, and research careers.

In these proceedings, we present the latest research and developments as well as new applications, practices, and products in the educational robotics domain. The book offers insights into the recent results related to curricula, activities, and their evaluation. The presented applications cover the whole educative range, from nursery and elementary school to high school, college, university, and beyond, for continuing education and possible outreach and workforce development. In total, 56 papers were submitted and 39 papers are now part of these proceedings after a careful peer review process. We would like to express our thanks to all authors who submitted papers to RiE 2019 and our congratulations to those whose papers were accepted.

This publication would not have been possible without the support of the RiE International Program Committee and the conference Co-Chairs. We also wish to express our gratitude to the volunteer students and staff of the partner organizations, which significantly contributed to the success of the conference. All of them deserve many thanks for having helped to attain the goal of providing a balanced event with a high level of scientific exchange and a pleasant environment. RiE 2019 was supported by the Austrian Federal Ministry of Education, Science and Research, for which we thankfully express our gratitude. We acknowledge the use of the EasyChair conference system for the paper submission and review process. We would also like to thank Dr. Thomas Ditzinger and Mr. Arumugam Deivasigamani from Springer for providing continuous assistance and advice whenever needed.

Organization

Committee

Co-chairpersons

Richard Balogh	Slovak University of Technology in Bratislava, Slovakia
Wilfried Lepuschitz	Practical Robotics Institute Austria, Austria
David Obdržálek	Charles University, Czech Republic

International Program Committee

Alimisis Dimitris	EDUMOTIVA-European Lab for Educational Technology, Greece
Angel-Fernandez Julian	Vienna University of Technology, Austria
Bredenfeld Ansgar	Dr. Bredenfeld UG, Germany
Carter Jenny	De Montfort University in Leicester, UK
Catlin Dave	Valiant Technology, UK
Demo G. Barbara	Dipartimento Informatica, Universita Torino, Italy
Dessimoz Jean-Daniel	Western Switzerland University of Applied Sciences and Arts, Switzerland
Eteokleous Nikleia	Department of Educational Sciences, Frederick University Cyprus, Cyprus
Ferreira Hugo	Instituto Superior de Engenharia do Porto, Portugal
Fiorini Paolo	University of Verona, Italy
Fislake Martin	Department of Technology Education, University of Koblenz and Landau, Germany
Gerndt Reinhard	Ostfalia University of Applied Sciences, Germany
Girvan Carina	Cardiff University, UK

Gonçalves José	Instituto Politécnico de Bragança (IPB), Portugal
Granosik Grzegorz	Lodz University of Technology, Poland
Grizioti Marianthi	University of Athens, Greece
Gueorguiev Ivaylo	European Software Institute – Center Eastern Europe, Bulgaria
Hofmann Alexander	University of Applied Sciences Technikum Wien, Austria
Kandlhofer Martin	Graz University of Technology, Austria
Kazed Boualem	University of Blida, Algeria
Koppensteiner Gottfried	Practical Robotics Institute Austria, Austria
Krajník Tomáš	Czech Technical University in Prague, Czech Republic
Kulich Miroslav	Czech Technical University in Prague, Czech Republic
Kynigos Chronis	University of Athens, Greece
Lammer Lara	Vienna University of Technology, Austria
Lúčny Andrej	Comenius University in Bratislava, Slovakia
Mellado Martin	Instituto ai2, Universitat Politècnica de València, Spain
Merdan Munir	Practical Robotics Institute Austria, Austria
Moro Michele	University of Padova, Italy
Papakostas A. George	Eastern Macedonia and Thrace Institute of Technology, Greece
Pedaste Margus	University of Tartu, Estonia
Petrovič Pavel	Comenius University in Bratislava, Slovakia
Pina Alfredo	Public University of Navarra, Spain
Santos João Machado	University of Lincoln, UK
Schmöllebeck Fritz	University of Applied Sciences Technikum Wien, Austria
Schreiner Dietmar	Vienna University of Technology, Austria
Solitro Ugo	University of Verona, Italy
Stelzer Roland	INNOC – Austrian Society for Innovative Computer Sciences, Austria
Varbanov Pavel	European Software Institute – Center Eastern Europe, Bulgaria
Verner M. Igor	Technion – Israel Institute of Technology, Israel
Vincze Markus	Vienna University of Technology, Austria
Wyffels Francis	Ghent University, Belgium
Yiannoutsou Nikoleta	University of Athens, Greece

Local Conference Organization

Koppensteiner Gottfried	Practical Robotics Institute Austria, Austria
Merdan Munir	Practical Robotics Institute Austria, Austria
Tomitsch Tanja	Practical Robotics Institute Austria, Austria

Sponsor

Austrian Federal Ministry of Education, Science and Research

═ Federal Ministry
  Education, Science
  and Research

Contents

Comprehensive View on Educational Robotics

On Measuring Engagement Level During Child-Robot Interaction in Education

Chris Lytridis, Christos Bazinas, George A. Papakostas$^{(\boxtimes)}$,
and Vassilis Kaburlasos

HUman-MAchines INteraction (HUMAIN) Lab,
International Hellenic University, Agios Loukas, Kavala 65404, Greece
{lytridic, chrbazi, gpapak, vgkabs}@teiemt.gr

Abstract. The introduction of social robots in education has been a major theme in robotics research in recent years. Various studies have been conducted that demonstrate the merits of using robots as teachers or teacher assistants. These studies are mainly focused on activities where the child interacts with the robot to achieve a certain educational or therapeutic goal. The principal reason that robots in education are observed to have a positive effect, is that children appear to be more engaged during the educational process when a robot is involved. This paper reviews the current literature on the subject of using social robots in education for the purposes of identifying the most appropriate methodologies in measuring the engagement levels of children during child-robot interactions, specifically focusing on interactions occurring in an educational or therapeutic setting.

Keywords: Social robots · Child-robot interaction · Engagement levels · Education

1 Introduction

During the last few years there have seen significant advances in robotics both in terms of hardware capabilities as well as in terms of artificial intelligence. These advances have given robots the ability to operate with more autonomy and perform tasks more effectively. Consequently, robots have been introduced in areas where humans have traditionally operated alone. One such area is education, and there has been much research on how robots can be used in the classroom in order to facilitate, improve and support the learning process [1].

The field of educational robotics aims at promoting collective knowledge, developing skills and stimulating students through manipulation and interaction with robots. The robots that are used in educational activities vary in complexity depending on their assigned task. For example, educational activities can be achieved with LEGO for simple problem-solving tasks, or more complex robots that can be used in more advanced learning activities such as learning mathematics, a second language, social skills or learning through theatrical plays [2, 3]. The nature of these latter activities suggests that the robot is capable to have simple or complex interactions with the students. For this reason, robots have been also introduced in the specific area of

© Springer Nature Switzerland AG 2020
M. Merdan et al. (Eds.): RiE 2019, AISC 1023, pp. 3–13, 2020.
https://doi.org/10.1007/978-3-030-26945-6_1

special education where activities based on repeated interactions between the child and the teacher (in this case, a robot) are critical to the effective treatment of learning disabilities or developmental disorders [4].

A series of interaction games designed for children with autism have been proposed in [5]. The games deal with improving skills such as social interactions, memory association and motor response. Also, a system that allows a robot to recognize the emotional state of children and to select an appropriate behavior has been developed [6]. In another study, a social robot was used in a school setting as a didactic mediator to improve school children's arithmetic skills [7].

One of the main reasons behind the apparent improvements in academic performance when robots are used in the classroom is the fact that robots are especially appealing to children because of their physical form and novelty [8]. As a result, the children are more engaged and more motivated to participate in the educational activities. This paper summarizes the different ways in which the engagement level of children during child-robot interactions can be measured automatically. The paper is organized as follows: Sect. 2 reviews the research studies investigate engagement measurements in applications for typical and special education. In Sect. 3 the findings of the review are discussed. Finally, Sect. 4 summarizes the conclusions and future directions for research.

2 Measuring Engagement Levels

Human-Computer Interaction (HCI) is a vast and complex field in which many disciplines and approaches are involved. A variety of metrics has been developed to calculate engagement levels during human-computer interaction. While earlier studies were focused on measuring activity from standard input devices like the mouse or the keyboard, new measuring methods have emerged due to more advanced input methods. For example, in [9] pupil size change is treated as an input command, and therefore was used as a metric for engagement, while in [10] gaze direction and facial muscle activations are observed as HCI. Multi-modal interactions (combination of body pose, gaze and gestures) have also been used to determine intention to engage [11].

With the introduction of robots that can undertake a large variety of complex human-like tasks such as navigation, environment perception, action management, object manipulation and social interactions (e.g. visual, auditory, tactile etc.), Human-Robot Interaction (HRI) has been established as a separate field of study. Various metrics have been proposed to evaluate the performance of human-robot interaction [12, 13]. In the particular case of social robotics however, these metrics depend on the objective, which can be either measuring the quality and effectiveness of the interaction in terms of outcome, or measuring the engagement level during the social interaction between the robot and the human. As is the case in social interactions between humans, human-robot social interactions are usually complex and multi-modal. Therefore, the task of measuring the engagement in a given setting is a non-trivial task and involves the measurement of quantities that are not necessarily quantifiable e.g. emotional states or social behaviors.

The review that follows examines the literature to determine the available methods of measuring the engagement levels of humans when interacting with educational robots. Since educational activities in typical and special education have different objectives, engagement measures in these categories will be examined separately.

2.1 Typical Education

Various studies propose that Educational Robotics (ER) can be an effective tool for teaching and learning, mainly in the field of Science, Technology, Engineering, and Mathematics (STEM). Others propose integration of art and design into STEM education (i.e., STEAM) [14, 15]. Due to the variety of human-robot interaction scenarios in the studies, there is not a standard method measuring the participants' engagement. Researchers are trying to determine the quality of the HRI using different metrics, depending on the nature of the interaction.

In many studies, robots have been used in traditional learning subjects such as mathematics or language learning. Arithmetic and mathematical tasks, from learning about angles by programming a robot to draw them [16], to teaching/learning numeracy [7] are proposed. In the first study, a teacher observed the children's behavior and noted that they were collaborating to find the solution. In the latter study, an evaluation questionnaire showed increased motivation and better understanding of mathematical concepts. A questionnaire together with interaction time analysis using timestamps collected by a robot was used to estimate long-term engagement in [17]. The participants of this study were divided in two groups: the first group was using a standard iPod-LEGO robot, while the second group's robots were customized according to the students' topics of interest (hobbies, music, etc.). The duration of these tactical sessions was 1 month and included math, language, problem solving and others. The questionnaire revealed higher engagement levels when the customized robot was used. Long-term engagement was addressed in a four-week study in English learning lessons with a robotic agent as an assistant [18]. The robotic agent was able to record logs from its touch screen, used a RFID reader to capture mobility of the children, and together with video recordings and interviews that were analyzed, the study revealed that the children remained highly engaged due to the use of the robot. Mobility of participants was also used as an engagement metric in [19] More specifically, the distance between the participants and the robot, the posture of the children, the time and number of users per session were studied via video recordings and interviews in a study where the children had free access to a robot during their free playing time. During this free time, the students were able to apply their previously gained knowledge from robotics lessons.

Other studies are examining the use of *robots as peers*. In this scenario the students are trying to teach something to the robot. In [20], in an interactive word learning task in which children are trying to teach new words to robots, real time gaze detection and speech detection is used as a metric for engagement. In similar educational activities, HRI is measured via the annotation of video recordings [21], the observation of video recordings combined with speech analysis [22], or tablet usage logs showing the time spent by demonstration and the duration of sessions in a study where a tablet was added in the process of teaching handwriting to the peer [23].

Storytelling is also a common learning approach. In [24], video, audio and contextual analysis is proposed to evaluate engagement but also to predict disengagement. Szafir and Mutlu use electroencephalography (EEG) together with a questionnaire to measure interaction [25]. Interactive story-making in collaboration with a robot is proposed in [26]. In this study, post-experimental analysis of video and audio recordings revealed that children engagement of children was increased for specific behavioral states of the robots.

In *educational games* with robots, gaze detection is considered a very important feature to estimate engagement. Gaze direction is proposed as a metric in a selection game with a robot, in which the robot instructs the participants to select one object from a group of available objects [27]. Researchers observed the participant's gaze while playing, or while being instructed by the robot. Multi-modal observation using video, the Kinect motion sensing system and audio sensors is used in a mix-and-match game, introducing a "side-kick" robot along with the main robot, that intervenes when necessary [28]. Post-experimental analysis showed that engagement was increased when the side-kick robot was involved.

The use of robots in *theatrical games* for educational purposes have also been proposed [29]. In a study involving a Projector robot for augmented children's play, video observations were used to show higher attention levels of the children and enhanced educational effects [15]. In [30], multiple robots were used in theatrical acting, dancing, singing and drawing. In this study, a questionnaire was the instrument that was used to show that motivation, attention, interaction and overall engagement was significantly increased when compared to art lessons without robots.

2.2   Special Education

Robots have also been used in special education to deal with a variety of conditions such as learning disabilities, behavioral disorders and developmental disabilities. In the case of special education, eliciting response and maintaining a high degree of engagement is crucial for the successful treatment of such conditions, even more so than in the case of typical education. Given the fact that the presence of robots is universally accepted to have a positive impact on attracting the attention of students, it follows that Socially Assistive Robots (SARs) have a significant role to play as pedagogical rehabilitation tools [8], especially when the educational objectives coincide with the improvement of social skills.

Researchers have proposed various methods in attempting to deal with the various disabilities. For example, robot-assisted tasks for understanding directions and making choices have been proposed for children with severe intellectual disabilities [31], imitation games for children with cerebral palsy [32], simple construction or programming tasks for children with learning disabilities [33, 34], or robot-mediated teaching of sign language for children with hearing impairments [35]. The high levels of engagement reported in these studies is primarily based on observations made by specialists manually, either during the sessions or using the video recordings of the sessions. The specialists use annotations to describe the various behaviors which are adopted by the children during the educational activities. In other studies, specialized questionnaires are completed by the teachers or parents and are used to assess qualitatively and quantitively the children's performance and/or their observed engagement.

A significant amount of work in the field of SARs in special education has been carried out in therapeutic environments for children with autism and more specifically children with high functioning autism spectrum disorder (ASD) [36–38]. This is because of the robot's suitability to improve social skills that are not particularly developed in individuals with this condition. Due to the diverse nature of the educational activities undertaken in this type of studies, different ways to measure engagement have been attempted i.e. the type of engagement measurements used depend on the particular feature under study. A portion of the studies focuses on detecting and assessing the social behaviors which result from engaging in game scenarios such as collaborative games [39, 40], or individual activities in the form of games such as cooking [41], specialized space-themed games [42], or imitation games [43–45]. Other more general social tasks that are evaluated include emotion recognition [46] assessment of verbal communications [38, 47], image processing [48] and general social interaction activities [49]. In these studies, the social behaviors that are under examination usually have complexities and subtleties which are very difficult to be automatically detected and annotated by a machine with the required accuracy. Moreover, it is difficult to assess their characteristics qualitatively. For this reason, video recordings are used by experts, usually in conjunction with questionnaires and interviews, in order to extract the required engagement data i.e. the behavioral traits that indicate engagement. For example, in [50] the children's affective states were recorded based on their responses when interacting with robots of mechanical and humanoid appearances. In the dance therapy proposed in [51] the authors employ social response time (SRT) which is defined as the time it took to the children to perform the target tasks. In [52], external cameras and the robot's own vision system is proposed to be used as the observation material used by psychologists to monitor the child' initial response to stimulus and its behavior, by measuring the frequency of the child looking at the robot and the duration of each interaction.

Other studies are examining more specific features that are revealing the engagement level during a certain task. Due to the specificity of these features, their detection by measurement devices is more accurate and reliable than the automatic annotation of complex behaviors. In general, the main engagement-related features that are measured in this research area are speech, bodily cues (which includes gestures, body pose) and gaze. To measure engagement, these features are considered either separately (i.e. the examination of a single modality to reveal the engagement level) or in combination (i.e. multi-modal engagement measurements).

One of the most common engagement cues is eye contact [20]. In [53] an eye tracker is used to index the child's gaze toward the robot during the task, in order to assess improvements in joint attention skills. In another study, a speech recognition system was used to assess the correct learning of words in [54]. Movement recognition in order to compare human-robot poses during an imitation game has been used in [55]. In [56], apart from video recordings which were used to detect where the child is facing, a custom wearable device was used to monitor the facial muscles involved in smiling. Motion analysis and vocal reaction sensing were used in conjunction using RGB-D sensors and a tablet device in order to assess engagement and modulate the robot's behavior [57]. Similarly, in [58] a framework which incorporates eye contact, gestures and utterances for ASD evaluation is proposed.

3 Discussion

From the literature review presented in the previous section It becomes obvious that a variety of engagement measures in educational activities have been proposed. It can be observed that there are significant differences on each study's used metrics. Table 1 summarizes the methods used to measure engagement in typical and special education.

Table 1. Summary of engagement measurement methods depending on educational activity

	Educational activity	Measurement method
Typical education	Robot as a peer	Gaze and speech detection, device logs (duration, number of interactions)
	Storytelling	Audio, visual, contextual analysis, electroencephalography (EEG)
	Games	Gaze detection, video observation, motion detection, audio analysis
	Arts (theater, dance, singing, drawing)	Video observation, questionnaire
	Arithmetic - mathematical tasks	Questionnaire, teacher observation
	Free interaction - application of curriculum	Interviews, video analysis (distance, posture, time per session, number of participants per session)
	Math, language, problem solving sessions	Questionnaire, interaction time analysis, touch screen logs, measuring mobility, video observation, interviews
Special education	Games	Video observation, smile and facing behaviors, intonation, bodily appearance
	General interaction	Video observation, interaction duration, biomarkers to check stress levels, key behavioral traits
	Imitation	Video observation, motion sensors, interaction initiations, eye gaze, gaze shifting, smiling/laughter, questionnaire, attention measurement
	Dance	Questionnaire, social response time, video observation (gaze, distance, touch, imitation)
	Attention exercises	Observation by clinicians, eye tracking
	Conversation, questions	Video observation, annotating behavioral traits
	Others	Video observation, rate of correct responses, speech recognition, motion analysis, vocal reaction sensing, interviews

Engagement metrics in each study are generally selected according to the educational activity and the learning objectives. However, it can be observed that even in studies that have dealt with similar educational methods and scenarios, the features that were used to measure student engagement were not the same.

There are two main reasons for this: firstly, the robots that are used in these studies have different hardware characteristics [59], and therefore their processing power, and sensing and data collection capabilities vary. In some cases, the robots are able to utilize their on-board equipment to extract the necessary features. This is especially important in the cases when the level of engagement has to be measured in real-time so that the robot can adjust its behavior in order to increase the student's engagement level. However, it is not always possible to measure engagement using the robot's on-board devices, since the robots are selected primarily based on their suitability to the task and not on their measurement capabilities. To remedy this, external measurement devices (such as motion detectors, electroencephalographs and eye trackers) are used for the reliable measurement of features such as motion, posture, gaze and speech characteristics. These features can be used separately or in conjunction in order to evaluate engagement.

Secondly, behavioral engagement cues can be more complex and cannot be reliably detected by typical measurement equipment. Therefore, these require trained specialists to detect and assess. In this case, many studies employ video recordings of the educational activities which are then used by experts in order to observe certain behaviors or linguistic features that indicate engagement, such as face expressions, arms approaching the robot, inspection of the robot by the child, verbal communications etc. Expected engagement behaviors are annotated and subsequent statistical analysis reveals the engagement levels during the child-robot interaction. Questionnaires and interviews are also used in post-experimental analysis for the measurement of engagement levels during the activities.

4 Conclusions

This paper presented a thorough review of the methods of measuring the engagement levels of children during child-robot interactions in an educational setting. The review has revealed the various methodologies employed in detecting and measuring the features which indicate engagement, either directly or indirectly, in educational activities both in typical as well as in special education.

The results of this review can help towards the design of therapeutic interventions for children with certain types of learning disabilities and children with ASD. In fact, the suitability of the engagement metrics examined in this paper will guide our research in the field of social robots in education.

Acknowledgment. This research has been co-financed by the European Union and Greek national funds through the Operational Program Competitiveness, Entrepreneurship and Innovation, under the call RESEARCH – CREATE – INNOVATE (project code: T1EDK-00929).

References

1. Benitti, F.B.V.: Exploring the educational potential of robotics in schools: a systematic review. Comput. Educ. **58**, 978–988 (2012)
2. Toh, L.P.E., Causo, A., Tzuo, P.-W., Chen, I.-M., Yeo, S.H.: A review on the use of robots in education and young children. J. Educ. Technol. Soc. **19**, 148–163 (2016)
3. Alimisis, D.: Educational robotics: open questions and new challenges. Themes Sci. Technol. Educ. **6**, 63–71 (2013)
4. Dimitrova, M., Wagatsuma, H.: Designing humanoid robots with novel roles and social abilities. Lovotics **3**, 2 (2015)
5. Lytridis, C., Vrochidou, E., Chatzistamatis, S., Kaburlasos, V.: Social engagement interaction games between children with autism and humanoid robot NAO. In: Graña, M., López-Guede, J.M., Etxaniz, O., Herrero, Á., Sáez, J.A., Quintián, H., Corchado, E. (eds.) Advances in Intelligent Systems and Computing, pp. 562–570. Springer, Cham (2018)
6. Lytridis, C., Vrochidou, E., Kaburlasos, V.: Emotional speech recognition toward modulating the behavior of a social robot. In: Proceedings of the 2018 JSME Conference on Robotics and Mechatronics, Kitakyushu, Japan (2018)
7. Vrochidou, E., Najoua, A., Lytridis, C., Salonidis, M., Ferelis, V., Papakostas, G.A.: Social robot NAO as a self-regulating didactic mediator: a case study of teaching/learning numeracy. In: Proceedings of the 26th International Conference on Software, Telecommunications and Computer Networks (SoftCOM 2018), Symposium on: Robotic and ICT Assisted Wellbeing, Split, Croatia (2018)
8. Kaburlasos, V., Vrochidou, E.: Social robots for pedagogical rehabilitation: trends and novel modeling principles. In: Dimitrova, M., Wagatsuma, H. (eds.) Cyber-Physical Systems for Social Applications. Advances in Systems Analysis, Software Engineering, and High Performance Computing (ASASEHPC), pp. 1–21. IGI Global (2019)
9. Ehlers, J., Strauch, C., Huckauf, A.: A view to a click: pupil size changes as input command in eyes-only human-computer interaction. Int. J. Hum. Comput. Stud. **119**, 28–34 (2018)
10. Tuisku, O., Surakka, V., Vanhala, T., Rantanen, V., Lekkala, J.: Wireless face interface: using voluntary gaze direction and facial muscle activations for human-computer interaction. Interact. Comput. **24**(1), 1–9 (2012)
11. Schwarz, J., Marais, C.C., Leyvand, T., Hudson, S.E., Mankoff, J.: Combining body pose, gaze, and gesture to determine intention to interact in vision-based interfaces. In: Proceedings of the 32nd Annual ACM Conference on Human Factors in Computing Systems - CHI 2014, pp. 3443–3452 (2014)
12. Steinfeld, A., Fong, T., Kaber, D., Lewis, M., Scholtz, J., Schultz, A., Goodrich, M.: Common metrics for human-robot interaction. In: HRI 2006: Proceedings of the 2006 ACM Conference on Human-Robot Interaction, pp. 33–40. ACM, New York (2006)
13. Murphy, R.R., Schreckenghost, D.: Survey of metrics for human-robot interaction. In: ACM/IEEE International Conference Human-Robot Interact, pp. 197–198 (2013)
14. Jeon, M., Fakhrhosseini, M., Barnes, J., Duford, Z., Zhang, R., Ryan, J., Vasey, E.: Making live theatre with multiple robots as actors. In: Human-Robot Interaction, pp. 445–446 (2016)
15. Ahn, J., Yang, H., Kim, G.J., Kim, N., Choi, K., Yeon, H., Hyun, E., Jo, M., Han, J.: Projector robot for augmented children's play. In: Proceedings of the 6th International Conference on Human-Robot Interaction - HRI 2011 (2011)
16. Kao, M.C., Chen, C.L.A., Ko, C.M.: Work in progress - using robot in developing the concept of angle for elementary school children. In: Proceedings - Frontiers in Education Conference, FIE (2007)

17. Barco, A., Albo-Canals, J., Garriga, C.: Engagement based on a customization of an iPod-LEGO robot for a long-term interaction for an educational purpose. In: Proceedings of the 2014 ACM/IEEE International Conference on Human-Robot Interaction - HRI 2014 (2014)
18. Park, S.J., Han, J.H., Kang, B.H., Shin, K.C.: Teaching assistant robot, ROBOSEM, in English class and practical issues for its diffusion. In: Proceedings of IEEE Workshop on Advanced Robotics and its Social Impacts, ARSO (2011)
19. Hyun, E., Yoon, H.: Characteristics of young children's utilization of a robot during play time: a case study. In: Proceedings - IEEE International Workshop on Robot and Human Interactive Communication (2009)
20. Xu, T.L., Zhang, H., Yu, C.: See you see me: the role of eye contact in multimodal human-robot interaction. ACM Trans. Interact. Intell. Syst. **6**, 1–22 (2016)
21. Tanaka, F., Ghosh, M.: The implementation of care-receiving robot at an english learning school for children. In: Proceedings of the 6th International Conference on Human-Robot Interaction - HRI 2011 (2011)
22. Yu, C., Scheutz, M., Schermerhorn, P.: Investigating multimodal real-time patterns of joint attention in an HRI word learning task. In: 2010 5th ACM/IEEE International Conference on Human-Robot Interaction (HRI) (2010)
23. Jacq, A., Lemaignan, S., Garcia, F., Dillenbourg, P., Paiva, A.: Building successful long child-robot interactions in a learning context. In: ACM/IEEE International Conference on Human-Robot Interaction (2016)
24. Leite, I., McCoy, M., Ullman, D., Salomons, N., Scassellati, B.: Comparing models of disengagement in individual and group interactions. In: Proceedings of the Tenth Annual ACM/IEEE International Conference on Human-Robot Interaction - HRI 2015 (2015)
25. Szafir, D., Mutlu, B.: Pay attention!: designing adaptive agents that monitor and improve user engagement. In: Proceedings of the SIGCHI Conference on Human Factors in Computing Systems (2012)
26. Gordon, G., Breazeal, C., Engel, S.: Can children catch curiosity from a social robot? In: Proceedings of the Tenth Annual ACM/IEEE International Conference on Human-Robot Interaction - HRI 2015 (2015)
27. Admoni, H., Datsikas, C., Scassellati, B.: Speech and gaze conflicts in collaborative human-robot interactions. In: Proceedings of the 36th Annual Conference on Cognitive Science Society (CogSci 2014) (2014)
28. Vázquez, M., Steinfeld, A., Hudson, S.E., Forlizzi, J.: Spatial and other social engagement cues in a child-robot interaction. In: Proceedings of the 2014 ACM/IEEE International Conference on Human-Robot Interaction - HRI 2014 (2014)
29. Verner, I.M., Polishuk, A., Krayner, N.: Science class with RoboThespian: using a robot teacher to make science fun and engage students. IEEE Robot. Autom. Mag. **23**(2), 74–80 (2016)
30. Jeon, M., Barnes, J., Fakhrhosseini, M., Vasey, E., Duford, Z., Zheng, Z., Dare, E.: Robot opera: a modularized afterschool program for STEAM education at local elementary school. In: 14th International Conference on Ubiquitous Robots and Ambient Intelligence, URAI 2017, pp. 935–936 (2017)
31. Aslam, S., Shopland, N., Standen, P.J., Burton, A., Brown, D.: A comparison of humanoid and non humanoid robots in supporting the learning of pupils with severe intellectual disabilities. In: Proceedings - 2016 International Conference on Interactive Technologies and Games EduRob Conjunction with iTAG 2016, iTAG 2016, pp. 7–12 (2016)
32. Malik, N.A., Yussof, H., Hanapiah, F.A., Anne, S.J.: Human Robot Interaction (HRI) between a humanoid robot and children with cerebral palsy: experimental framework and measure of engagement. In: IECBES 2014, Conference Proceedings - 2014 IEEE Conference Biomedical Engineering and Sciences "Miri, Where Engineering in Medicine and Biology and Humanity Meeting", pp. 430–435 (2014)

33. Kärnä-Lin, E., Pihlainen-Bednarik, K., Sutinen, E., Virnes, M.: Can robots teach? Preliminary results on educational robotics in special education. In: Proceedings of the 6th IEEE International Conference on Advanced Learning Technologies (ICALT 2006), Kerkrade, The Netherlands, pp. 319–321 (2006)
34. Bugmann, J., Karsenti, T.: Learning to program a humanoid robot: impact on special education students. In: Mikropoulos, T.A. (ed.) Research on e-Learning and ICT in Education: Technological, Pedagogical and Instructional Perspectives, pp. 323–337. Springer, Cham (2018)
35. Akalin, N., Uluer, P., Kose, H., Ince, G.: Humanoid robots communication with participants using sign language: an interaction based sign language game. In: Proceedings of IEEE Workshop on Advanced Robotics and its Social Impacts, ARSO, pp. 181–186 (2013)
36. Scassellati, B., Admoni, H., Matarić, M.: Robots for use in autism research. Annu. Rev. Biomed. Eng. **14**, 275–294 (2012)
37. Ricks, D.J., Colton, M.B.: Trends and considerations in robot-assisted autism therapy. In: Proceedings - IEEE International Conference on Robotics and Automation, pp. 4354–4359 (2010)
38. Lewis, L., Charron, N., Clamp, C., Craig, M.: Soft systems methodology as a tool to aid a pilot study in robot-assisted therapy. In: ACM/IEEE International Conference on Human-Robot Interaction, April 2016, pp. 467–468 (2016)
39. Barakova, E.I., Vanderelst, D.: From spreading of behavior to dyadic interaction—a robot learns what to imitate. Int. J. Intell. Syst. **29**, 495–524 (2014)
40. Huskens, B., Palmen, A., Van der Werff, M., Lourens, T., Barakova, E.I.: Improving collaborative play between children with autism spectrum disorders and their siblings: the effectiveness of a robot-mediated intervention based on Lego® therapy. J. Autism Dev. Disord. **45**, 3746–3755 (2015)
41. Simut, R.E., Vanderfaeillie, J., Peca, A., Van de Perre, G., Vanderborght, B.: Children with autism spectrum disorders make a fruit salad with probo, the social robot: an interaction study. J. Autism Dev. Disord. **46**, 113–126 (2016)
42. Clabaugh, C., Becerra, D., Deng, E., Ragusa, G., Matarić, M.: Month-long, in-home case study of a socially assistive robot for children with autism spectrum disorder. In: Companion 2018 ACM/IEEE International Conference on Human-Robot Interaction - HRI 2018, pp. 87–88 (2018)
43. Costa, S., Lehmann, H., Dautenhahn, K., Robins, B., Soares, F.: Using a humanoid robot to elicit body awareness and appropriate physical interaction in children with autism. Int. J. Soc. Robot. **7**, 265–278 (2015)
44. Tapus, A., Peca, A., Aly, A., Pop, C., Jisa, L., Pintea, S., Rusu, A.S., David, D.O., Hammouda, O., Borbély, G.: Children with autism social engagement in interaction with Nao, an imitative robot: a series of single case experiments. Interact. Stud. **13**, 315–347 (2012)
45. Iacono, I., Lehmann, H., Marti, P., Robins, B., Dautenhahn, K.: Robots as social mediators for children with Autism - a preliminary analysis comparing two different robotic platforms. In: 2011 IEEE International Conference on Development and Learning (ICDL), pp. 1–6 (2011)
46. English, B.A., Coates, A., Howard, A.: Recognition of gestural behaviors expressed by humanoid robotic platforms for teaching affect recognition to children with autism - a healthy subjects pilot study. In: Kheddar, A., Yoshida, E., Sam, S., Suzuki, G., Cabibihan, J.-J., Eyssel, F., and He, H. (eds.) Social Robotics, ICSR 2017. Lecture Notes in Computer Science, pp. 567–576. Springer, Cham (2017)

47. Mavadati, S.M., Feng, H., Salvador, M., Silver, S., Gutierrez, A., Mahoor, M.H.: Robot-based therapeutic protocol for training children with Autism. In: 25th IEEE International Symposium on Robot and Human Interactive Communication, RO-MAN 2016, pp. 855–860 (2016)

48. Kajopoulos, J., Wong, A.H.Y., Yuen, A.W.C., Dung, T.A., Kee, T.Y., Wykowska, A.: Robot-assisted training of joint attention skills in children diagnosed with autism. In: Tapus, A., André, E., Martin, J.-C., Ferland, F., Ammi, M. (eds.) International Conference on Social Robotics, pp. 296–305. Springer, Cham (2015)

49. Robins, B., Dautenhahn, K., Dickerson, P.: From isolation to communication: a case study evaluation of robot assisted play for children with autism with a minimally expressive humanoid robot. In: Proceedings of 2nd International Conferences on Advances in Computer-Human Interactions, ACHI 2009, pp. 205–211 (2009)

50. van Straten, C.L., Smeekens, I., Barakova, E.I., Glennon, J., Buitelaar, J., Chen, A.: Effects of robots' intonation and bodily appearance on robot-mediated communicative treatment outcomes for children with autism spectrum disorder. Pers. Ubiquitous Comput. **22**, 379–390 (2018)

51. Suzuki, R., Lee, J., Rudovic, O.: NAO-dance therapy for children with ASD. In: Proceedings of the Companion 2017 ACM/IEEE International Conference on Human-Robot Interaction - HRI 2017, pp. 295–296 (2017)

52. Shamsuddin, S., Yussof, H., Ismail, L., Hanapiah, F.A., Mohamed, S., Piah, H.A., Zahari, N. I.: Initial response of autistic children in human-robot interaction therapy with humanoid robot NAO. In: 2012 IEEE 8th International Colloquium on Signal Processing and its Applications, pp. 188–193 (2012)

53. Warren, Z.E., Zheng, Z., Swanson, A.R., Bekele, E., Zhang, L., Crittendon, J.A., Weitlauf, A.F., Sarkar, N.: Can robotic interaction improve joint attention skills? J. Autism Dev. Disord. **45**, 3726–3734 (2015)

54. Saadatzi, M.N., Pennington, R.C., Welch, K.C., Graham, J.H.: Small-group technology-assisted instruction: virtual teacher and robot peer for individuals with autism spectrum disorder. J. Autism Dev. Disord. **48**, 3816–3830 (2018)

55. Greczek, J., Kaszubski, E., Atrash, A., Matarić, M.: Graded cueing feedback in robot-mediated imitation practice for children with autism spectrum disorders. In: The 23rd IEEE International Symposium on Robot and Human Interactive Communication, pp. 561–566 (2014)

56. Hirokawa, M., Funahashi, A., Pan, Y., Itoh, Y., Suzuki, K.: Design of a robotic agent that measures smile and facing behavior of children with autism spectrum disorder. In: 25th IEEE International Symposium on Robot and Human Interactive Communication, RO-MAN 2016, pp. 843–848 (2016)

57. Bevill, R., Park, C.H., Kim, H.J., Lee, J.W., Rennie, A., Jeon, M., Howard, A.M.: Interactive robotic framework for multi-sensory therapy for children with autism spectrum disorder. In: 2016 11th ACM/IEEE International Conference on Human-Robot Interaction (HRI), pp. 421–422 (2016)

58. Petric, F., Miklić, D., Kovačić, Z.: POMDP-based coding of child-robot interaction within a robot-assisted ASD diagnostic protocol. Int. J. Humanoid Robot. **15**, 2–6 (2018)

59. Papakostas, G.A., Strolis, A.K., Panagiotopoulos, F., Aitsidis, C.N.: Social robot selection: a case study in education. In: 2018 26th International Conference on Software, Telecommunications and Computer Networks (SoftCOM), Split, pp. 1–4 (2018)

The Effect of Commercially Available Educational Robotics: A Systematic Review

Bjarke Kristian Maigaard Kjær Pedersen$^{(\boxtimes)}$,
Jørgen Christian Larsen , and Jacob Nielsen

University of Southern Denmark, 5230 Odense M, Denmark
bkp@mmmi.sdu.dk

Abstract. Educational institutions planning to invest in Educational Robotics are faced with a wide selection of products. Yet, we have not been able to find any review studies on the effect of these products, to guide the institutions to get the most out of their investments. For this review, 29 Educational Robotics products were therefore selected, and eight major databases were searched for effect studies involving these. The search yielded 301 results, of which 17 were selected for synthesizing. The studies and their respective findings are discussed in the review. Unfortunately, there were not enough studies to compare the effect of the products and more research is therefore needed. In addition, the studies methodologies and design have been analyzed, and a series of recommendations for how future experimental/quasi-experimental studies within the field can be design and conducted, have been established.

Keywords: Review · Robot · Education · Educational robotics ·
Experimental · Quasi-experimental · Effect · Computational thinking · CT ·
STEM · K-12 · Primary education · Higher education · Tertiary education ·
LEGO mindstorms

1 Introduction

Educational Robotics (ER) [1–3] is a popular way of supporting and exemplifying Computational Thinking (CT) [4] and teaching of STEM (Science, Technology, Engineering and Mathematics). When educational institutions plan to invest in ER, they are faced with wide a selection of products. However, we have not been able to find any review studies examining the effect of different commercially available ER products, which we believe to be instrumental in order to let the educational institutions take informed choices based on research, in regard to getting the most out of their investments. Recent review studies examining how educational robotics have been used to support STEM teaching in traditional topics in K-12 and tertiary educational institutions [5–7], have found that there is a lack of experimental or quasi-experimental studies on the topic. We believe this lack might be prescribed to the complex scenarios of educational research, in which parameters and conditions cannot justifiably be isolated from the contexts they exist in, as e.g. in a medicinal study [8].

The purposes of this paper are therefore: First, to examine and map the extent of experimental/quasi-experimental research involving commercially available ER

© Springer Nature Switzerland AG 2020
M. Merdan et al. (Eds.): RiE 2019, AISC 1023, pp. 14–27, 2020.
https://doi.org/10.1007/978-3-030-26945-6_2

products (effect studies) in order to evaluate and compare their effect, if any effect have been measured. Second, to examine how these studies have been conducted, to identify specific traits they might have in common as a guide for other researches on how to plan and conduct future experimental/quasi-experimental studies on the topic.

For this reason, we have selected 29 commercially available ER products to be examined, with our main research questions being:

- To what extent have experimental/quasi-experimental research been conducted on the respectable products, what was the focus and findings of the studies?
- How were the experimental/quasi-experimental research designed and conducted?

2   Method

This study is based on the eight stages in the process of conducting a systematic review [9] and have covered the eight stages involved herein as seen below (with the exception of the quality assessment, which has been merged with the exclusion criteria as we are solely focusing on experimental/quasi-experimental research studies involving control/comparison groups):

Planning the review:
Identification of the need for a review, development of a review protocol.
Conducting the review:
Identification of research, selection of primary studies, data extraction, data synthesis.
Reporting the review:
Communicating the results.

2.1   Planning the Review

As established in the introduction, we believe there is a need for a systematic review on how experimental/quasi-experimental studies are carried out within the field of ER, as well as a need for a review which measures the effect of commercially available ER products, to enable educational institutions taking more informed choices regarding which ER products to invest in. The protocol for selecting the primary studies to be included in the review and the data to be extracted, can be seen in Table 1.

Table 1. Exclusion criteria and data to be extracted.

Exclusion criteria		Data to be extracted	
EC1	The publication is a duplet	DE1	Author, title, and publishing year
EC2	The publication is not written in English	DE2	The methodology
EC3	The publication is not an article published at a conference or in a journal	DE3	How the experiment was designed and conducted

(*continued*)

Table 1. (*continued*)

Exclusion criteria		Data to be extracted	
EC4	The publication does not revolve around teaching with or through the searched for technology	DE4	The timespan of the study, the sample size and number of groups included in the study
EC5	The publication does not make use of the searched for technology	DE5	The age/level and gender of the participants
EC6	The publication does not mention the usage of control/comparison groups in the abstract, title or keywords	DE6	The purpose of the study and the results
EC7	The publication does not suit the research question		
EC8	The publication cannot be retrieved		

2.2   Conduction of the Review

A total of 29 commercially available ER products were selected for the review: Bee-Bot/Blue-Bot [10], Codey Rocky [11], Cozmo [12], Cubelets [13], Dash/Que [14], Edison [15], Hummingbird [16], Kibo [17], Kubo [18], LEGO Boost [19], LEGO Mindstorms NXT/EV3 [20] - [21], LEGO WeDo 2.0 [22], LittleBits [23], mBot [24], micro:bit [25], MiP [26], MU Spacebot [27], NAO [28], Neuron [29], Ozobot [30], Primo/Cubetto [31], Pro-Bot [32], Robot Mouse [33], Romibo [34], Scribbler [35], Sphero/Ollie [36], Thymio [37], Ultimate 2.0 [38], Vex [39]. The products were selected based on the researchers' initial knowledge as well as on a short initial study into existing ER products. Furthermore, the products had to have been developed for educational purposes.

After the selection, eight databases were likewise selected and searched in order to collect the data. The search was when possible restricted to the title, abstract and keywords, as well as works published in the years 2010–2018. The search took place in November-December 2018. Table 2 presents an overview of the searched databases as well as the queries used in the search, while Table 3 presents the substrings in use[1]. The purposes of the individual substrings are as following:

Technology: To ensure that the returned results make use of the searched for products.
Study: To ensure that the returned results are either experimental or quasi-experimental studies.
Education: To ensure that the returned results involves learning or teaching through the searched for technology.
Purpose: To ensure that the returned results has a focus on the purpose of active learning through working with the searched for products.

---

[1] Please note that the technology substring itself consist of several smaller substrings, which were used individually throughout the search. Likewise, please note that the IEEE Xplore search query differed depending on the technology substring in use.

Table 2. The databases and search queries.

Database	Search query
Academic Search Premier (EBSCO) (DB1)	(AB ((technology substring) AND ((study substring)) OR TI ((technology substring) AND ((study substring)) OR KW ((technology substring) AND ((study substring))) AND ((education substring)) AND ((purpose substring))
ACM Digital Library (Full Text Collection) (DB2)	(recordAbstract:((technology substring) AND (study substring)) OR acmdlTitle:((technology substring) AND (study substring)) OR keywords.author.keyword: ((technology substring) AND (study substring))) AND (education substring) AND (purpose substring)
Engineering Village (DB3)	(((((technology substring) AND (study substring)) WN KY) AND (((education substring)) WN All fields)) AND (((purpose substring)) WN All fields))
ERIC (EBSCO) (DB4)	(AB ((technology substring) AND ((study substring)) OR TI ((technology substring) AND ((study substring)) OR KW ((technology substring) AND ((study substring))) AND ((education substring)) AND ((purpose substring))
IEEE Xplore (DB5)	(((("Abstract":"technology substring") OR "Publication Title":"technology substring") OR "Author Keywords":"technology substring") AND (("Abstract":"Experimental") OR ("Abstract":"Group") OR ("Abstract":"Groups") OR ("Abstract":"Compar*") OR ("Publication Title":"Experimental") OR ("Publication Title":"Group") OR ("Publication Title":"Group") OR ("Publication Title":"Compar*") OR ("Author Keywords":"Experimental") OR ("Author Keywords":"Group") OR ("Author Keywords":"Group") OR ("Author Keywords":"Compar*")) AND ("Education" OR "Educational" OR "Learn" OR "Learning" OR "Teach" OR "Teaching") AND ("programming" OR "coding" OR "Computational Thinking" OR "Science, Technology, Engineering and Mathematics" OR "STEM")
Science Direct (DB6)	Find articles with these terms: (education substring) Title, abstract, keywords: (technology substring) AND (study substring)
Scopus (DB7)	(TITLE-ABS-KEY ((technology substring) AND (study substring)) AND ALL ((education substring)) AND ALL ((purpose substring)))
Web of Science (DB8)	ALL FIELDS: ((technology substring)) *AND* ALL FIELDS:((study substring)) *AND* ALL FIELDS: ((education substring)) *AND* ALL FIELDS: ((purpose substring))

Table 3. An overview of the substrings in use.

Substrings	
Technology	("Bee-Bot" OR "Blue-Bot"), ("Codey Rocky"), ("Cozmo" AND "Robot"), ("Cubelets"), (("Dash" OR "Que") AND "Wonder Workshop"), ("Edison" AND ("robot" OR "Microbric")), ("Hummingbird" AND ("BirdBrain" OR "Robotics")), ("KIBO" AND "Robot"), ("Kubo" AND "robot"), ("LEGO" AND "Boots"), ("LEGO" AND "Mindstorms"), ("LEGO" AND "WeDo"), ("LittleBits"), ("mBot" AND "Makeblock"), ("micro:bit"), ("WowWee" AND "MiP"), ("MU SpaceBot"), ("NAO" AND ("Aldebaran Robotics" OR "Softbank Robotics")), ("Neuron" AND "makeblock"), ("Osmo" AND "Tangible Play"), ("Ozobot"), (("Primo" OR "Cubetto") AND "robot"), ("Pro-Bot"), ("Roboblock"), ("Robot Mouse"), ("Romibo"), ("Scribbler" AND "robot"), (("Sphero" OR "Ollie") AND "Robot"), ("Thymio"), ("Ultimate 2.0" AND "Makeblock"), ("VEX" AND "Robotics")
Study	("group*" OR "compar*" OR "experimental")
Education	("education" OR "educational" OR "learn" OR "learning" OR "teach" OR "teaching")
Purpose	("programming" OR "coding" OR "Computational Thinking" OR "Science, Technology, Engineering and Mathematics" OR "STEM")

3   Results

The search yielded a total of 301 results, divided between the individual technologies and databases as seen in Table 4, with the following products returning zero results: Codey Rocky, Cozmo, Hummingbird, Kubo, mBot, MU Spacebot, Neuron, Ozobot, Robot Mouse, Ultimate 2.0. Of the 301 articles, 284 were excluded based on the established exclusion criteria: EC1 (40.8%), EC2 (1%), EC3 (3.9%), EC4 (14.2%), EC5 (2.3%), EC6 (25.6%), EC7 (5.9%), EC8 (0.3%). The remaining 17 articles are presented in Table 5.

Table 4. Search results per database/technology.

	DB1	DB2	DB3	DB4	DB5	DB6	DB7	DB8	Total
Bee-Bot, Blue-Bot	2	0	3	1	0	0	2	2	10
Cubelets	0	0	0	0	0	0	2	0	2
Dash, Que	1	0	0	0	0	0	0	0	1
Edison	1	0	0	0	0	0	0	0	1
Kibo	0	0	0	0	0	0	1	0	1
LEGO Boost	3	0	0	1	0	1	1	0	6
LEGO Mindstorms	33	6	31	11	15	17	59	35	207
LEGO WeDo	5	0	1	4	0	0	3	2	15
LittleBits	2	0	0	0	0	0	1	0	3
micro:bit	1	0	3	0	1	0	2	0	7

(*continued*)

Table 4. (*continued*)

	DB1	DB2	DB3	DB4	DB5	DB6	DB7	DB8	Total
MiP	2	0	0	0	0	0	0	0	2
NAO	1	1	0	0	0	0	2	0	4
Primo, Cubetto	0	0	2	0	1	0	1	1	5
Pro-Bot	0	0	0	0	0	0	0	1	1
Romibo	0	1	0	0	0	0	0	0	1
Scribbler	0	1	0	0	0	0	3	0	4
Sphero, Ollie	1	0	0	0	0	1	3	0	5
Thymio	1	1	1	0	2	0	4	3	12
Vex	1	1	3	2	1	1	3	2	14
Total	54	11	40	19	20	20	87	46	301

Table 5. An overview of the selected articles.

ID	Title	Technology	Compares/d with	Demographics[a]
#1 [40]	*Teaching computer science to 5–7 year-Olds: An initial study with scratch, Cubelets and unplugged computing*	Cubelets	Scratch and unplugged computing	Primary school
#2 [41]	*Contextualized learning tools: animations and robots*	Scribbler	Greenfoot animations	University
#3 [42]	*The effect of the programming interfaces of robots in teaching computer languages*	Thymio	Textual and visual programming interfaces	High school/college
#4 [43]	*Students Learn Programming Faster through Robotic Simulation*	VEX	Physical robotics vs. simulations	High school
#5 [44]	*"I want my robot to look for food": Comparing Kindergartner's programming comprehension using tangible, graphic, and hybrid user interfaces*	LEGO WeDo 2.0	Tangible, visual and hybrid interfaces	Kindergarten
#6 [45]	*Introducing Computer Programming to Children through Robotic and Wearable Devices*	LEGO Mind-storms NXT	Lilypad and Scratch	Secondary education school

(*continued*)

Table 5. (*continued*)

ID	Title	Technology	Compares/d with	Demographics[a]
#7 [46]	*Textual vs. visual programming languages in programming education for primary schoolchildren*	LEGO Mindstorms EV3	Textual and visual programming interfaces	Primary school
#8 [47]	*Learning to program with lego mindstorms – difference between K-12 students and adults*	LEGO Mindstorms EV3	The participants understanding and interest, based on their age	Mixed
#9 [48]	*Grouping matters in computational robotic activities*	LEGO Mindstorms EV3	Different approaches to group roles and gender divided groups	Elementary school
#10 [49]	*Creativity and contextualization activities in educational robotics to improve engineering and computational thinking*	LEGO Mindstorms EV3	Scratch and different approaches to teaching	Secondary education
#11 [50]	*From LEGO to Arduino: Enhancement of ECE freshman design with practical applications*	LEGO Mindstorms	Arduino	University
#12 [51]	*The Effect of Lego Mindstorms Ev3 Based Design Activities on Students' Attitudes towards Learning Computer Programming, Self-Efficacy Beliefs and Levels of Academic Achievement*	LEGO Mindstorms EV3	A C++ editor	University
#13 [52]	*The Effect of Scratch- and Lego Mindstorms Ev3-Based Programming Activities on Academic Achievement, Problem-Solving Skills and Logical-Mathematical Thinking Skills of Students*	LEGO Mindstorms EV3	Scratch and a C++ editor	University
#14 [53]	*The Effect of Reflective Strategies on Students' Problem Solving in Robotics Learning*	LEGO Mindstorms NXT	The usage of reflective strategies	Junior high school

(*continued*)

Table 5. (continued)

ID	Title	Technology	Compares/d with	Demographics[a]
#15 [54]	Motivating programming students by Problem Based Learning and LEGO robots	LEGO Mindstorms	Different learning designs and a C++ editor	University
#16 [55]	Distractions in programming environments	LEGO Mindstorms NXT	A full- and a limited interface	High school
#17 [56]	The Effect of Tangible Artifacts, Gender and Subjective Technical Competence on Teaching Programming to Seventh Graders	LEGO Mindstorms NXT	Physical robotics vs. simulations	Elementary school

[a]Please note that even though the demographics are taken directly from the specific articles, the age of participants at the same educational level may vary due to national difference in the classification of the individual levels within the educational systems.

4 Discussion

In this section we will go over what the selected studies aimed at measuring the effect of, their findings, as well as how the studies were designed, the methodologies in use and how the effect was measured.

4.1 The Effect of the Different Educational Robotics Products

The limited number of studies on the effect of ER, have proved insufficient in regard to concluding anything on the effect of teaching or learning through the searched for Educational Robotics products or on ER in general. Furthermore, to our surprise none of the studies included in this review have sought to measure the effect of ER as a facilitator for teaching traditional topics: mathematics, physics, etc. Likewise, only two studies aimed at comparing ER with another tangible technology: #6 Lilypad [57] and #11 Arduino [58]. However, when synthesizing the selected articles, we identified three major areas as being the focus of the studies included in this review. The areas revolve around the technology itself, ways of teaching as well as demographics. A breakdown of the areas in addition to a highlight of the most important findings, can be seen below:

1. Technology - comparing the ER with:

   Another tangible technology: #6, #11
   A screen-based technology: #1, #2, #4, #6, #13, #17
   Traditional media: #1, #12, #13, #15
   Itself (different interfaces): #3, #5, #7, #16

2. Teaching: #9, #10, #14, #15
3. Demographics: #6, #8, #9

Technology: Studies #12-13 have found that university students achieved higher scores in an introduction course to programming, when working with a combination of ER (LEGO Mindstorms) and a traditional C++ editor, compared to the editor alone. In study #15 university students working with both ER (LEGO Mindstorms) and a C++ editor, were reported as being happier than those working solely with the editor. Study #2 compared university students working with Greenfoot Animations [59], and animations in combination with ER (Scribbler, Finch [60]). While no increase in scores were registered, the instructors reported that the students working with ER were more engaged in the studies and working more collaboratively than those working solely with the animations. The students in both study #2 and #15, did however report working with ER to be cumbersome and to some degree frustrating. This were prescribed partly to the unpacking/packing or the ER, as well as limited space to work on. Study #6 also supports the notion of pupils being more engaged when working with tangible technologies (LEGO Mindstorms and Lilypad) as opposed to screen-based technologies (Scratch [61]). Like study #2, study #4 did not find any difference in the learning outcomes, when working with ER (VEX) and screen-based technologies (VEX simulation), however the pupils working with the screen-based technology finished the course at a faster rate.

Study #3 have found that pupils being introduced to programming and ER (Thymio) though a Visual Programming Language (VPL) achieved higher scores, than those working with a Textual Programming Language (TPL). Yet, pupils working with a VPL first, then switching over to a TPL, scored very close to the VPL only group. Study #16 found that pupils using a subset of the LEGO Mindstorms NXT interface, as compared to the entire interface, likewise scored better and afterwards even proved better at transferring the obtained knowledge to the Alice program [62].

Teaching: Acontextualized/creative approach as well as Problem Based Learning (PBL), have been found more effective than a more traditional and strict approach (#10, #15). Assigning individual group members with either fixed or rotating roles, have been found to yield more effective results than a no roles approach, with a small favor towards fixed roles (#9). Pupils keeping a journal of their project, finished the projects goal at a faster rate than those who did not keep a journal (#14). This is prescribed to the students reflecting deeper upon the assigned tasks and possible solutions, due to filling out the journal – the effect could possibly be related to the effect of rubber duck debugging [63].

Demographics: LEGO Mindstorms have been found to be suitable for teaching adults as well as children (#8). Study #9 did not find any difference when measuring the achievements of gender divided groups, the lack of a difference might be due to the contextualization of the assignment, which revolved around the creation of music. While not the focus of the study, study #2 found that female students were more likely to pass the course if it included working with ER as compared to their male counterparts. In study #17 however, female pupils had a lower subjective technical competence score, and performed worse when tasked with writing new programs, while study #6 found no difference in interest between the genders, female pupils performed better in all programming categories.

4.2   How the Studies Were Designed and Conducted

Context, Conduction and Sample Sizes: Three main contexts in which the studies have been conducted have been identified: Universities (regular teaching: #2, #11–13, #15), Workshops (workshop course: #7, #8, #16) and K-12 institutions (regular teachings: #10, elective courses: #4, shorter studies: #1, #3, #5–6, #9, #14, #17). Likewise, four main trends regarding who is conducting the teaching in the studies have also been identified: The author whom is also the regular teacher (#2, #11–13, #15), the author whom is not the regular teacher (#3, #6–8, #10, #14, #16–17)[2], the author in combination with the regular teacher (#1, #5, #9) and the regular teacher (#4). If we look at the twelve studies set outside the universities, we find that very few of these have been set in the context of regular teachings or have been conducted by or in combination with the regular teacher. This poses a serious problem as we cannot expect the findings to hold, when the task of teaching afterwards resides with teachers whom may not be as well versed in technology, as university researchers must be expected to be. Furthermore, eight of these studies have an average sample size per group of sixteen or fewer participants (#1, #3–5, #7, #16–17). This could indicate that researchers in general have a hard time gaining access to subjects. In addition, studies with a between-groups design, have been found to be the most popular choice (#2–5, #7–17) as compared to a within-groups design (#1, #6).

Pre-, Mid- and Post-tests: The most common setup make use of both a pre-/post-test which have been used in ten studies (#4, #6, #8–9, #12–14, #16–17), while a mid-/post-test and a post-test only setup have been used in respective five (#1, #3, #5, #7, #10) and three studies (#2, #11, #15)[3]. While a pre-test is relatively effortless to include, it is of fundamental value to evaluate the effect regarding the post-test. In relation to #2 and #15, we cannot know if there truly has been no effect on the grades from working with ER, or if there already were a difference in the students' emotions from the beginning and not as a result of the ER intervention.

Quantitative and Qualitative Data: Qualitative data is an efficient method for casting a light on the reasons behind the acquired quantitative data, i.e. as seen in #14 were it effectively describes the reasons for the added benefits of having the students keep a project journal, or as in #2 where the qualitative data highlighted differences not found in the quantitative data. Yet, only six studies (#1–2, #5, #11, #14–15) have focused on collecting qualitative data in addition to the quantitative data (See footnote 3).

What have been Measured and How: The focus of the measured can roughly be divided into two main trends: The learning outcome (#1–6, #8–14, #16–17) and the emotions/motivation (#2, #6–8, #12, #15–17).The learning outcome for the studies carried out at the universities have mainly made use of the course grades (#2, #11) and

---

[2] This have been assumed in the instances, were nothing indicating otherwise have been specified.

[3] These numbers are based on the authors' evaluation of the selected studies, since studies (#1–3, #7–8, #10–11, #15, #17) does not explicitly mention if the studies include pre-, mid- or post-tests, while studies (#3–4, #6–10, #14, #16–17) does not explicitly mention if the data is quantitative and or qualitative.

multiple-choice tests (#12–13) to evaluate the outcome. Outside the universities, performance tests have also been a popular way to measure the learning outcome. These have often involved questionnaires and multiple-choice tests (#4, #6, #8, #14, #16), and programming tasks revolving around solving a specific problem, or series of problems (#1, #3, #5, #17). The emotions and motivation have mainly been measured through questionnaires (#6–8, #12, #15–17) and interviews (#2, #15).

4.3   Recommendations

Based on the findings above, we recommend building a closer relationship between the universities and local educational institutions. This to provide researchers with easier access to participants, ex. parallel classes (K-12) often averaging between 20–30 students per class. Parallel classes would also be ideal for studies, which implements a between-groups design. In addition, a closer collaboration would also enable researchers to let the regular teachers conduct the studies, either alone or in combination with the researchers - as part of the regular teaching - thereby strengthening the validity of the studies. Likewise, we also recommend that future studies always conduct a pre-test in order to strengthen the validity of the post-test, as well as collecting both quantitative and qualitative data, to strengthen the understanding of the results.

With the popularity of questionnaires and multiple-choice, we recommend the development of a standardized test for measuring the learning outcome (sequences, loops, conditionals, functions). This would make it possible to compare the learning outcome across and between platforms.

Furthermore, for the purpose of clarity we recommend that future studies explicitly state if the tests included are pre-, mid- or post-tests, as well as if the data is quantitative or qualitative. In addition to this, we also recommend that future studies state both the educational level and the age of the participants, since the classification of the former can vary from country to country.

4.4   Future Directions and Limitations

With the limited amount of studies found on the effect of Educational Robotics, we feel that there is a serious lack of research on the field – not only in regard to comparing the effect of available ER products, but even more importantly in regard to comparing ER to traditional ways of teaching. Regarding limitations, we acknowledge that an expanded list for ER products and /or different search strategies might have yielded a different result.

5   Conclusion

A search in eight major databases for studies measuring the effect of Educational Robotics - spread out on 29 products - yielded 301 results out of which 17 articles were selected based on the established exclusion criteria. The limited number of articles have not made it possible to conclude anything on the effect of the different ER products or on ER in general – further studies on the topic is needed.

If we should recommend some general methodologies on behalf of our study, these would be: When conducting experimental/quasi-experimental studies, using parallel classes (K-12) or dividing the students of a university course into two groups, could prove an ideal way of naturally dividing the participants in a between subject design. When collecting the data, we recommend using both quantitative and qualitative methods, with the possibility of quantifying tests to measure the learning outcome (multiple-choice, performance tests), followed up by qualitative data to measure the emotions (questionnaires and interviews). Regarding this, we also recommend performing both pre- and post-tests. In addition, we recommend setting the experiment in a context as close as possible to the one the results are expected to be used in afterwards, i.e. during the regular teaching and by or in combination with the regular teacher. For the purpose of clarity, we also recommend that future studies explicitly state if the tests included are pre-, mid- or post-tests, if the data is quantitative or qualitative, as well as both the current educational level and age of the participants.

References

1. Atmatzidou, S., Demetriadis, S.: Advancing students' computational thinking skills through educational robotics: a study on age and gender relevant differences. Rob. Auton. Syst. **75**, 661–670 (2016)
2. Blanchard, S., Freiman, V., Lirrete-Pitre, N.: Strategies used by elementary schoolchildren solving robotics-based complex tasks: innovative potential of technology. Procedia - Soc. Behav. Sci. **2**(2), 2851–2857 (2010)
3. Bers, M.U., et al.: Computational thinking and tinkering: exploration of an early childhood robotics curriculum. Comput. Educ. **72**, 145–157 (2014)
4. Wing, J.M.: Computational thinking and thinking about computing. Philos. Trans. A. Math. Phys. Eng. Sci. **366**(1881), 3717–3725 (2008)
5. Benitti, F.B.V.: Exploring the educational potential of robotics in schools: a systematic review. Comput. Educ. **58**(3), 978–988 (2012)
6. Benitti, F.B.V., Spolaôr, N.: How have robots supported STEM teaching? In: Robotics in STEM Education, pp. 103–129. Springer, Cham (2017)
7. Spolaôr, N., Benitti, F.B.V.: Robotics applications grounded in learning theories on tertiary education: a systematic review. Comput. Educ. **112**, 97–107 (2017)
8. Barab, S., Squire, K.: Design-based research: putting a stake in the ground. J. Learn. Sci. **13**(1), 1–14 (2004)
9. Kitchenham, B.: Procedures for Performing Systematic Review. Keele University, Keele, UK (2004)
10. Bee-Bot/Blue-Bot. https://www.bee-bot.us/bee-bot.html. Accessed 4 Jan 2019
11. Codey Rocky. https://www.makeblock.com/steam-kits/codey-rocky. Accessed 4 Jan 2019
12. Cozmo. https://www.anki.com/en-us/cozmo. Accessed 4 Jan 2019
13. Cubelets. https://www.modrobotics.com/cubelets/. Accessed 4 Jan 2019
14. Dash/Que. https://uk.makewonder.com/?. Accessed 4 Jan 2019
15. Edison. https://meetedison.com/. Accessed 4 Jan 2019
16. Hummingbird. https://www.birdbraintechnologies.com/hummingbirdbit/. Accessed 4 Jan 2019
17. Kibo. http://kinderlabrobotics.com/kibo/. Accessed 4 Jan 2019
18. Kubo. https://kubo.education/. Accessed 4 Jan 2019

19. LEGO Boost. https://www.lego.com/da-dk/themes/boost. Accessed 4 Jan 2019
20. LEGO Mindstorms NXT. https://shop.lego.com/en-US/LEGO-MINDSTORMS-NXT-2-0-8547. Accessed 4 Jan 2019
21. LEGO Mindstorms EV3. https://education.lego.com/en-us/shop/mindstorms-ev3. Accessed 4 Jan 2019
22. LEGO WeDo 2.0. https://education.lego.com/en-us/shop/wedo-2. Accessed 4 Jan 2019
23. LittleBits https://littlebits.com/. Accessed 4 Jan 2019
24. mBot. https://www.makeblock.com/steam-kits/mbot. Accessed 4 Jan 2019
25. micro:bit. https://microbit.org/. Accessed 4 Jan 2019
26. MiP. https://wowwee.com/mip/. Accessed 4 Jan 2019
27. MU Spacebot. http://www.morpx.com/. Accessed 4 Jan 2019
28. NAO. https://www.softbankrobotics.com/us/NAO. Accessed 4 Jan 2019
29. Neuron. http://neuron.makeblock.com/en/. Accessed 4 Jan 2019
30. Ozobot. https://ozobot.com/. Accessed 4 Jan 2019
31. Primo/Cubetto. https://www.primotoys.com/. Accessed 4 Jan 2019
32. Pro-Bot. https://www.bee-bot.us/probot.html. Accessed 4 Jan 2019
33. Robot Mouse. https://www.generationrobots.com/en/403207-robot-mouse.html. Accessed 4 Jan 2019
34. Romibo. https://www.makeblock.com/kid-coding/210584.html. Accessed 4 Jan 2019
35. Scribbler. https://www.robotshop.com/uk/parallax-scribbler-3-robot.html. Accessed 4 Jan 2019
36. Sphero/Ollie. https://www.sphero.com/. Accessed 4 Jan 2019
37. Thymio. https://www.thymio.org/en:thymio. Accessed 4 Jan 2019
38. Ultimate 2.0. https://www.makeblock.com/steam-kits/mbot-ultimate. Accessed 4 Jan 2019
39. VEX. https://www.vexrobotics.com/vexiq. Accessed 4 Jan 2019
40. Wohl, B., Porter, B., Clinch, S.: Teaching computer science to 5–7 year-Olds: an initial study with scratch, Cubelets and unplugged computing. In: WiPSCE, London, United Kingdom (2015)
41. Remshagen, A., Rolka, C.: Contextualized learning tools: animations and robots. In: ACM Southeast Regional Conference, Kennesaw, Georgia, USA (2014)
42. Bağcı, B.B., Kamasak, M., Ince, G.: The effect of the programming interfaces of robots in teaching computer languages. In: International Conference on Robotics and Education, Sofia, Bulgaria (2017)
43. Allison, L., et al.: Students learn programming faster through robotic simulation. Tech Dir. **8** (72), 16–19 (2013)
44. Strawhacker, A., Bers, M.U.: "I want my robot to look for food": comparing Kindergartner's programming comprehension using tangible, graphic, and hybrid user interfaces. Int. J. Technol. Des. Educ. **25**(3), 293–319 (2015)
45. Merkouris, A., Chorianopoulos, K.: Introducing computer programming to children through robotic and wearable devices. In: Workshop in Primary and Secondary Computing Education, London, England (2015)
46. Tsukamoto, H., et al.: Textual vs. visual programming languages in programming education for primary schoolchildren. In: Proceedings - Frontiers in Education Conference, Erie, PA, USA (2016)
47. Umbleja, K.: Learning to program with lego mindstorms – difference between K-12 students and adults. In: International Conference on Interactive Collaborative Learning, Budapest, Hungary (2017)
48. Taylor, K., Baek, Y.: Grouping matters in computational robotic activities. Comput. Hum. Behav. **93**, 99–105 (2019)

49. Valls, A., Albó-Canals, J., Canaleta, X.: Creativity and contextualization activities in educational robotics to improve engineering and computational thinking. In: 8th International Conference on Robotics in Education, Sofia, Bulgaria (2017)
50. Berry, C.A.: From LEGO to Arduino: enhancement of ECE freshman design with practical applications. In: ASEE Annual Conference and Exposition, New Orleans; United States (2016)
51. Korkmaz, Ö.: The effect of Lego Mindstorms Ev3 based design activities on students' attitudes towards learning computer programming, self-efficacy beliefs and levels of academic achievement. Baltic J. Modern Comput. 4(4), 994–1007 (2016)
52. Korkmaz, Ö.: The effect of scratch- and Lego Mindstorms Ev3-based programming activities on academic achievement, problem-solving skills and logical-mathematical thinking skills of students. Malays. Online J. Educ. Sci. 4(3), 73–88 (2016)
53. Lin, C.H., Liu, E.Z.F.: The effect of reflective strategies on students' problem solving in robotics learning. In: 4th International Conference on Digital Game and Intelligent Toy Enhanced Learning, Takamatsu, Japan (2012)
54. Lykke, M., et al.: Motivating programming students by problem based learning and LEGO robots. In: Global Engineering Education Conference, Istanbul, Turkey (2014)
55. Mason, R., Cooper, G.: Distractions in programming environments. In: Conferences in Research and Practice in Information Technology Series, Adelaide, Australia (2013)
56. Brauner, P., et al.: The effect of tangible artifacts, gender and subjective technical competence on teaching programming to seventh graders. In: 4th International Conference on Informatics in Secondary Schools - Evolution and Perspectives, Zurich, Switzerland (2010)
57. Lilypad Arduino. https://store.arduino.cc/lilypad-arduino-simple. Accessed 16 Jan 2019
58. Arduino. https://www.arduino.cc/. Accessed 16 Jan 2019
59. Greenfoot. https://www.greenfoot.org/home. Accessed 16 Jan 2019
60. Finch. https://www.birdbraintechnologies.com/finch/. Accessed 16 Jan 2019
61. Scratch. https://scratch.mit.edu/. Accessed 16 Jan 2019
62. Alice. https://www.alice.org/. Accessed 16 Jan 2019
63. Rubber Duck Debugging. https://rubberduckdebugging.com/. Accessed 16 Jan 2019

Workshops, Curricula and Related Aspects

On the Use of Robotics for the Development of Computational Thinking in Kindergarten: Educational Intervention and Evaluation

Evgenia Roussou[1] and Maria Rangoussi[2]($\boxtimes$) (iD)

[1] MSc "ICT for Education", National and Kapodistrian University of Athens,
Athens, Greece
[2] Department of Electrical and Electronics Engineering,
University of West Attica, Athens, Greece
mariar@uniwa.gr

Abstract. Computational thinking is a relatively new concept that has attracted research interest during the last decade and has gained popularity in the field of applied education as well. This paper attempts to shed light upon the development of computational thinking in early childhood through an applied, experimental approach aiming to clarify the connection between computational thinking and programming with emphasis on visual and tangible programming of educational robots. The research is based on a case study that investigates the impact of robotics on the cultivation of computational thinking skills in early childhood through an educational intervention implemented in a typical public kindergarten in Athens, Greece. The intervention focuses on the development of particular aspects of computational thinking with the use of a programmable floor robot. The implementation follows a detailed lesson plan that encompasses all the robotic activities involved. Pupils' activities, behavior and exchange are monitored for evaluation purposes. Data collected during the intervention are analyzed in an attempt to answer research questions related to the potential of robotics in the development of computational thinking in early childhood. Results indicate that the use of a robot in a playful way, suitable for the stage of development of kindergarten pupils, leads to notable enhancement of the kindergartners' computational skills – a finding that is consistent to those of existing research studies on similar questions. Given the limited scale of this study, the positive results obtained may serve as a basis for the design of further, more extensive research on this issue.

Keywords: Computational thinking · Kindergarten · Educational robot · Haptic programming · Robot programming · Educational intervention · Evaluation

© Springer Nature Switzerland AG 2020
M. Merdan et al. (Eds.): RiE 2019, AISC 1023, pp. 31–44, 2020.
https://doi.org/10.1007/978-3-030-26945-6_3

1 Introduction - Computational Thinking: A 21st Century Skill

Thinking, communication and collaboration form a triplet that holds a prominent position among the various skills and competences modern Education seeks to develop in young learners. *"We are currently preparing students for jobs and technologies that don't yet exist... in order to solve problems we don't even know are problems yet"* – this is how the nature of the challenges Education is faced with in the 21st century was summarized in [1]. Earlier than that, Einstein had called for a new way of thinking for problem solving: *"We cannot solve our problems with the same thinking we used when we created them"*.

Computational thinking (CT) has recently received much attention and has become the focus of many research studies aiming to establish its merits and the feasibility of its development through designed activities at various grades of education. There is no definition of CT that enjoys a common consensus; it is widely understood as an analytic and algorithmic way of thinking for problem solving. The term was first used by Papert, [2], to describe the combination of critical thinking and the potential of computing – a combination that encompasses and enhances skills for problem solving, communication, collaboration, creativity and computation. These skills are considered essential for the 21st century; interestingly enough, the introduction and use of Information and Communication Technologies (ICT) in Education aims at the development of the same set of 'horizontal' skills as well as at the cultivation of multiple intelligences. Wing, whose pioneering research in Carnegie-Mellon University, USA was instrumental in drawing attention on CT, has claimed that *"CT involves solving problems, designing systems and understanding human behavior, by drawing on the concepts fundamental to computer science"*, [3] – a definition that has raised much discussion, and that was later modified into *"CT is the thought processes involved in formulating problems and their solutions so that the solutions are represented in a form that can be effectively carried out by an information-processing agent"*, [4]. As it has been pointed out in [5], this definition paves the way for the introduction of CT in compulsory education as it maintains that (a) being a thought process, CT is not necessarily tied to a specific technology, while (b) *"it is a specific type of problem-solving that entails distinct abilities, e.g. being able to design solutions that can be executed by a computer, a human, or a combination of both"*. In that case, all three key dimensions of CT, namely *"computational concepts, computational practices and computational perspectives"*, should be taken into account, [6].

Research on CT by Wing and many other groups has drawn the attention of academic and education communities to the potential benefits of integrating CT in the Primary and Secondary School Curricula. A plethora of research papers published indicates that programming and robotics are the most suitable tools to promote CT skills. Over 500 articles of scientific and grey literature have already appeared on the issue; and yet, experts agree that further research and data from actual school settings are vital to the comprehension of all aspects of CT.

As CT is finding its way to the curricula of K-12 and academic education around the world, [7–9], researchers realize that *'...this area of research is still in its infancy,'*

[10], while '... *lots of countries are still in the process of, or have not yet started, introducing CT into curriculums in all levels of education,*' [11], – possibly because of our as yet incomplete understanding of its meaning and the confusion to computer programming, [5], or the mere use of computer technology, [12]. In particular, the appropriate pedagogical framework for the incorporation of CT in education remains an open issue – see, e.g., [13].

Although CT is closely related to computer programming, the latter being recognized today as an efficient tool for the development of CT, it may be taught independently of computers or programming. It focuses on problem solving through strategies that rely on five basic concepts, namely, problem identification and decomposition, abstraction, logical thinking, algorithms and debugging, [14]. In fact, CT is related to a variety of other ways of thinking, e.g., mathematical, engineering, creative, critical or design thinking; it builds on the skills each of them develops and extends them beyond specific subjects or disciplines, specific ages or needs.

Conventional instruction of programming, however, is not well suited to younger learners, e.g., kindergarten pupils, who are not yet skilled in reading and writing. Alternative approaches that involve *optical* or *haptic* programming and robotic devices of different forms have been employed and investigated for these ages, [15–23]. These studies have established the advantages of robot programming in early childhood and have reported a positive impact on cognitive skills related to STEM subjects, on critical thinking and on social skills such as communication and collaboration, [24–28]. The enhancement of individual CT constituent skills, such as debugging, algorithms, problem-solving, sequencing and understanding causality, through the use of robotics and developmentally appropriate materials and approaches, is reported in [19, 25, 29–32], among others. Moreover, the relative advantage of haptic over optical programming, in particular, has been concluded in [23], in a KIBO versus ScratchJr experiment.

Inspired by the pioneering research work by Bers and her group (DevTech Research Group, Tufts University, USA) on the use of educational robots like KIBO by preschool and kindergarten pupils, the present study aims to investigate the feasibility, the efficiency and the potential of haptic programming of a floor robot for the development of CT in kindergarten pupils. Moreover, our aim is to study the impact of programming activities on particular CT skills. The major research question investigated is whether (and to what extend) participation of the kindergarten pupils in tangible programming of a floor robot enhances (i) their ability to understand cause and effect relationships, (ii) their ability to formulate hypotheses, (iii) their understanding of logical order and sequencing, and (iv) the development of problem solving skills through the spotting and correction of code errors (debugging). Other research questions address the quality of the pupils' experience with robot programming (involvement, well-being) and the development of social skills (communication, collaboration).

The novelty of the present approach lies in (i) its very objective which is not to teach elementary robotics as a subject per se or to use robotics as a teaching aid to other related subjects (or even as an entertainment or distraction), but rather to employ a (suitable) programmable robotic device as the centerpiece of an intervention aimed to develop CT skills in preschool and kindergarten pupils, (ii) its duration of eleven (11) weeks, which is longer than most of the reported interventions, and (iii) its context of an authentic classroom environment rather than a custom setup lab.

2   Materials and Methods

The main objective of this research is the investigation of the impact of robot programming in the development of specific CT skills in kindergartners, through an
experimental, hands-on approach. According to the CT framework set in [6], the subset
of CT-related skills compatible with the developmental stage of children in the preschool and kindergarten ages includes

- Sequencing: the execution of a series of steps in the proper order,
- Looping: a mechanism to repeat a sequence of steps for a number of times,
- Hypothesis: decision making upon conditions or facts, and
- Debugging: detection and alleviation/correction of error points in a code.

Debugging, in particular, may prove useful to anyone facing a tough problem that
cannot be solved straightforwardly. It may be decomposed into (i) recognition of a
malfunction; (b) spotting of the point where the malfunctioning occurs; (iii) formulation of hypotheses on possible causes; (d) checking of all alternative hypotheses until
the malfunction is alleviated. Through debugging, pupils become able to recognize
cause-and-effect relations and to formulate and check alternative hypotheses.

The ability to recognize cause-and-effect relations is fundamental for programming.
Since any program 'written' by a programmer and 'executed' by a computer or processor is essentially a sequence of commands, each of which causes a specific activity
or response of the computer or one of its components, the cause-and-effect notion is
clearly 'wired in' any programming task. Successful programming presupposes from
the part of the programmer (i) an understanding of the notion of causality and (ii) good
knowledge of each command (cause) and its results (effects). Sequencing, on the other
hand, is fundamental to the development of the notion of time evolution, of reading
skills and of language and speech skills in preschoolers.

The present research focuses on a slightly modified set of CT elements, namely, on
sequencing, hypothesis testing and debugging. The later is further broken into cause-
and-effect relations and problem solving, to be treated separately. Looping is not
employed because of practical limitations that have to do with the robotic device
selected. This is a mouse-like educational floor robot, namely, the Colby robotic mouse
of Learning Resources ™ that was purchased and used for this study, complete with its
Code & Go Robot Mouse Activity Set, i.e., with maze grids, maze walls, tunnels,
double-sided coding cards and double-sided activity cards (Fig. 1). It is an attractive

Fig. 1. Photos of the Code & Go Robot Mouse Activity set (Learning Resources ™), as used in
the classroom: components (left); the Colby robotic mouse with haptic programming self-
explanatory buttons on top (middle); a desktop path setup with Colby and cheese piece (right).

and inexpensive solution that is functionally equivalent to, e.g., the BeeBot educational robot that has already been used in research studies with preschoolers, [29].

Methodologically, this research is a field experiment in an existing kindergarten class with a population consisting of pupils aged 4.5 to 6. The researcher (first author) is the class teacher that participates in the experiment and at the same time observes the behavior of the individuals of the sample, [33] – in that aspect this study has action research elements. Video recordings, questionnaires (pre- and post-tests), observation sheets completed by the teacher and semi-structured interviews of the pupils held by the teacher are the tools used to evaluate the intervention and its outcomes and to collect data for further analysis. The small size of the population taken into account, a mixed-methods approach is opted for: the intervention outcomes (impact of robotic programming on the development of specific CT skills) are evaluated quantitatively through questionnaires (pre- and post-tests), while the intervention per se, its social aspects and the overall experience of pupils are evaluated qualitatively, through interviews with the teacher, observation sheets and video recordings. The involvement and well-being of pupils during the intervention are measured by the class teacher using the 5-levels Leuven scale, [34]. Due to the small population size (see next section), results cannot be generalized: they are to be interpreted as indications rather than proofs.

3 Design and Implementation of the Educational Intervention

3.1 Objectives, Background, Outline and Sample

The design and implementation of an educational intervention for kindergarten pupils, built on haptic programming of an educational robot, is the centerpiece of the present research. The intervention follows a detailed educational scenario designed in stages:

- The *introductory, experiential ('unplugged') stage* that lasts for 2 weeks and includes story-telling, role-playing games and dramatization. Pupils in pairs play in turn the role of the programmer or the robot and give or execute commands to navigate on a path constructed by blocks on the class floor.
- The *technological stage* that lasts for 7 weeks and includes haptic robot programming activities and testing of the robot on floor paths of increasing difficulty. Pupils work either all in one group, or in random groups (4 members) and in groups of friends (3–4 members).

The major objective of the educational scenario is to investigate the impact of robot programming activities on the four (4) CT skills mentioned earlier, namely, (i) understanding of causality relations, (ii) formulation and testing of hypotheses, (iii) sequencing, and (iv) problem solving. Secondary objectives include progress in the kids' space awareness, speech communication, decision making and reflection on their choices, collaboration with mutual respect and respect for the rules, and logging of their actions for subsequent discussion. Side objectives are self-expression through art and the development of imagination and creativity. All these objectives are in strong correspondence to those set by the national curriculum for kindergarten.

The project takes place in a public kindergarten in a municipality near Athens, Greece – a working-class area of low income and medium education level. Kids have access to their parents' devices (smart phones, tablets) for gaming and communication purposes. None of the pupils has had any previous experience with robotics or programming, though.

The research extends from late March to mid-June 2018. One week before the introductory stage, the class teacher gives the pupils a diagnostic pre-test to assess their existing CT skills. The main activities of the two stages of the intervention follow. A post-test is delivered by the teacher one week after the intervention. Individual interviews with the teacher follow; *additional* evaluation activities during the interviews include a test with cards delivered by the teacher that requires combined use of all the CT skills and skill transfer tests (haptical programming of a different robot – BeeBot – and optical programming in ScratchJr, on a tablet). Finally the teacher holds interviews with the pupils' parents, to cross-validate results with the pupils' behavior at home.

The intervention is implemented in a class of eighteen (18) pupils, all between 4.5 and 6 years of age. Thirteen (13) pupils out of them, six (6) boys and seven (7) girls, form the *experimental group* of the intervention; the other five (5) pupils are excluded from data analysis for various reasons (frequent absences, illnesses, non-consenting parents). The experimental group is used for the quantitative analysis part relative to the CT skills development. For the qualitative analysis part (interviews and additional evaluation activities), however, the teacher has selected a sub-group of six (6) pupils (three (3) boys and three (3) girls), who, according to her observations during the intervention, participated actively and consistently and who intentionally tried to get better in programming the robot. This selection was dictated by practical limitations brought about by the end of the school term, given that the interviews and additional activities are time-consuming for the class teacher. Finally, seven (7) pupils (three (3) boys and four (4) girls) of another class in the same kindergarten, who had no contact or interaction with the robot, were used as the *control group* for quantitative analysis.

3.2   The Pre- and Post-tests to Assess the Level of CT Skills

The pre- and post-test is essentially the same test, delivered one week before the beginning of the activities as a pre-test and one week after the completion of all activities as a post-test, to the experimental and the control group. It is based on pictures and it is custom-designed by the class teacher who took into account the general abilities and cognitive constraints of the preschool age and her experience with the specific class of pupils. The size of the test (36 questions) is adjusted to take 15 min; this is the maximum concentration period for this age. It is given to each pupil individually by the teacher, who shows the card, reads the question in neutral voice and avoids any sign that would guide the child to the correct answer – and, of course, does not comment on the child's choices. The test aims to measure the level of the CT skills of the pupils – in particular, of the CT skills targeted in this study (sequencing, hypothesis formulation, cause-and-effect relations, problem solving). It consists of four (4) distinct units, one for each of the CT skills:

(a) Unit 1 (10 tasks) checks whether pupils understand the cause of a problem and whether they can identify the likely outcomes (Unit 1 sample task in Fig. 2).

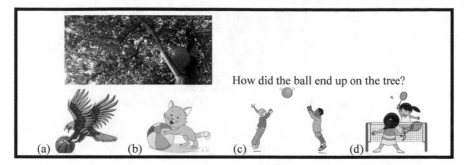

Fig. 2. Unit 1 sample task card.

(b) Unit 2 (10 tasks) investigates whether or not and how pupils can speculate on a likely connection between two seemingly unconnected, randomly chosen items. There is no unique correct answer here; any answer that formulates a hypothesis is considered as satisfactory; this is judged by the teacher (Unit 2 sample task in Fig. 3).

Fig. 3. Unit 2 sample task card.

(c) Unit 3 (8 tasks) focuses on sequencing: pupils are challenged to arrange a set of pictures in a logical order so as to 'tell' a short story. The number of pictures in each set varies from 3 to 9 for increasing difficulty, and stories are based on ordinary experiences and daily routines which 5 year-old pupils should be familiar with, (Unit 3 sample task in Fig. 4).

Fig. 4. Unit 3 sample task card.

(d) Unit 4 (8 tasks) aims to measure problem solving skills by presenting pupils with a set of items and asking them to select the appropriate one to complete a tough task, (Unit 4 sample task in Fig. 5).

What can the boy use to reach the book on the top shelf?

Fig. 5. Unit 4 sample task card.

3.3   The Two Central Stages of the Intervention

The underlying pedagogical principles of this intervention are *Learning-by-doing*, [35] and *Collaborative learning*, [36]. In both stages of the intervention, pupils are encouraged to actively participate, to work in small groups and to take initiatives on the action plan. The classroom used for the intervention is rearranged to host a 'Mouse Corner', where Colby the robot mouse and related materials (programming cards, floor boards etc.) are kept. Pupils' desks are arranged to accommodate small groups of 3–4 kids each.

In Week 1 of the introduction stage, the '*Winnie's Big Bad Robot*' book for kids by Australian author Valerie Thomas (Greek translation) is read to the class by the teacher, followed by discussion on the existence, nature and role of robots; kids make drawings of robots, name them and post them in class; they are prompted to move or dance like robots to music. In Week 2, the book '*Me and my robot*' by Greek author Manos Kontoleon is read in class, followed by discussion on 'programming' a robot to achieve a goal. The kids are encouraged to play a game in pairs: the 'programmer' kid uses a limited set of verbal 'commands' (forward, backward, left, right) to motion the 'robot' kid, blind-folded for the role, on a path formed by white A3-sized stepping blocks on the floor, safely around obstacles and to the terminal (goal). During the week, all kids play with enthusiasm on paths of increasing difficulty and repeatedly assume both roles until they master the notion of 'programming' a 'robot' by 'commands', (Fig. 6).

Fig. 6. Snapshots of the role-playing game in class: the 'programmer' kid uses verbal commands to safely motion the 'robot' kid on the path, around obstacles and to the goal.

During the seven (7) weeks of the technological stage, activities employ Colby, the robotic mouse. Colby is introduced by the teacher who reads aloud a letter Colby addresses to the kids. After taking some time to get to know Colby and its buttons, the pupils are asked to achieve a simple yet demanding goal: to guide Colby to its prize (a yellow plastic pyramid resembling a piece of cheese) by pressing the correct buttons. For two weeks, they all work together as one team, waiting patiently for their turn to program the robot and guide it the cheese. This phase offers pupils the possibility and the time to experiment, learn how to operate the robot and get familiar with the coding cards. For the next two weeks (phase 2), they work in small randomly formed groups of four (4) in the Mouse Corner: each group is free to 'play' with Colby for 15 min, and try different command sets to lead it to the cheese block. The routes are becoming more complex for increasing difficulty; yet, all pupils remain attracted, involved and focused on the task. Phase 3 differs in that the groups are no longer random but formed of 3-4 friends who share a common dream: to win the 'Robot Olympics' scheduled for the last week of the intervention. Each group takes on a name for the Olympics. They are allowed to practice in the Mouse Corner for 20 min daily on routes of increased difficulty. The 'Robot Olympics' constitute the last phase of the intervention: during the week, all groups contest each other in three (3) matches daily, trying to complete faster a very 'tough' route. On the last school day of the term, all competitors are awarded medals and certificates in a ceremony, (Fig. 7).

Fig. 7. Snapshots of the robot programming activities of the 2nd stage: phase 1 (top left & top middle); phase 2 (top right); phase 3 (bottom left); Robot Olympics (bottom middle); award conferring ceremony in the classroom (bottom right).

The post-test, the interviews of the test group of pupils with the teacher and the additional evaluation activities follow. Interviews the teacher holds with parents for verification purposes conclude the intervention.

Pupils are video-recorded continuously during all the activities. After class, the teacher watches the day's video, registers actions, interactions and performance and completes observation sheets concerning the pupils' interactions with the robot and with each other, their programming efforts and their motivation and participation levels on a daily basis. This systematic observation is multi-purpose: it serves as a feedback to the teacher for self-improvement (along the action research lines) and it therefore provides a formative evaluation of the intervention while at the same time it documents the pupils' progress on the cognitive and social axes and it therefore provides a final evaluation of the intervention and its outcomes. (*Detailed data on the qualitative analysis activities and results are omitted due to lack of space; they are available and can be provided by the authors on request*).

4   Evaluation Results

The pre- and post-test consists of a total of thirty six (36) questions or tasks. Results are shown comparatively for the pre- and the post-test, for the experimental group in Table 1 and for the control group in Table 2. The 'improvement' columns in both tables show that the overall impact of the intervention was positive: all thirteen (13) pupils of the experimental group did improve their scores, their progress ranging from 11% (4 tasks) to 42% (15 tasks) additional correct answers in the post-test. Comparably, the control group shows very limited improvement: 0 to 1 additional correct answers in the post-test.

Table 1. Pre-test versus Post-test results for the *experimental* group: correct answers over 36 questions (absolute values and percentages).

Pupil ID	Pre-test	Post-test	Improvement (Post – Pre)
A1	24 (66%)	31 (86%)	7 (20%)
A2	25 (69%)	33 (91%)	8 (22%)
A3	24 (66%)	34 (94%)	10 (28%)
A4	22 (61%)	32 (88%)	10 (27%)
A5	22 (61%)	34 (94%)	12 (33%)
A6	22 (61%)	27 (75%)	5 (14%)
A7	19 (52%)	30 (83%)	11 (31%)
A8	12 (33%)	27 (75%)	15 (42%)
A9	15 (42%)	24 (66%)	9 (25%)
A10	25 (69%)	30 (83%)	5 (14%)
A11	22 (61%)	31 (86%)	9 (25%)
A12	25 (69%)	29 (80%)	4 (11%)
A13	22 (61%)	30 (83%)	8 (22%)

Table 2. Pre-test versus Post-test results for the *control group*: correct answers over 36 questions (absolute values and percentages).

Pupil ID	Pre-test	Post-test	Improvement (Post − Pre)
E1	26 (72%)	27 (75%)	1 (3%)
E2	28 (77%)	29 (80%)	1 (3%)
E3	30 (83%)	31 (86%)	1 (3%)
E4	28 (77%)	28 (77%)	0 (0%)
E5	28 (77%)	29 (80%)	1 (3%)
E6	33 (91%)	33 (91%)	0 (0%)
E7	30 (83%)	31 (86%)	1 (3%)

Further comparative analysis of the results per Unit (CT skill) makes clear that the pupils' participation in the programming activities increased their understanding of causal relationships: in the post-test, correct answers in Unit 1 range from 8 to 10 over 10 questions in total. Moreover, in their verbal communication with the teacher during the tests, pupils show an increased ability to reason (explain the thoughts behind their choices) – even when choices are erroneous. They all try to justify their choices and in doing so they avoid random answers or irrational explanations.

For the second research question on the ability to formulate hypotheses, a quantitative look reveals very little improvement: correct answers in the post-test vary between 5 and 10 whereas in the pre-test they ranged between 4 and 9. However, a more qualitative approach, focusing on the verbal communication and the phrasing of the pupils' answers, reveals that there has been noticeable improvement in their use of full conditional sentences. Moreover, most pupils actually attempted to form more realistic and complex hypotheses. The majority refrained from answering "*I don't know*" – a common answer in the pre-test. For instance, in the question about the elephant and the lion, common pre-test answers included "*I don't know*" or "*They'll fight*" or "*They'll play*". In the post-test, however, the answers included "*If the elephant sees the lion, they may become friends and play together*" or "*If the lion is hungry, it will eat the elephant but if it can't, because the elephant is big, it will go away*". It is also worth mentioning that the pupils who demonstrated the greatest improvement (moving from 4 correct answers in the pre-test to 10 correct answers in the post-test) were the ones who worked together almost throughout the program and had the most vivid arguments while programming the floor robot: they usually disagreed on the right moves to be selected and applied to the robot and they were often heard saying things like: "*No, no! If you press the orange button, the robot will fall off the 'road'! If you press the purple one, it will take the right turn! Press the purple one!*".

For the third research question on sequencing, the progress of all pupils has been noticeable: they managed to sequence correctly 6–8 card sets (out of 8) in the post-test as opposed to 2–5 in the pre-test. It is noteworthy that even those pupils who in the pre-test had difficulty in sequencing short sets of 4 cards and either gave up or ordered the longer sets of 6 or 9 cards at random, were successful in the post-test: they managed to sequence correctly all the 'easy' sets and were also partially successful in identifying the correct order of longer sets. This skill is further illustrated by their ability to utilize Colby's coding cards efficiently towards the end of the intervention.

The lowest improvement is registered in the fourth research question on problem solving (6–8 correct answers in the post-test out of 8 tasks); this is due to the rather high scores in the pre-test, however (6–7 correct answers). Observation data on the kids' interactions reveal that they were daily faced with failure to guide the mouse to its prize. They consistently solved this problem by applying specific strategies: (a) they observed the route to spot the block where Colby strayed off-route, (b) they checked the order of their coding cards and spotted the erroneous one(s) and then replaced it accordingly, (c) they tried the new sequence of cards (commands), (d) they watched Colby execute the new sequence, ready to cheer at success or redo stages (a) to (c) until they got it right. In fact, these are the steps of the actual debugging process; it is therefore valid to claim that by mastering them, kindergartners demonstrate improved problem solving skills.

5 Conclusions – Further Research

On the basis of the data collected and analyzed during the educational intervention, it may be claimed that the systematic use of a programmable floor robot in a pedagogically suitable, playful way offers kindergarten pupils the opportunity to develop important aspects of their computational thinking skills, namely (i) to understand causality relations, (ii) to formulate and test hypotheses, (iii) to sequence objects, and (iv) to develop strategies for problem solving. Improvements in language skills, especially in the use of conditional sentences, and in social skills (communication and collaboration towards a common goal) were also observed. While high post-test scores attest to the cognitive benefits of the intervention, high scores in involvement and well-being, measured by the teacher to vary between 4 and 5 in the Leuven scale depending on the stage and the phase of the intervention, verify that all pupils did enjoy participating in the robotic activities. The positive attitude assumed towards programming a robot has been clearly expressed in the drawings that pupils made in class after the intervention; it has also been verified by their parents, who reported their behavior and relative exchange at home. Such findings are in good agreement to existing studies, e.g., [16–19, 21, 23, 29, 30], both as to the cognitive and as to the social outcomes. Clearly, further experimentation on a larger sample is needed in order (a) to draw conclusions that may be generalized and tested for statistical significance, and (b) to fully comprehend how programming a robot helps develop the computational thinking as well as the social/collaborative and the language/expression skills of kids in the kindergarten age.

References

1. Fisch, K., McLeod, S.: Did you know? https://www.youtube.com/watch?v=pMcfrLYDm2U. Accessed 10 Oct 2018
2. Papert, S.: Mindstorms: Children, Computers, and Powerful Ideas. Basic Books, New York (1980)
3. Wing, J.M.: Computational thinking. Commun. ACM **49**(3), 33–35 (2006)
4. Wing, J.M.: Research Notebook: Computational Thinking - What and Why? Carnegie Mellon University Magazine 'The Link', Pittsburg (2011)
5. Bocconi, S., Chioccariello, A., Dettori, G., Ferrari, A., Engelhardt, K.: Developing computational thinking in compulsory education – Implications for policy and practice. EUR 28295 EN (2016). http://www.eun.org/resources/detail?publicationID=861. Accessed 10 Oct 2018
6. Brennan, K., Resnick, M.: Using artifact-based interviews to study the development of computational thinking in interactive media design. In: Proceedings Annual American Educational Research Association meeting, Vancouver, BC, Canada (2012)
7. Lye, S.Y., Koh, J.H.L.: Review on teaching and learning of computational thinking through programming: what is next for K-12? Comput. Hum. Behav. **41**, 51–61 (2014)
8. Mannila, L., Dagiene, V., Demo, B., Grgurina, N., Mirolo, C., Rolandsson, L., Settle, A.: Computational thinking in K-9 education. In: Proceedings ITiCSE-WGR 2014, Uppsala, Sweden (2014)
9. Chuang, H.-C., Hu, C.-F., Wu, C.-C., Lin, Y.-T.: Computational thinking curriculum for K-12 education - a Delphi survey. In: Proceedings International Conference on Learning and Teaching in Computing and Engineering, Taipei, Taiwan. IEEE (2015)
10. Angeli, C., Voogt, J., Fluck, A., Webb, M., Cox, M., Malyn-Smith, J., Zagami, J.: A K-6 computational thinking curriculum framework: implications for teacher knowledge. Educ. Technol. Soc. **19**(3), 47–57 (2016)
11. Lockwood, J., Mooney, A.: Computational thinking in secondary education: where does it fit? A systematic literary review. Int. J. Comput. Sci. Educ. Sch. **2**(1), 41–60 (2018)
12. Yadav, A., Stephenson, C., Hong, H.: Computational thinking for teacher education. Commun. ACM **60**(4), 55–62 (2017)
13. Kotsopoulos, D., Floyd, L., Khan, S., Namukasa, J.K., Somanath, S., Weber, J., Yiu, C.: A pedagogical framework for computational thinking. Digit. Exp. Math. Educ. **3**(2), 154–171 (2017)
14. Yadav, A., Zhou, N., Mayfield, C., Hambrusch, S., Korb, J. T.: Introducing computational thinking in education courses. In: Proceedings of ACM Special Interest Group on Computer Science Education, Dallas, TX, USA (2011)
15. Scharf, F., Winkler, T., Herczeg, M.: Tangicons: algorithmic reasoning in a collaborative game for children in kindergarten and first class. In: Proceedings 7th International Conference on Interaction Design and Children (IDC 2008), Chicago, Illinois, USA, pp. 242–249. ACM (2008)
16. Stoeckelmayr, K., Tesar, M., Hofmann, A.: Kindergarten children programming robots: a first attempt. In: Proceedings 2nd International Conference on Robotics in Education (RiE 2011), Vienna, Austria, pp. 185–192 (2011)
17. Gordon, M., Rivera, E., Ackermann, E., Breazeal, C.: Designing a relational social robot toolkit for preschool children to explore computational concepts. In: Proceedings 14th International Conference on Interaction Design and Children (IDC 2015), pp. 355–358. ACM, New York (2015)

18. Kazakoff, E., Bers, M.U.: Programming in a robotics context in the kindergarten classroom: the impact on sequencing skills. J. Educ. Multimedia Hypermedia **21**(4), 371–391 (2012)
19. Bers, M.U., Flannery, E.L., Kazakoff, E.R., Sullivan, A.: Computational thinking and tinkering: exploration of an early childhood robotics curriculum. Comput. Educ. **72**, 145–157 (2014)
20. Sullivan, A., Bers, M.U.: Robotics in the early childhood classroom: learning outcomes from an 8-week robotics curriculum in pre-kindergarten through second grade. Int. J. Technol. Des. Educ. **26**(1), 3–20 (2015)
21. Elkin, M., Sullivan, A., Bers, M.U.: Programming with the KIBO robotics kit in preschool classrooms. Comput. Sch. **33**(3), 169–186 (2016)
22. Toh, L.P.E., Causo, A., Tzuo, P.W., Chen, I.M., Yeo, S.H.: A review on the use of robots in education and young children. Educ. Technol. Soc. **19**(2), 148–163 (2016)
23. Pugnali, A., Sullivan, A., Bers, M.U.: The impact of user interface on young children's computational thinking. J. Inf. Technol. Educ. Innovations Pract. **16**, 171–193 (2017)
24. João-Monteiro, M., Cristóvão-Morgado, R., Bulas-Cruz, M., Morgado, L.: A robot in kindergarten. In: Proceedings Eurologo'2003 - Re-inventing Technology on Education, Porto, Portugal (2003)
25. Highfield, K., Mulligan, J., Hedberg, J.: Early mathematics learning through exploration with programmable toys. In: Figueras, O., Cortina, J.L., Alatorre, S., Rojano, T., Sepulveda, A. (eds.) Proceedings Joint Meeting PME 32 and PME-Na, Morelia, Mexico, vol. 3, pp. 169–176. Cinvestav-UMSNH (2008)
26. Grover, S., Pea, R.: Computational thinking in K-12: a review of the state of the field. Educ. Res. **42**(1), 38–43 (2013)
27. Fessakis, G., Gouli, E., Mavroudi, E.: Problem solving by 5–6 year old kindergarten children in a computer programming environment: a case study. Comput. Educ. **63**, 87–97 (2013)
28. Strawhacker, A., Lee, M., Caine, C., Bers, M-U.: ScratchJr demo: a coding language for kindergarten. In: Proceedings 14th International Conference on Interaction Design and Children (IDC 2015), Boston, MA, USA. ACM (2015)
29. Highfield, K.: Robotic toys as a catalyst for mathematical problem solving. Aust. Primary Math. Classroom **15**(2), 22–27 (2010)
30. Kazakoff, E., Sullivan, A., Bers, M.: The effect of a classroom-based intensive robotics and programming workshop on sequencing ability in early childhood. Early Childhood Educ. J. **41**(4), 245–255 (2013)
31. Komis, V., Misirli A.: Étude des processus de construction d'algorithmes et de programmes par les petits enfants à l'aide de jouets programmables. Dans Sciences et technologies de l'information et de la communication (STIC) en milieu éducatif: Objets et méthodes d'enseignement et d'apprentissage, de la maternelle à l'université. Clermont-Ferrand, France (2013)
32. Ohlson, T., Monroe-Ossi, H., Fountain, C., McLemore, B., Carlson, D., Wehry, S.: Exploring programming and robotics in early childhood classrooms. In: Proceedings International Society for Technology in Education, Denver, Colorado, USA. ISTE (2016)
33. Cohen, L., Manion, L., Morrison, K.: Research Methods in Education. Routledge, New York (2007)
34. Laevers, F. (ed.): Well-being and Involvement in Care Settings. A Process-oriented Self-evaluation Instrument. Research Centre for Experiential Education, Leuven University, Belgium (2011)
35. Dewey, J.: Experience and education. In: Garforth, F.W. (ed.) John Dewey: Selected Educational Writings. Heinemann, London (1966)
36. Vygotsky, L.S.: Mind in Society: The Development of Higher Psychological Processes. Harvard University Press, Cambridge (1978)

"CREA": An Inquiry-Based Methodology to Teach Robotics to Children

Maria Blancas[1,4], Cristina Valero[1], Anna Mura[1], Vasiliki Vouloutsi[1], and Paul F. M. J. Verschure[1,2,3(✉)]

[1] Synthetic Perceptive Emotive Cognitive Systems (SPECS) Group,
Institute for Bioengineering of Catalonia (IBEC),
Eduard Maristany Av., 10-14, Barcelona, Spain
{mblancas, cvalero, amura, vvouloutsi,
pverschure}@ibecbarcelona.eu
[2] Catalan Institution for Research and Advanced Studies (ICREA),
Lluis Companys Av., 23, Barcelona, Spain
[3] Barcelona Institute of Science and Technology (BIST), Barcelona, Spain
[4] Pompeu Fabra University (UPF), Roc Boronat St., 138, Barcelona, Spain

Abstract. Learning programming and robotics offers the opportunity to practice problem-solving, creativity, and team-work and it provides important competencies to train for the 21st century. However, programming can be challenging, and children may encounter difficulties in learning the syntax or using the coding environment. To address this issue, we have developed a methodology for teaching programming, design and robotics based on inquiry-based learning and hands-on oriented activities together with visual programming. We have applied and evaluated this new methodology within the extracurricular activity of an international elementary school in Barcelona. Our findings showed acquisition and learning of technical language, understanding of electronics devices, understanding the mapping of coding into action via the robot's behavior. This suggests that our approach is a valid and effective teaching methodology for the instructional design of robotics and programming.

Keywords: Educational technology · Instructional design · Robotics

1 Introduction

Teaching computer science (CS) and robotics in primary school should have the goal to provide students with learning tools and skills other than programming or building a robot, to enable them practicing problem-solving, creativity and team-work [1]. This would allow students to reflect on their solutions [2], make confident decisions and develop social, emotional and cognitive skills [3]. Thus, these learning tools are consistent with the so-called Four Cs of the 21st Century skills: Communication, Critical thinking, Collaboration, and Creativity. It is not surprising, then, that CS and robotics are gaining popularity in school curricula [4–6]. However, programming can be complicated for young students [7], as they may face difficulties in learning the

---

M. Blancas and C. Valero—These authors contributed equally to the study.

M. Merdan et al. (Eds.): RiE 2019, AISC 1023, pp. 45–51, 2020.
https://doi.org/10.1007/978-3-030-26945-6_4

syntaxes, coding environments and commands of a programming language [8]. Learning Computational Thinking (CT), defined as the set of thinking skills and approaches essential to solving complex problems using computers, before being introduced to formal programming, would help ease students into the formal languages [9, 10]. Additionally, using tangible environments may help to bypass the need for familiarity with computers [11] and get rid of distraction derived from learning how to use them.

During our experience with primary school children (10–12 years old), we have identified several critical points that need optimization to improve teaching robotics/coding courses. Here we address three of them: (1) excessive focus on theory, (2) preference for personalized hands-on and discovery-oriented projects, and (3) dependency on formal programming languages instead of CT. We thus propose "CREA" as a methodology for teaching CS to this age-group.

2   The Framework

Our primary focus is that students learn how to build and program a robot that performs a certain task. Our methodology, "CREA" (accounting for "Coding Robots through Exploring their Affordances"), proposes collaborative inquiry-based activities for learning CS and robotics where the students are at the center of the learning process and where the teacher's role is to guide rather than to provide information. To do so, the process of building and programming a robot can be decomposed into smaller, manageable components. These components, which are shown to benefit students when learning to program [12], make use of visual representations, require verbal explanations and promote discovering on their own. In the following sections, we present the technology and the robotic platform employed in our study, as well as a description of the methodology employed to teach young children how to build and program a robot.

2.1   Tools

The basic component of our robotic platform is an Arduino Uno[1] with a Grove[2] board (for safety and ease of use) on top of which children can connect a variety of sensors and actuators, such as potentiometers, buttons, light sensors and LEDs (Fig. 1(1–6)). To program the Arduino, students used Visualino, a multiplatform visual programming environment similar to Scratch. Visualino not only provides graphical programming blocks that can create a program when assembled (Fig. 1(5)) and promote tinkering but also generates native code for the Arduino, thus no computer is required for the Arduino to work (https://www.arduino.cc/).

Students were also provided with worksheets to support some sessions. The purpose of these worksheets was for children to write down the robot's behavior and later use them to identify the specific components of the robot. After that, these worksheets were used as bases for the robot's programming. In each session, children received a different robot (and its related worksheet, filled in by their peers) than the one they

---

[1] https://www.arduino.cc/

[2] http://wiki.seeedstudio.com/Grove_System/

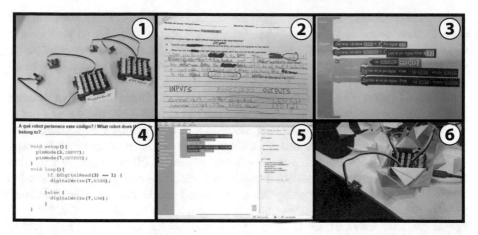

Fig. 1. Examples of the six sessions of "CREA". (1) Robot Observation (RO), (2) I Instruction Analysis (IA), (3) Paper Blocks (PB), (4) Arduino Pairing (AP), (5) Code in Visualino (CV), (6) Robot Construction (RC).

worked with the previous session. This exchange was to make sure all the groups get to know all the robots used and to promote collaboration between groups.

2.2 "CREA" - The Methodology to Teach Programming to Children

Our Program consisted of six sessions (units) lasting one hour and a half each. In the **first session** (*Robot Observation*, RO), children formed groups of two or three people and remained in the same group throughout all sessions. Here, each group was randomly introduced to a robotic platform and was asked to observe it, interact with it, identify its behaviors and write them down in a worksheet. The first session served as an introduction to the robots and the following sessions guided the students from initial robotic components observations to the programming and construction of the robot. Thus, students were not only presented with a possible outcome (a fully built robot) before engaging in the task, but they also could decompose it in discrete parts and reflect on how these parts worked in isolation.

For the **second session** (*Identification and Analysis*, IA), The aim was to identify the key elements that compose a robot (inputs, outputs, and pins) and to relate them to the components of programming (functions). To do that, students were organized in groups and each group's worksheet was transcribed in a printed version and randomly assigned to another group. Students were then asked to analyze the information on the sheets and to identify and classify the inputs, outputs, functions, variables, and pins (the position of the board where an item is connected) and color-code them similarly to the colors used in Visualino (Fig. 1(2)). Finally, students were asked to relate the inputs with the corresponding functions and outputs (for example: Button -input- → when pressed, turns on -function- → LED -output-).

In the **third session** (*Paper Blocks*, PB), the worksheets generated in the previous session were again randomly distributed to the groups along with a paper version of the blocks used in Visualino to program each robot. Groups had to first read the instructions from the worksheets, find the corresponding blocks of Visualino and connect them accordingly, almost like assembling a puzzle (Fig. 1(3)). This session allowed students to familiarize with the Visualino blocks and the programming units, without being exposed to the actual interface. At the end of sessions IA and PB, groups received feedback on their work separately.

In **session four** (*Arduino Pairing*, AP), students familiarize with the software interface. The groups were given papers with code in C (Fig. 1(4)) and groups had to pair each paper with the corresponding robot. In this way, students are introduced to formal programming and how code is translated into action or behavior. Afterward, each group was given a random picture of the assignment of PB (an image of Visualino blocks) and were asked to translate it into actual Visualino code. In the **fifth session** (*Code in Visualino*, CV), each group chose a robot and decide on a new design and task to be implemented in the robot, by making changes in the Visualino code. This is in agreement with previous work showing that the ability to choose one's project increases the motivation to perform a task [13]. Finally, in the **last session** (*Robot Construction*, RC), each group applied what they had learned in the previous sessions to program their proposed robot.

3 Methodology

We applied our methodology in an extracurricular course of "Robotics and Programming" in an international elementary school in Barcelona. Our participants, the whole class, consisted of 10 students with 9 boys and 1 girl (age: 11.11 ± 0.60). To evaluate the effectiveness of our methodology, before the first session the students were given a pre-test and at the end of the last session a post-test for their knowledge about CS and robotics. Both tests contained 12 questions of multiple-choice (four answer options) regarding hardware and coding elements that would be covered during the sessions: inputs and outputs used with the Arduino board and coding with Visualino. Additionally, at the end of the last session, we performed a semi-structured interview with each student. The goal of the interview was to acquire insights on how the students evaluated the class. More specifically, we presented the students with images (see Fig. 1) that depicted each of the six sessions and asked them to evaluate them based on the following criteria: likeability (if they enjoyed each session), perceived difficulty, if they would do it again, if they worked alone or in collaboration and if they would remove any part. The evaluation was done by ordering each image according to each question (1–6 scale). The perceived difficulty of the exercises served not only for the students to assess their performance, but also identify what they found challenging [14]. All content of the class was presented using the native language of the students (Spanish).

4 Results

To quantify the learning progress between the different sessions, we used a Wilcoxon signed-rank test. Our results show improvement in programming and robotics-related knowledge after each session ($p = 0.016$), with a mean performance of $33.33\% \pm 9.32$

in the pre-test and a mean performance of 52.08% ± 13.01 in the post-test. Regarding the short interview, children highlighted the discovery-oriented approach of the RO session; additionally, they reported enjoying guessing the robot's behavior. Children's perception of the IA session was conditioned by their writing skills. Nonetheless, it is worth noting that in their descriptions they have used most technical words learned during the session.

The PB session seemed to generate confusion, as the setup could be easily dismantled. Here, we found a negative correlation between how much the students enjoyed the session (likeability) and the perceived difficulty ($r = -0.93, p < 0.001$). AP was well received by the students, as it allowed them to move freely around the class and promoted exploratory behavior. However, students encountered difficulties in understanding the Arduino code. Students also enjoyed the CV session as it allowed them to use the computer. Finally, the last session (RC), had the highest score in likeability as it involved crafting, creativity and allowed them to choose their project. A summary of children's evaluation of the sessions is provided in Table 1 (Fig. 2).

Table 1. RO: Robot Observation, IA: Instruction Analysis, PB: Paper blocks, AP: Arduino Pairing, CV: Code in Visualino, RC: Robot Construction. Perc. Diff.: Perceived Difficulty, Again: if they would do the task again. Asterisk represents significant correlation between likeability and perceived difficulty.

Unit	Likeability (1–6)	Perc. Diff. (1–6)	Again (%)
1. RO	4.28 ± 0.75	4.39 ± 1.36	100
2. IA	2.50 ± 1.37	3.00 ± 1.22	22.22
3. PB*	2.54 ± 1.59	2.55 ± 1.67	44.44
4. AP	3.72 ± 1.30	3.39 ± 1.11	66.66
5. CV	2.83 ± 1.12	3.11 ± 1.96	77.78
6. RC	5.22 ± 0.97	4.56 ± 1.67	100

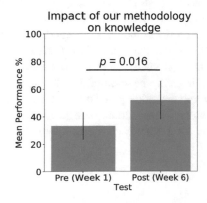

Fig. 2. Mean performance at the pre- and post-tests represent improvement in programming and robotics knowledge after the intervention. Lines represent standard deviation.

5 Discussion and Conclusion

This study proposes a methodology, "CREA", to teach programming and robotics to primary school children and it consists of six learning stages that aim at scaffolding the students' learning process. We applied and evaluated the proposed methodology in an extracurricular course for primary school students (10–12 years old). Our results show an increase in performance between the first stage and the last stage of the course suggesting that our method was effective as a teaching strategy. Nevertheless, due to the relatively small sample and having no control condition to compare with, we cannot make any general claims.

Regarding the students' evaluations of the learning activities in the course (programming, design and building), children enjoyed more those that involved crafting, creativity and the usage of computers, as well as those that promoted teamwork. When referring to the RO and IA activities, children used verbs like "discover" and "guess"; however, they found not interesting or more challenging the writing part of the IA. When comparing programming using paper blocks or the computers, opinions were divided: some children preferred the constrained possibilities offered by paper blocks while others enjoyed exploring their options using the Visualino software. For future versions of the methodology we will consider a larger and harder version of the pieces in the PB part (even stickers or magnets), so they are easier to assemble.

In comparison, other methodological studies to teach CS and robotics to an age range similar to ours include more (up to 28 sessions in [15]) or longer sessions (up to 150 min in [16]). In the first case, the results show a significant increase in knowledge on Science Engineering technical concepts in the experimental condition compared to a control one. In the second case, results suggest a variability among students' conceptual development. In other cases, though, learners were older than in our study (for example, between 10 and 15 years old [17]). Moreover, both of these studies contained sessions focused on learning theory first, something we have avoided to follow an inquiry-based learning approach.

Overall, our results suggest that our inquiry-based methodology can provide the first steps to effectively teaching instructional design of robotics and programming. For example, as children seemed to prefer crafting-related activities and computer-mediated environments, the course activities could be more hands-on oriented but include the use of computers. Additionally, given that students reported having difficulties in understanding the Arduino syntax (C code), to propose alternative ways to teach CS, for instance visual programming may be preferable for such age groups. In our future steps, we will strengthen the methodology to use it in robotics with more components and more complex behaviors. More studies will be needed to validate our methodology, including controlling for factors such as multilanguage background, gender and different learning needs.

References

1. Chaudhary, V., Agrawal, V., Sureka, P., Sureka, A.: An experience report on teaching programming and computational thinking to elementary level children using lego robotics education kit. In: Proceedings - IEEE 8th International Conference on Technology for Education, T4E 2016, pp. 38–41 (2017)
2. Kalelioğlu, F.: A new way of teaching programming skills to K-12 students: Code.org. Comput. Hum. Behav. **52**, 200–210 (2015)
3. Bers, M.U.: Coding as a Playground. Routledge, Abingdon (2017)
4. Petre, M., Price, B.: Using Robotics to Motivate 'Back Door' Learning. Kluwer Academic Publishers, Dordrecht (2004)
5. Tanja, K., et al.: Generation NXT: building young engineers with LEGOs. IEEE Trans. Educ. **53**(1), 80–87 (2010)
6. Varney, M.W., Janoudi, A., Aslam, D.M., Graham, D.: Building young engineers: TASEM for third graders in woodcreek magnet elementary school. IEEE Trans. Educ. **55**, 78–82 (2012)
7. Kelleher, C., Pausch, R.: Lowering the barriers to programming: a survey of programming environments and languages for novice programmers. ACM Comput. Surv. (CSUR) **37**(2), 83–137 (2003)
8. Cockburn, A., Bryant, A.: Leogo: an equal opportunity user interface for programming. J. Vis. Lang. Comput. **8**(5–6), 601–619 (1997)
9. Wing, J.M.: Computational thinking and thinking about computing. Philos. Trans. R. Soc. A Math. Phys. Eng. Sci. **366**(1881), 3717–3725 (2008)
10. Kazimoglu, C., Kiernan, M., Bacon, L., MacKinnon, L.: Understanding computational thinking before programming. Int. J. Game-Based Learn. **1**(3), 30–52 (2011)
11. Smith, A.C.: Using magnets in physical blocks that behave as programming objects. In: Proceedings of the 1st International Conference on Tangible Embedded Interaction - TEI 2007, p. 147 (2007)
12. Zhang, J.X., Liu, L., de Pablos, P.O., She, J.H.: The auxiliary role of information technology in teaching: enhancing programming course using alice. Int. J. Eng. Educ. **30**(3), 560–565 (2014)
13. Radenski, A.: Freedom of choice as motivational factor for active learning. ACM SIGCSE Bull. **41**(3), 21 (2009)
14. Guzdial, M.: Paving the way for computational thinking. Commun. ACM **51**(8), 25 (2008)
15. Barker, B.S., Ansorge, J.: Robotics as means to increase achievement scores in an informal learning environment robotics as means to increase achievement scores in an informal learning environment. J. Res. Technol. Educ. **1523**(June), 229–243 (2017)
16. Chambers, J., Carbonaro, M., Murray, H.: Developing conceptual understanding of mechanical advantage through the use of Lego robotic technology. Australas. J. Educ. Technol. **24**(4), 387–401 (2008)
17. Karahoca, D., Karahoca, A., Uzunboylu, H.: Robotics teaching in primary school education by project based learning for supporting science and technology courses. Procedia Comput. Sci. **3**, 1425–1431 (2011)

Educational Robotics in Kindergarten, a Case Study

Garyfalia Mantzanidou[✉]

Kindergarten of Drepano, Achaia, Greece
falydemi@gmail.com

Abstract. There has been considerable interest recently in trying to address the future with regards to STEAM (Science, Technology, Engineering, Arts and Mathematics). STEAM has been proved to be highly beneficial for the students as a means of development of essential skills and aptitudes, enhancement of their learning process and improvement in problem-solving.

In this work we examine a case study and the activities that took place in order to study the use of the educational robot in teaching History as well as the learning of music using the robot within the framework of the Erasmus+ KA2 program entitled "Be a Master think Creatively".

Through the use of STEAM, the students manage to express themselves creatively, they were offered a "playground" on which to experiment on, clear signs of inspiration can be identified, and they take advantage of opportunities for collaboration that open and are offered. By integrating STEAM in a pleasing and fulfilling manner, the means are offered for escaping more traditional teaching methods. Overall, they were led to learning through action and interaction.

Keywords: STEAM · Educational robotics · BEEBOT · Erasmus · Kindergarten

1 Introduction

Robotics is a relatively new scientific branch—especially in Greece—that has recently become pretty widespread. Robotics supports STEAM education from kindergarten to high school according to Benniti's study (2012).

Educational robotics is an educational tool that arouses the interest and the curiosity of the students using enjoyable activities in an attractive learning environment (Eguchi 2010).

As a pedagogical approach it is subsumed within the framework of constructivism (Piaget 1972), or rather according to Papert (Papert 1993) within constructionism. Piaget (Piaget 1974) supports that when children handle objects is given them the potential to construct their knowledge. Constructivist approach about learning advocates that the learning environment should provide original activities as a part of the open-ended problem-solving process of the real world, to encourage expression and personal involvement in the learning process and to support social interaction. The objective is to adequately supply the children with the appropriate objects in order to learn in practice more than before (Papert 1980). Its main tool is the construction and the function of a

© Springer Nature Switzerland AG 2020
M. Merdan et al. (Eds.): RiE 2019, AISC 1023, pp. 52–58, 2020.
https://doi.org/10.1007/978-3-030-26945-6_5

robot and its aim is to instill technological literacy, problem-solving skills and creativity into students already in their first education stages. Moreover, the activities that are selected by the pedagogues have to face the meaningful problems and setup real-world connections (Chalmers 2012) because this way the students are motivated to be involved and to promote social interaction. McNamara considers that the existing robotic systems are designed to develop together with the students (McNamara 2014).

Robotics provides students with a unique learning opportunity to design, build and program an important construction (Bers et al. 2002). Furthermore, they have the potential to learn in practice and to improve their development of metacognitive ability. Programming leads students to think independently.

The engagement with ER in the classroom was an inducement as it is widely accepted that children are interested in robots. ER manages to combine learning with play, and thus education easily turns into an entertaining activity.

The chance was given by participating in the Erasmus+ KA2 program entitled Be a Master think Creatively. The school was responsible for organizing History and Music activities and used an educational robot to this end.

The aim was to compare the performances of the Kindergarten pupils in these two domains before and after the activities with the educational robot.

Hypothesis 1. The performance of the pupils has improved in the post-test stage compared to those of the pre-test stage after the use of the educational robot for the learning of History.

Hypothesis 2. The performance of the pupils has improved after the use of the educational robot for learning musical notes and the tutelage of the robot in the traditional song "Feggaraki".

1.1 BeeBot

BEEBOTS follow the principles of the LOGO programming language and are user-controlled for setting the execution of motion commands. As Pekárová (Pekárová 2008) argues, this game has been awarded as the most impressive material in the global market of educational technology.

They are programmed using colored keys located on them and they can also be programmed in order to move forward, backwards, turning left and right. BeeBot is ideal for teaching directionality, distance, sequencing, understanding of algorithms and their application as programs in digital devices, as well as for programming language for young children. Young students using BEEBOT are driven to solve open-ended problems that require modeling, discovery and correction of any program problems, comprehension and use of algorithms in a playful way.

Research has shown the benefits of using Bee-Bot to children, as it can help them to develop not only literacy and numeracy skills, but also math, geography, science, and even history skills (Pekárová 2008).

1.2 School Background

The activities described in this part took place in an urban, public school in Patra, Greece, serving students in prekindergarten through Kindergarten grade. In the project participated 18 pupils 12 boys and 6 girls and a teacher. All the students were Greeks. Neither the students nor the teacher had been previously exposed to the BEEBOT robotics kit. Participation and engagement with New Technologies are boosted in the day-to-day school program.

1.3 Educational Approaches and Methodology

The activities that were carried out were integrated into the Interdisciplinary Framework of Educational Programs of the Kindergarten. The Experiential, Collaborative Learning and Interdisciplinary pedagogical method was used. Each participant group was small and the BEEBOT was used as a learning support tool in a research-based experimental way and in contact with educational robotics (learning by doing).

1.4 Teaching Experiment

As a methodology, it develops the need to study the learning process itself rather than just its results. For pedagogues, the teaching experiment is research in the classroom and, by extension, research of the learning process to improve learning in the classroom itself (Czarnocha 2006). The "Teaching Experiment" in the classroom is an experimental framework where the subject involved in the research finds itself in a structured environment, which can change according to the situation that has a specific orientation and aim (Chronaki 2008).

2 Experimental Programming Activities

Before the robot was presented, a conversation by teacher was held to discover children's previous opinions/knowledge.

What is a robot and what it is not: Students were asked to answer what they think that a robot is and why they have that opinion.

The following activities included programming concepts such as sequence and repetition. In particular, the children were asked to instruct a classmate to arrive in the morning at school. In practice, they found out that when they had not given the right instructions, they guided their classmate to reach school sometimes without shoes, thus they had to "reprogram" them (ICT, Algorithmic and Mathematical Thinking). The students were taught the basic concepts of the programming language in a playful way, their interest and curiosity were stimulated and established correlations between programming and everyday life. (Time taken 30 min).

2.1 Meeting BeeBot

The aim of BeeBot is to stimulate pupils' interest and to familiarize them with its operation in order to be introduced to robotics through play.

The robot was presented to the students, their views on its operation mode were discussed, and the students played a pantomime game during which they had to imitate the robot's movements. (Time taken 15 min).

Afterwards the students became acquainted and experimented with the BEEBOT symbols/keys in order to learn how to give the correct commands, properly program the robot and understand its steady pace (15-centimeter range of motion). They then freely experimented on a ready-made BEEBOT track mat. (Time taken 30 min). Twisting the BEEBOT constituted the biggest challenge as they had to understand that "turn right or left" means that it literally turns right or left and does not move forward or sideways (learning from experimentation and mistakes). To help them, they prior had plotted the route on paper to record their thoughts and then in practice to see if they understood the route they planned (see Fig. 1).

Fig. 1. Plot the route on paper.

Devoting time to explore the robot's function, the children will be able to solve problems through programming solutions by recording commands and correct sequence of commands (Algorithms). It helps students to understand that algorithms are the appropriate steps to make something happen (Language, ICT, Mathematics). The learning objectives are not only the students' understanding of robotic systems and of the concept of programming but also their learning how to run a program. Optimal route.

2.2   STEM Activities with the Traditional Song Feggaraki

The traditional song was an opportunity to explain children the Greek Folklore Tradition as the children's song has some common ground with the popular legend about the "krifo scholio". According to some Greek scholars' writings after the Greek War of Independence in 1821, Krifo Scholio was supposed to be an underground school. The oral tradition also says that the children were marching and singing that song in the night while going to the underground school that was forbidden by the Ottoman law. It is considered that the aforementioned picture helped children to raise awareness of Greek language (Terzopoulou 2007).

The students sang the song and put their own interpretation on the words of the song but most of them gave the same interpretation with the popular legend. They also played pantomime game and mimicked the children going to "krifo scholio" (Hidden school) having the moon as a guide. Then the teacher proposed them to try to involve

BEEBOT (the students called it Mellou) in this game. The children suggested creating their own floor/track for Mellou using the history of the song. (Time taken 15 min).The BEEBOT walked provided that they would have given it the right commands (understanding the correct sequence of commands). (Language, Environment, Creation-Expression, Mathematics, ICT).

2.3   Creation of Floor/Track Inspired by the Krifo Scholio Story

Having been helped by the teacher, the pupils designed the first square of the track. Next, they created alone the entire track that was easy for them, in which they glued images that had chosen online while searching about Krifo Scholio. (Shapes).

2.4   Routes

The students attempted to lead Mellou to the Krifo Scholio trying to use the right instructions, choosing the most appropriate route and avoiding dangers and obstacles. Each student chose their own route while trying to give the right instructions. Pupils cooperated and helped each other. After some attempts and having understood the orientation and the right-to-left rotation by giving the right commands, the children managed to properly program the robot (understanding of programming, sequence and prediction that lead to the development of problem-solving skills and improving their critical thinking (Time taken 45 min). The learning goals are the students' familiarization with the robotic systems, their understanding of the concept of programming, running the program and choosing the best route (see Fig. 2).

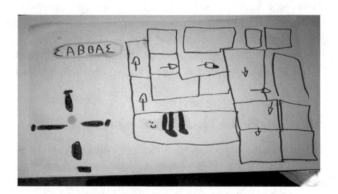

Fig. 2. Plot the route on paper.

2.5   Playing Music with Boomwhackers

The Boomwhackers are colorful plastic sound tubes that help children to learn the pentagram easily. These tubes are tuned and durable and each of them produces a sound that corresponds to a note. The sound is produced by hitting the tube at a fixed piece of equipment such as a chair, floor, etc. The students learn the notes easily and play rhythms and melodies when a song is combined with Boomwhackers tubes. They also learn how to make and feel music through play.

In these activities the musical tubes were used to introduce students into the musical instruments, to learn the notes, to improve their memory, to learn coding to understand algorithms. (Creation – Expression, Language, Mathematics).

Due to the colors the students learned easily not only the musical notes but also the notes of the song so they sang Feggaraki using the notes.Then they played the song with Boomwhackers. (Time taken 20 min).Each student held their tube and tried to play their note in the correct order (sequence).

2.6   Creation of Coded Track and Routes with BEEBOT Following the Notes of the Song

The pupils created a coded floor with the notes in the colors of the sound tubes of "Feggaraki" song, the choice of the layout of the colors with the notes was upon pupils. While the students were singing, they had to show the corresponding notes on the coded floor. They drew the notes of the song according to the colors of the musical tubes in the interactive board. They designed the route that BEEBOT would follow according to the notes of the song on paper and then programmed the robot to move on the floor following the notes (Music, Educational Robotics). (Time taken 40 min).

The students developed their imagination, creativity, critical thinking and learned how to direct the robot into the notes of the song (ICT, Algorithmic thought, Mathematics, Music).

3   Discussion - Conclusions

The pre-test and post-test results illustrate the students' learning performance when using educational robots for learning History and Musical Notes.

The results of post-test skills of children show significant improvements in childrens' performances in the area of learning History using the robot. It was found that the students' interest was undiminished and the integration of the educational robotics in the teaching of History helped to improve their knowledge, skills and abilities. During the learning procedure, the pupils applied and tested programming within a creative and enjoyable environment that promotes both mathematical skills and computational thinking. The results of post test skills established that using educational robots for learning music can be an effective tool not only for the development of students' knowledge and skills, but also for their facilitation with the notes while playing the song in a motivational and enthusiastic way within the learning process. In summary, having a well-designed educational activity based on modern learning theories, turns the educational robot into a practical educational tool in the kindergarten classroom.

BEEBOT is ideal for the students' first contact with the world of educational robotics and constitutes the perfect means to show students, starting with the kindergarteners, that engineering and computer science can be very entertaining. Working in small groups is ideal for students to get in touch with robotics in a playful way. At first, the students encountered problems to understand the directional change but soon realized that they are in control of the robot's movements. The pupils expressed themselves freely, developed their opinions, were actively involved, received visual feedback and improved their imagination and creativity. They comprehended mathematical concepts, developed their motor skills and computational thinking.

While programming the robot, the children came into contact with technology, science, mathematics and engineering (STEAM) and at the same time they are expressing themselves creatively.

References

Benniti, F.B.V.: Exploring the educational potential of robotics in schools: a systematic review. Comput. Educ. **58**, 978–988 (2012)

Bers, M., Ponte, I., Juelich, C., Viera, A., Schenkerm, J.: Teachers as designers: integrating robotics in early childhood education. Inf. Technol. Child. Educ. Annu. **2002**(1), 123–145 (2002)

http://makepuppet.org/stem/research/item1_earlychildhood_designcourse_BersITCE.pdf

Chalmers, C., Chandra, V., Hudson, S.M., Hudson, P.B.: Preservice teachers teaching technology with robotics (2012). https://eprints.qut.edu.au/51472/2/51472.pdf

Chronaki, A.: The teaching experiment. Studying learning and teaching process. In: Svolopoulos, B. (ed.) Connection of Educational Research and Practice, Athens, Atrapos (2008)

Czarnocha, B., Prabhu, V.: Teaching-Research and Design Experiment: Two Methodologies of Integrating Research and Classroom Practice (2006)

Eguchi, A.: What is educational robotics? Theories behind it and practical implementation. In: Society for Information Technology and Teacher Education International Conference, San Diego, USA (2010)

McNamara, S., Cyr, M., Rogers, C., Bratzel, B.: LEGO brick sculptures and robotics in education. In: Proceedings of the ASEE Annual Conference. Session #3353, 1999 (2014)

Papert, S.: Mindstorms – Children, Computers and Powerful Ideas. Basic Books, New York (1980). http://www.papert.org/articles/SituatingConstructionism.html

Papert, S.: The Children's Machine – Rethinking School in the Age of the Computer. Basic Books, New York (1993)

Piaget, J.: The Principles of Genetic Epistemology (Trans. W. Mays) Routledge and Kegan Paul, London (1972)

Piaget, J., Inhelder, B.: The Child's Construction of Quantities. Routledge and Kegan Paul (1974)

Pekárová, J.: Using a Programmable Toy at Preschool Age: Why and How? (2008)

Terzopoulou, M.: Discussing Europe on the Banks of Maritza River: Thrace as a European Cultural Space (2007)

Comparison of LEGO WeDo 2.0 Robotic Models in Two Different Grades of Elementary School

Michaela Veselovská[⊠], Zuzana Kubincová, and Karolína Mayerová

Faculty of Mathematics, Physics and Informatics,
Comenius University in Bratislava, Mlynská dolina, 84248 Bratislava, Slovakia
{veselovska, kubincova, mayerova}@fmph. uniba. sk

Abstract. In this paper, we are dealing with the implementation of activities with the LEGO WeDo 2.0 robotic kit into primary education. In particular, we are reporting on an introductory activity for third and fourth grade pupils. On the basis of qualitative data analysis, we have discovered several differences in the creation of their first robotic models when comparing pupils in different grades. We have identified differences in terms of building, programming and presenting models. We plan to compare our findings from the introductory activity and also other activities with the findings of the Taiwanese researchers, with whom we collaborate in testing the activities with the robotic kit. We plan to compare the results of the analysis from a different cultural point of view and adapt the activities to both national curricula so that they can be part of the ordinary primary education.

Keywords: Educational robotics · Children · Programming education · LEGO WeDo 2.0

1 Introduction

Constructivism [1, 2], constructionism [3] and social constructivism [4] are well known and accepted learning theories. For the implementation of their methods in teaching, activities using robotic kits are very well suited [5]. Typically, they are designed as project-oriented activities in which pupils work in smaller teams where they jointly build a robotic model and program its behavior. In addition to learning how to apply the knowledge and skills acquired in other subjects, interpersonal competences, teamwork skills, communication skills, and so on, are developed in pupils during their collaborative work. The educational robotics is challenging for pupils [6]. Its advantage is also in its suitability to be used in various subjects [7–9]. As we consider the educational robotics as a very promising area of education since the first grade of elementary school, we started introducing activities with the LEGO WeDo robotic kit to the teaching already several years ago [10–13]. The results were very encouraging. In the meantime, a new version of this robotic kit featuring several new characteristics has been issued. The differences prevent using the created activities with this new robotic kit without any modifications. That is why we are currently developing a new

© Springer Nature Switzerland AG 2020
M. Merdan et al. (Eds.): RiE 2019, AISC 1023, pp. 59–64, 2020.
https://doi.org/10.1007/978-3-030-26945-6_6

series of LEGO WeDo 2.0 activities, so far for elementary school. As part of a joint project with the National Tsing Hua University in Hsinchu, Taiwan, we plan to validate these activities in both countries. In the pilot study conducted in this project, we tested an introductory activity with a small sample of children [14]. Based on the acquired results, we have modified the activity to make it easy to use in the classroom.

A few weeks ago we started testing of a series of LEGO WeDo 2.0 activities in the classroom. This paper presents the results of the initial activity testing. Its aim is to describe the differences in the process of creation of first own robotic models by pupils in third and fourth grade of elementary school who did not have any previous experience of programming and the design of robotic models.

2 Methodology

In our research, we asked the following research question: *What are the differences between 3rd and 4th grade elementary school pupils when creating their first robotic models built using their fantasy with the LEGO WeDo 2.0 robotic kit?*

We used various qualitative methods to collect and analyze the data [15], including unstructured observation and field notes method, semi-structured interviews, creating and analyzing videos of the working procedure of each pair of pupils as well as of the whole class, and creating photos of the work of each pair of pupils.

The activity was tested during the regular teaching of Informatics which is a compulsory subject at elementary schools in Slovakia. The testing was conducted at a private elementary school in Bratislava in one third grade class and one fourth grade class (pupils aged eight to ten years). Besides pupils and their teacher, there were three researchers in the classroom, who were collecting data. The third grade pupils were taught by one of the researchers, while their regular teacher supervised the course. The fourth grade pupils were taught by their regular informatics teacher, and three researchers were collecting data. The teachers were provided a methodical material in the form of a website, which was created by the researcher who managed the activity testing in the third grade class. In the third grade class, there were six pupils (one girl and five boys) and in the fourth grade class there were seven pupils (one girl and six boys). Pupils worked in pairs during the activity, but in the fourth grade class, the girl worked alone. She has had previous experience with working with the robotic kit, so her work will not be taken into account in this study.

3 Hello Robot – First Activity with LEGO WeDo 2.0

3.1 Activity Goals

Our goal was to implement the activities primarily in a constructionistic way. However, based on the results of the pilot study [14], we have decided to include also instructional elements of teaching already in the first activity. The initial activity dealt with the concept robot and took one teaching hour (45 min).

The goal of this activity was, in the course of an interview, to organize the current informal pupils' knowledge about robots in certain way. The findings from similar activities carried out with the previous version of the LEGO WeDo kit were published in our previous paper [12]. Another goal of this activity was to familiarize the pupils with the LEGO WeDo 2.0 robotic kit, and to contribute not only to the development of fine motor skills but also to teamwork and communication skills, by collaborative building and programming a model.

3.2   The Course of the Activity

The opening activity was divided into four parts. In the first part, the teacher conducted an interview with the pupils in order to find out their previous experience and their notion of robots. The teacher asked them questions such as: *Do you know the word robot? What do the robots consist of? What do they need to be able to move? Can they think by themselves? Can they be good or bad?*

The next part contained instructive elements of teaching. The teacher showed the pupils some electronic and electro-mechanical components like a hub and an engine. She showed the pupils how to connect the engine to the hub and she also showed the connection of two LEGO parts to the engine. She demonstrated the pairing of the kit with the application in order to control the kit by Bluetooth, as well as the creation and launching of the first robotic model control program using command to set the direction of engine rotation. In the third part, the teacher asked the pupils to divide into pairs, to build their own model of the robot, and to present it at the end of the activity. At the beginning of this section, the teacher gave each pair of pupils a robotic kit and a tablet. She also explained to each pair of pupils the pairing of the kit with a tablet application via Bluetooth. In the fourth part of the activity, both the pupils and the teacher sat in a circle on the carpet and presented robotic models they built. The results of the data analysis of the third and fourth parts of the activity are described in the next chapter.

4   Comparison of First Robotic Models Creation in 3rd and 4th Grade

On the basis of data analysis, we have divided the creation of the first own robotic model designed according to the pupils' fantasy into three parts: construction, programming, and presentation of the robotic model.

4.1   Construction of a Robotic Model According to the Pupils' Fantasy

The first idea, all pupils came with in the construction of their first robotic model, was the car. In the third grade two pairs out of three built a certain type of car (a car with a propeller and a car with a drill - drill). In the fourth grade, one pair built a model of a mountain car with laser.

All the pairs of third grade pupils and one pair of fourth grade pupils tried to attach the wheels to the model so the model moved. However, until the end of the activity, no pair managed to fasten the wheels to the engine so that the car moves.

Most of the pairs of **third grade** pupils **have** completely **dismantled** and **rebuild** their **models three times**. However, the designs of the models have not changed significantly. The final models were built in 11 min. When creating the third model, most of the third grade pupils acted as follows: *they first constructed a model* (construction of the final model took a maximum of three minutes), *then they tested the behavior of the model using the program they created* (the testing came to pass within five minutes), *finally, they finished the model construction* (it lasted at most three minutes). Most pairs of **fourth grade** pupils **dismounted** their **models once** during the building process however, they did not disassemble the models completely (except for the pair who built the car). The other pairs kept some part of the previous model, e.g. a propeller, and used them to build another model. Then they built the final model within 5 min (before the presentation started). The pair who built a blender did not disassemble the model and built it in 20 min.

Pupils of both grades mostly have not mastered the cooperation in pairs. The model was mostly built by one dominant member of the pair.

The robotic models constructions of the fourth grade pupils were somewhat more complicated than those of the third grade pupils. However, this complexity may be in particular due to the fact that the fourth grade pupils devoted themselves to the construction of one model for a longer time than the younger pupils.

4.2 Programming of the Robotic Model

All the pairs in the fourth grade focused mostly on building the model. They dealt with the model programming only a fraction of the time compared to building the model. Testing of the model behavior was not clearly separated from constructing the model (in the third grade these phases were apart). Pupils used only one command icon to draw up their programs. Only one group created program to test movement of the model, but they were unsuccessful.

All pairs of third grade pupils examined multiple commands icons and connected them together creating a program (mainly command icons for engine). Two pairs of third grade pupils created programs with multiple command icons, using predominantly commands to set the direction of engine rotation and the speed of engine rotation. One pair of third grade pupils has not managed to create a program in time. This was a case of a mixed pair where the boy did not want to cooperate with the girl during the whole activity.

In the course of the activity, we noticed that the third grade pupils had been studying the commands in the program longer and testing the behavior of their model more than the fourth grade pupils. Most pairs of fourth grade pupils attempted to add multiple command icons to their programs immediately before the presentation or during the presentation of their models. All pairs in the fourth grade presented programs containing two command icons - turning the engine to the right or left and setting the engine speed. These programs were created just before the presentation.

4.3   Presentation of the Robotic Model

During the presentation of the models, all the pairs of pupils first spoke about what they were planning to build but failed to do so. Pupils talked, for example, about planning: *to build a car or a helicopter (helicopter image was on the box from the kit), to attach the wheels to the car (afterwards they attached the skis instead of it), to make the wheels of the car to spin, but only the propeller was spinning, and so on.*

Two pairs in fourth grade described the design of their models by attributing fictive features to some LEGO bricks, e.g. density sensors were placed on the blender, and the laser was placed on the mountain car. No pair of pupils (neither in the third grade nor in the fourth grade) explained their model control program. When presenting the model they ran the program without a comment. As a result, we cannot clearly judge to what extent the pupils understood the programs they created and whether they did not randomly link several commands to the program.

5   Discussion

The differences detected in the work of the two observed grades of pupils, could also arise from the different approach of researcher and computer science teacher.

6   Conclusion

In this paper, we present the introductory activity with the LEGO WeDo 2.0 robotic kit designed for pupils of the third and fourth grade of elementary school. We discuss the work of pupils in the creation of their first robotic model based on their fantasy, focusing on the main differences in the work of pupils from the different grades. Based on a qualitative data analysis, we found several differences in terms of construction, programming and presentation of the robotic model:

- The third grade pupils built the final model on the third attempt, while completely dismantling their previous models. Nearly all pairs of fourth grade pupils built the final model on the second attempt, but they did not disassemble the previous model completely.
- The third grade pupils explored multiple command icons, they assembled programs with multiple command icons, and tested the behavior of the model when controlling it by the created program. Fourth grade pupils focused more on building the model and they did not study command icons too much. They created the programs with multiple command icons (more than two) only during the presentation.
- At the presentation, the fourth grade pupils attributed fictional features (density and laser sensors) to LEGO bricks. The third grade pupils also attributed their models fictional behavior, but they did not ascribe the imaginative features to the design LEGO bricks.

The results of the data analysis from the initial activity and also the other activities will be compared with the data collected in cooperation with Taiwanese researchers.

We also plan to prepare a comparative study taking to account the different cultural backgrounds and adapt the activities to both national curricula so that they can be part of ordinary primary education.

Acknowledgement. This research was supported from the Slovak national VEGA project no. 1/0797/18 as well as from the bilateral APVV project no. SK-TW-2017-0006.

References

1. Piaget, J.: The Principles of Genetic Epistemology. Basic Books, New York (1972)
2. Von Glasersfeld, E.: An introduction to constructivism. In: Watzlawick, P. (ed.) The Invented Reality: How Do We Know What We Believe We Know? (Contributions to Constructivism), pp. 17–40. Norton, New York (1984)
3. Papert, S.: The Children's Machine. Basic Books, New York (1993)
4. Palincsar, A.S.: Social constructivist perspectives on teaching and learning. Annu. Rev. Psychol. **49**, 345–375 (1998)
5. Alimisis, D., Moro, M., Arlegui, J., Pina, A., Frangou, S., Papanikolaou, K.: Robotics & constructivism in education: the TERECoP project. In: EuroLogo, vol. 40, pp. 19–24 (2007)
6. Eguchi, A.: Robotics as a learning tool for educational transformation. In: Proceeding of 4th International Workshop Teaching Robotics, Teaching with Robotics & 5th International Conference Robotics in Education Padova (Italy) (2014)
7. Mubin, O., Stevens, C.J., Shahid, S., Al Mahmud, A., Dong, J.J.: A review of the applicability of robots in education. J. Technol. Educ. Learn. **1**(209-0015), 13 (2013)
8. Han, J.H., Kim, D.H., Kim, J.W.: Physical learning activities with a teaching assistant robot in elementary school music class. J. Convergence Inf. Technol. **5**(5), 1406–1410 (2009)
9. Chang, C.W., Lee, J.H., Chao, P.Y., Wang, C.Y., et al.: Exploring the possibility of using humanoid robots as instructional tools for teaching a second language in primary school. Educ. Technol. Soc. **13**(2), 13–24 (2010)
10. Veselovská, M., Mayerová, K.: Assessing robotics learning at lower secondary school. Information and Communication Technology in Education 2015, pp. 240–247. Pedagogical Faculty, University of Ostrava, Ostrava (2015)
11. Mayerové, K., Veselovská, M.: How to teach with LEGO WeDo at primary school. In: Robotics in Education, pp. 55–62. Springer, Cham (2017)
12. Mayerová, K., Veselovská, M.: How we did introductory lessons about robot. In: Teaching Robotics, Teaching with Robotics, Padova, pp. 127–134 (2014)
13. Veselovská, M., Mayerová, K.: Pilot study: educational robotics at lower secondary school. In: Constructionism and Creativity Conference, Vienna (2014)
14. Veselovská, M., Mayerová, K., Kubincová, Z., Chiu, F.Y.: A pilot study: comparing the work of children with LEGO WeDo 2.0 in Slovakia and Taiwan. In: International Conference of Education, Research and Innovation (ICERI), pp. 6030–6037. IATED (2018)
15. Creswell, J.W.: Educational Research: Planning, Conducting, and Evaluating Quantitative. Upper Saddle River, New Jersey (2002)

STEAM Robotic Puzzles to Teach in Primary School. A Sustainable City Project Case

Francisco Ruiz Vicente[1], Alberto Zapatera[2], Nicolás Montes[3],
and Nuria Rosillo[3(✉)]

[1] Colegio de Fomento Aitana,
Ctra. Murcia-Alicante, km 69, 03320 Elche-Torrellano, Alicante, Spain
`aitn_faruiz@fomento.edu`
[2] Department of Educational Sciences, University CEU Cardenal Herrera,
C/Carmelitas 1, 03203 Elche, Alicante, Spain
`alberto.zapatera@uchceu.es`
[3] Department of Mathematics, Physics and Technological Sciences,
University CEU Cardenal Herrera,
C/San Bartolomé 55, 46115 Alfara del Patriarca, Valencia, Spain
`{nicolas.montes,nrosillo}@uchceu.es`

Abstract. In the present work, after conducting an analysis on the STEAM properties for the educational curriculum of the Law LOMCE 2013, an intervention proposal is developed and implemented through a STEAM learning project that uses educational robotics as a teaching tool incorporating different methodological elements from the flipped classroom, problem and project based learning, as well as cooperative learning. Of the 11 areas of opportunity detected the topic "sustainability" was chosen for the development of the presented project. The present work focuses on the range of students aged 9–12, which is, according to the knowledge of the authors, the least studied in the scientific literature.

The educational robotics kit developed ad hoc, which makes up the Sustainable City project, is composed of 25 tiles of 15 × 15 cm each. The board reproduces a city block which must be programmed. When the robot goes around the perimeter street, the different elements that make it a sustainable city are activated.

Keywords: STEAM · Educational robotics · Cooperative puzzle · Board · Active methodologies

1 Introduction

The arrival of the 21st century has provided an authentic technological revolution regarding information and knowledge systems. In developed countries the access to all types of knowledge is immediate, but, at the same time, its obsolescence has grown exponentially.

Education cannot ignore this new reality, in which most of the current students of secondary education will work in jobs that do not yet exist handling unimaginable knowledge. From this perspective, it will no longer be as important to accumulate

© Springer Nature Switzerland AG 2020
M. Merdan et al. (Eds.): RiE 2019, AISC 1023, pp. 65–76, 2020.
https://doi.org/10.1007/978-3-030-26945-6_7

specific knowledge as learning to search and select the right ideas, to discard false, outdated and obsolete information and especially to have the ability to learn autonomously throughout life. It is understood, therefore, the need to start introducing new educational methodologies and tools and new types of multidisciplinary teaching activities at the service of a global learning project.

With these forecasts emerges the need to train new generations whose STEM skills (Science, Technology, Engineering, Mathematics) are sufficiently developed so that students can adapt and develop technologies yet to be discovered. The educational policies of countries such as the United States or Korea, among others, and even the European Community, have begun to bet on the enhancement of the STEM disciplines, [1, 2]. However, in this process, doubts are emerging about how to strengthen these disciplines and the effectiveness of the models that have been implemented, [3, 4]. In this context, Art emerges as a new discipline to be added to STEM which becomes STEAM, generating much more interdisciplinary, creative and innovative learning situations [5].

On the other hand, the so-called active methodologies have emerged placing the students at the centre of their learning. These methodologies are more effective in generating meaningful learning used to train people to become critical, creative and prepared in order to face the unknown challenges of the future. One of these active methodologies is project-based learning, which based on an initial question or challenge, aims to generate a final product, generating learning through the tasks which are carried out to create it. If any of these tasks, in addition to being part of the project, pose a new challenge or problem to solve, we will need to overcome these with techniques of another methodology, the problem-based learning. This type of learning emerges from university education and is, in turn, an active methodology that focuses on the student and generates a good dose of meaningful learning. Both methodologies use the large methodological umbrella of cooperative learning and therefore to be implemented it is necessary a new organizational structure of the classroom, a different way of managing times and evaluation systems as well as changing the role of teachers and their training.

Educational robotics is presented as a creative process based on trial-error and as a technological process based on the interactions between science, society and technology, which are embodied in the construction, programming and manipulation of a robotic platform, [6, 7]. Therefore, educational robotics is a perfectly integrative tool in a STEAM learning environment which can deeply contribute to the motivation, interest and performance of students, [8]. Although it does not imply a learning methodology in itself, educational robotics can play the role of a manipulative tool in the teaching process and place the robot as a companion and co-learner of the student and even, as a tutor, [9]. Although it is still in the phase of empirical verification (where the range of students aged 9–12 is the least studied, [11–13]), the already numerous scientific studies developed envision a series of positive contributions that are often repeated: (1) academic performance, (2) interest and motivation, (3) social skills and cooperative work, (4) creativity and (5) problem solving skills [10].

1.1 Educational Robotics and Curricula

By relating educational robotics and curriculum, in [14] several questions in the air were left: "What part of the science, technology and mathematics curriculum can be transferred to robotics? The whole plan? half the plan?. It is not known, and there is clearly an important work to be done in response to that question. "One of the great challenges facing educational robotics is the official curriculum and how to adapt it so that it can be worked through robotics [7–9]. In [14] it is proposed to order this task to teachers since they are the ones who are trained to carry out the necessary adaptations, creating educational materials and tools. In the same line, in [9] it is emphasized that the greatest efforts should focus on the development of learning materials and on the identification of the appropriate curricula.

1.2 Educational Robotics and Low-Cost Platforms

Over the last few years, low-cost platforms are having a key role in the different levels of education worldwide. Many institutions are using them to improve the teaching experience in very different subjects, from primary and secondary schools to universities (specifically those offering engineering degrees), as well as private initiatives around the world. The most widely used low-cost devices are typically, [15]: Arduino, Raspberry Pi, and Kinect. In the specific case of robotics, numerous low-cost mobile robots have appeared in the last years, such as Adept, Moway, E-Puck, LEGO Mindstorm, etc. These devices do not provide the same accuracy as industrial robots but sufficient for educational purposes or even research tests and experiments.

1.3 Educational Robotics Board

One of the most usual ways to use robotics in education is through the use of boards that try to represent reality at scale. One of the best known examples is the First Lego League [16], where in each edition a thematic board is constructed where the participants have to solve missions. Another of the examples, already at the university level, is Duckietown.

The Duckietown project was conceived in 2016 as a graduate class at MIT. A group of over 15 Postdocs and 5 professors were involved in the initial development. The goal was to build a platform that was small-scale and cute yet still preserved the real scientific challenges inherent in a full-scale real autonomous robot platform, [17]. The autonomous vehicles are called "Duckiebots". Duckiebots live in "Duckietowns", colorful miniature environments that are assembled from modular tiles. In Duckietown, inhabitants (duckies) are transported via an autonomous mobility service (Duckiebots).

These examples favor the teaching-learning process since the students are more motivated and involved in this process.

1.4 Goal of the Paper

The main goal of this research work is the design of an intervention proposal that specifies a STEAM learning project for 4th, 5th and 6th grade students of Primary

Education (9–12 years old) in which, through the use of educational robotics as a tool, and the design of a robotic board that represents a concrete reality, active methodologies such as problem-based learning and cooperative learning are introduced along with flipped class-room sessions. The proposal must be intra-curricular, to be able to be worked within school hours.

Section 2 explains how to obtain STEAM projects based on the Spanish Education Law LOMCE. As a result, it is possible to group all the contents in 11 STEAM conceptual opportunity areas. In Sect. 3, one of them is selected, "Sustainability" and a proposal of intervention is developed giving the "Sustainable City" project. In Sect. 4, the methodology to be applied in class is developed, where the use of a robotic board in the form of a cooperative puzzle is the great umbrella that encompasses the bulk of the learning process. Section 5 defines in depth the robotic board designed and the functionality of each of the parts. The conclusions and future work are shown in Sect. 6.

2 Previous Works. Spanish STEAM Curricula

In relation to the Spanish curriculum, in 2013 there was a new change of educational law in the Spanish system and the Organic Law 8/2013, of 9 December, for the improvement of educational quality, LOMCE, [18], modifying several sections of the previous Education Law, LOE, [19].

In one of these modifications, LOMCE redefines the eight basic competences of the LOE, calling them "key competences" and summarizing them in seven. One of these key competences is called "mathematical competence and basic competences in science and technology", MCBST, whose definition in [20] is quite similar to what could be a competence in STEM learning, but in no case is there any reference to the art and the possibility of STEAM learning.

In our previous work, [21], a STEAM analysis of the LOMCE curriculum is carried out so that in the present work, an educational STEAM learning project can be designed to integrate educational robotics as a main didactic tool. The STEAM analysis of the curriculum was divided into three phases. Phase 1, STEAM learning in the curriculum, Phase 2, Content analysis and Phase 3, Opportunity areas.

As a result, two types of areas were detected with the greatest number of connections: those that develop conceptual contents and those that deal with attitudinal or procedural contents. The latter, which were called non-conceptual areas of opportunity, by not representing contents close to the students and being somewhat removed from their daily reality, were not considered as areas that could motivate and trigger the whole project, but neither did it seem advisable to leave them completely aside for the design of a project. The breakdowns of conceptual opportunity areas by discipline found are:

- Science:
  - The cell and living beings: Structure, classifications, relationships and ecosystems.
  - Economic and human activity: agriculture, industry and raw materials.
  - Sustainability.
  - The physical environment of Spain, Europe and the world.

- Technology:
  - Simple and compound machines and electrical appliances.
- Engineering:
  - Units of measure, comparison, measurements and measuring devices.
  - Matter, its properties, states and changes.
- Art:
  - The audio-visual message and the plastic art: function, message and culture.
  - Audio-visual and plastic composition: Elements, expressiveness and techniques.
- Mathematics:
  - Direct proportionality and percentages: calculation and variations.
  - Statistics: graphic charts, data type and centralization measures.

3 Design of Intervention Proposal

From the STEAM analysis of the curriculum developed in our previous work, [19], in the present work the area of "Sustainability" is selected as the main conceptual opportunity area and the "Scientific research" is chosen as the main non-conceptual area to develop the intervention proposal. Both areas belong to the discipline of science (S), which is the dominant discipline of the Project. The choice of these two areas as a starting point for the design is based on the fact that the STEAM analysis of the curriculum revealed that sciences are the discipline with the greatest number of opportunity areas and that sustainability was one of those that brought together the greatest number of contents (Table 1).

Table 1. Areas involved in the project.

Area. (NC (Non-Conceptual) C (Conceptual), STEAM)	Subject	Level (grades)
Comprehension of statements and data, (NC, E)	Mathematics	4th–6th
Creativity in problems and their solutions. (NC, A)	Mathematics Nature sciences	6th
Scales, maps, axes and representation of reality. (S, E)	Mathematics Nature sciences Plastic education	4th–6th
Skills for plastic and audio-visual creation. (NC, A)	Plastic education	4th–6th
Scientific research: Method, projects, culture and scientific activity. (NC, S)	Nature sciences	4th–6th
Simple and compound machines and electrical appliances. (S, T)	Nature sciences	4th–6th

(*continued*)

Table 1. (*continued*)

Area. (NC (Non-Conceptual) C (Conceptual), STEAM)	Subject	Level (grades)
Operations and mental calculation. (NC, M)	Mathematics	4th–6th
Direct proportionality and percentages. (C, M)	Mathematics	4th–6th
Resolution of real, closed, open and thematic problems. (NC, E)	Mathematics	4th–6th
Sustainability. (C, S)	Nature sciences	5th, 6th
Supervision, evaluation and improvement of products and projects. (NC, E)	Nature sciences Social sciences	4th–6th
Units of measure, comparison, measurements and measuring devices. (C, E)	Mathematics	4th–6th

Regarding conceptual opportunity areas, from the curricular elements that make up the thematic area of "Sustainability", four belong to the subject of Nature Sciences and three to Social Sciences. The absence of the discipline of art in the conceptual plane of design is significant. For this reason, the project incorporated the area of "Skill for plastic and audio-visual creation" so that the project is not just STEM learning. The design of the project incorporated this area, conditioning the final product to a certain degree of plastic personalization on the part of the students.

When the design approached its methodological definition through project-based learning, cooperative learning, problem-based learning and the flipped classroom, new non-conceptual areas were incorporated that ended up guaranteeing the integration of all the disciplines. The final project ends up adding the engineering areas of "Comprehension of statements and data", of "Problem solving" and of "Supervision, evaluation and improvement of products and projects", the mathematical area of "Operations and mental calculation" and the art area of "Creativity in problems and their solutions".

Table 2 collects the data derived from the areas involved in the project. It shows the contribution of each discipline and the influence of each subject on the project, as well as the connection with each course.

3.1   Contents

Once the STEAM areas that are included in the project design have been defined, it is possible to specify one by one the curricular items or contents included, as well as the rest of the elements of the curriculum: objectives, evaluation criteria, key competences and learning standards.

The contents of the project are grouped into conceptual and non-conceptual, specifying the subjects and the block to which they belong according to the LOMCE, see Table 2.

Table 2. Contents of the project.

Subject	Conceptual block	Non-conceptual block
Nature sciences	1. Matter and energy 2. Technology, objects and machines	1. Initiation to scientific activity
Social sciences	1. The world in which we live 2. Living in society	1. Common contents
Mathematics	1. Numbers 2. Geometry	1. Processes, methods and attitudes
Plastic education	1. Plastic expression 2. Geometric drawing	1. Plastic expression

4 Methodology

As a first step, the sequences of activities are divided into three: (1) presentation of the project, (2) completion of a cooperative puzzle and (3) presentation of the final product.

4.1 Presentation of the Project

The presentation of the project consists of an activity, which brings the subject to the students, motivates them and serves as an introduction to the presentation of the challenge to be solved. As a link activity to the cooperative puzzle, a flipped classroom session is set, whose objective is to know the basic conceptual framework of the project and for the teacher to have objective data to organize the cooperative groups.

4.2 Cooperative Puzzle

The cooperative puzzle serves as a great methodological umbrella that encompasses the learning process and subdivides it into four phases:

1. The organization of the puzzle, which starts with the training, by the teacher, with groups of 3–4 students whose task will be to know, analyze and distribute the individual challenges of which the project consist and which concludes with a new session of flipped classroom in which each student will specialize in the concepts of the part of the project that the student has to solve.
2. The "PBL (problem-based learning) research" starts when the teacher reorganizes the students in research groups or specialists formed by 3–4 students who, when distributing in the base group, has the same individual challenge and, a new group to be able to solve it is formed, "research PBL".
3. The reflection in groups, which starts with the disappearance of the research groups to form new base groups, which will carry out activities that imply individual and group self-evaluation of the previous phase, presentation to the rest of the group base of what was learned during the research and a peer evaluation that allows reflecting on the presentations made and ends with the third general session of flipped classroom in which each student will work the necessary concepts to go from the individual challenges to the challenge of the Final product.

4. The "PBL of execution, criticism and revision", whose objective is to address the design, execution, manufacture and/or assembly of the final product.

4.3 Final Product Presentation

Once the cooperative puzzle is finished, the final product is introduced, in which the base groups will continue to work together. In this case, the activities follow the cooperative scheme of "Numbered heads" that oblige the students of the group to carry out a general review of the product and the concepts learned, making sure that all the members of the group have learned and know how to present the product.

5 Materials and Resources

The materials necessary to carry out the project are essentially two, (1) a virtual learning environment necessary for the flipped classroom sessions which is also useful to support the individual work and evaluation activities and (2) the Sustainable City robotics kit, including its teaching material.

5.1 Virtual Learning Environment

The use of a virtual learning environment (VLE) becomes necessary at the moment when it is decided to have personalized data of each student after flipped classroom sessions. These sessions were designed so that after viewing a video the students had to answer a series of questions and the results of these sessions will be part of the continuous evaluation process during the project. Each kit of the Sustainable City has printed codes of access to the VLE, programmed in Moodle 3.0 and although the access is individual, the base group to which each student belongs is shown.

In addition to following the didactic activities that are indicated in class and that will appear in the section courses, each of the students can customize their virtual file on the platform, ask for help from the rest through the main forum and contact the teacher.

Within the platform and as the project progresses, the individual and group self-assessment activities are shown, as well as the final contents test.

5.2 Robotics Kit

In the robotics kit, specifically designed for the project, control boards, sensors and actuators compatible with Arduino Uno and PLA-type plastic parts printed with 3D printers were used. The use of these components makes it possible to provide the kit with flexibility since, with the same robotic components, it would be possible to generate any type of challenge, the didactic criteria being those that impose and define the design of the kit.

The kit preserves the three phases of work that define educational robotics based on interactive learning, construction, programming and manipulation, [7], guaranteeing in the project the promotion and use of the technology discipline (T) and allowing a first interdisciplinary connection.

The work kits are accompanied by a robot like the one in Fig. 1 fully assembled and programmed to follow a black line 5 cm wide. The robot is a version of the one designed by the BQ brand and called PrintBot (PrintBot Tadpole).

Fig. 1. BQ robot adapted for the project.

The Bitbloq online software, developed by BQ, is used to program it. The assembly and programming of the robot take place in the final phase of the cooperative puzzle while the manipulation takes place in the final phase of public presentation.

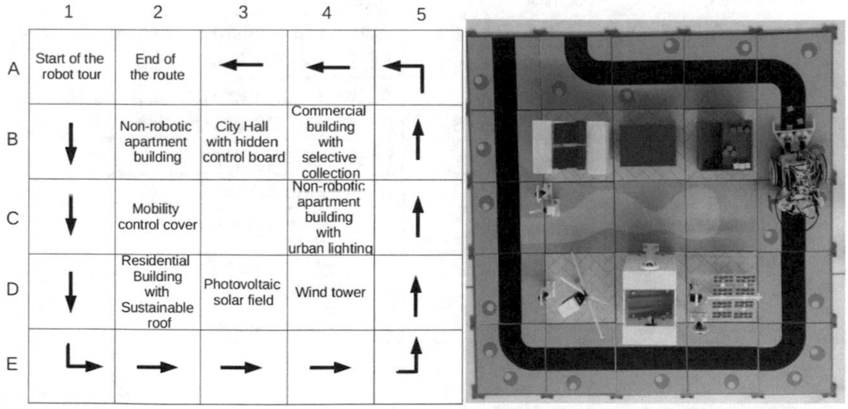

	1	2	3	4	5
A	Start of the robot tour	End of the route	←	←	←↑
B	↓	Non-robotic apartment building	City Hall with hidden control board	Commercial building with selective collection	↑
C	↓	Mobility control cover		Non-robotic apartment building with urban lighting	↑
D	↓	Residential Building with Sustainable roof	Photovoltaic solar field	Wind tower	↑
E	└→	→	→	→	↑┘

Fig. 2. Left: Organizational scheme of the board. Right: Full board image

The challenge board that makes up the Sustainable City, consists of 25 tiles of 15 × 15 cm each, is mounted on a structure of detachable cross-arms and all its parts are made of plastic PLA by 3D printing. Table 3 shows the function of each tile.

The board reproduces a city block which must be programmed so that when the robot goes around the perimeter street, the different elements that make it a sustainable city are activated. Figure 2, (left), shows the organizational scheme of the board that can be seen in the photograph of Fig. 2, right.

74   F. Ruiz Vicente et al.

Table 3. Tile function

Tile	Formalization	Components	Robotic challenge
A1	Outer street	None	Start of the robot tour
A2	Outer street	None	End of the route
A3	Outer street	None	Line following path
A4	Outer street	None	Line following path
A5	Outer street	None	Line following path
B1	Outer street	None	Line following path
B2	Building	None	None
B3	City hall	ZUM board	Connection wire board
B4	Commercial building	Micro-servo	Activate the micro-servo when IR sensor for the tile C4 is activated
B5	Outer street	None	None
C1	Outer street	None	None
C2	Mobility control cover	IR Led ZUM sensor	Blinking and brightness intensity of the led
C3	Inner street	None	None
C4	Apartment with urban lighting	IR Led ZUM sensor	Switch on the led when IR sensor is activated
C5	Outer street	None	Line following path
D1	Outer street	None	Line following path
D2	Building with solar roof	IR Led ZUM sensor	Switch on the led when IR sensor is activated
D3	Photovoltaic solar field	Micro-servo	Activate the micro-servo when IR sensor for the tile C2 is activated
D4	Wind tower	Continuous rotation servo	Activate the micro-servo when IR sensor for the tile D2 is activated
D5	Outer street	None	Line following path
E1	Outer street	None	Line following path
E2	Outer street	None	Line following path
E3	Outer street	None	Line following path
E4	Outer street	None	Line following path
E5	Outer street	None	Line following path

Of the nine tiles that make up the central core of the board, six are robotic tiles, which pose six challenges of sustainability and robotics: (1) wind power, (2) sustainable roofing, (3) photovoltaic field, (4) mobility control, (5) selective collection and (6) urban lighting.

The six tiles numbered B4, C2, C4, D2, D3 and D4 are those assigned two by two to the students of the base groups at the beginning of the cooperative puzzle, during the presentation and initial analysis of the board. Expert in biodiversity and energy; tiles D2 and D4. Expert in mobility and energy; tiles C2 and D3. Expert in efficiency and waste; tiles B4 and C4.

The students work their two tiles alternating the collaborative and the individual work, after which the assembly of the complete board and its programming begins. In this case the work is not distributed by tiles but by typology creating the roles of designer, assembler and programmer. A video of the project can be watch in [23].

6 Conclusions and Future Work

This paper presents, for the first time, how to carry out STEAM educational projects based on educational robotics taking into account the Spanish education law LOMCE. In particular, the work focuses on primary school students, aged 9–12. The methodology used can be generalized to other education laws in other countries, as well as to other age ranges.

As a result of the STEAM analysis of the LOMCE, 11 areas of opportunity that could be the main topic of a STEAM project were detected. Of all of them, the topic "sustainability" was chosen for the development of the presented project.

The educational robotics kit developed ad hoc, which makes up the Sustainable City, is composed of 25 tiles of 15 × 15 cm each. The board reproduces a city block which must be programmed so that when the robot goes around the perimeter street, the different elements that make it a sustainable city are activated.

The presented kit is a learning tool and not an end in itself because by incorporating the technology it can be customized according to the didactic criteria of the project and not vice versa, that is, the kit is put at the service of the didactic and it becomes the main learning tool without being an end in itself. In this same sense, it is an active learning tool since it respects the construction, programming and manipulation phases of active educational robotics, avoiding designs that only allow manipulation.

As a future work, we intend to apply the proposal to a real classroom, in order to demonstrate the effectiveness of the proposed robotics kit.

References

1. European Commission. http://ec.europa.eu/programmes/horizon2020/en/h2020-section/science-education. Accessed 30 Nov 2017
2. Yakman, G., Lee, Y.: Exploring the exemplary STEAM education in the U.S. as a practical educational framework for Korea. J. Korea Assoc. Sci. Educ. **32**(6), 1072–1086 (2012)
3. Pitt, J.: Blurring the boundaries - STEM education and education for sustainable development. Des. Technol. Educ. Int. J. **14**(1), 37–48 (2009)
4. Williams, J.: STEM education: proceed with caution. Des. Technol. Educ. Int. J. **16**(1), 26–35 (2011). Special edition: STEM-Underpinned by research
5. Yakman, G.: STΣ@M education: an overview of creating a model of integrative education. In: de Vries, M.J. (ed.) PATT-17 and PATT-19 Proceedings, pp. 335–358 (2008)
6. García, J.: Robótica Educativa. La programación como parte de un proceso educativo. RED. Revista de Educación a Distancia **14**(46), 88–99 (2015)
7. Karim, M.E., Lemaignan, S., Mondada, F.: A review: can robots reshape K-12 STEM education? In: IEEE International Workshop on Advanced Robotics and its Social Impacts, ARSO, Lyon (France) (2015)

8. Benitti, F.B.V.: Exploring the educational potential of robotics in schools: a systematic review. Comput. Educ. **58**(3), 978–988 (2012)
9. Mubin, O., Stevens, C.J., Shadid, S., Mahnud, A., Dong, J.J.: A review of the applicability of robots in education. J. Technol. Educ. Learn. **1**, 13 (2013)
10. Toh, L.P.E., Causo, A., Tzuo, P.W., Chen, I.M., Yeo, S.H.: A review on the use of robots in education and young children. Educ. Technol. Soc. **19**(2), 148–163 (2016)
11. Barker, B.S., Ansorge, J.: Robotics as means to increase achievement scores in an informal learning environment. J. Res. Technol. Educ. **39**(3), 229–243 (2007)
12. Liu, E.Z.F.: Early adolescents' perceptions of educational robots and learning of robotics. Br. J. Edu. Technol. **41**(3), 44–47 (2010)
13. Slangen, L., Keulen, H.V., Gravemeijer, K.: What pupils can learn from working with robotic direct manipulation environments. Int. J. Technol. Des. Educ. **21**, 449–469 (2011)
14. Johnson, J.: Children, robotics and education. In: Proceedings of 7th International Symposium on Artificial Life and Robotics, Oita, pp. 16–21 (2003)
15. Irigoyen, R.E., Larzabal, E.: Low-cost platforms used in control education: an educational case study. In: 10th IFAC Symposium Advances in Control Education, pp. 256–261 (2013)
16. http://www.firstlegoleague.org
17. Paull, L., et al.: Duckietown: an open, inexpensive and flexible platform for autonomy education and research. In: IEEE International conference on Robotics and Automation (ICRA), pp. 1497–1504 (2017)
18. LOMCE. https://www.boe.es/buscar/act.php?id=BOE-A-2013-12886. Accessed 30 Nov 2017
19. LOE. https://www.boe.es/buscar/doc.php?id=BOE-A-2006-7899. Accessed 30 Nov 2017
20. ECD/65/2015. https://www.boe.es/buscar/doc.php?id=BOE-A-2015-738. Accessed 30 Nov 2017
21. Ruiz-Vicente, F., Zapatera, A., Montés, N.: How to design STEAM projects for primary school in Spain. In: 12th Annual International Technology, Education and Development Conference, INTED, March 2018 (accepted paper)
22. Ruiz Bolívar, C.: Instrumentos de Investigación Educativa. Fedupel, Venezuela (2002)
23. https://www.youtube.com/watch?v=6mtdiKEtrtA

STEAM Approach to Autonomous Robotics Curriculum for High School Using the Robobo Robot

Francisco Bellas(✉), Alma Mallo, Martin Naya, Daniel Souto,
Alvaro Deibe, Abraham Prieto, and Richard J. Duro

Integrated Group for Engineering Research, Universidade da Coruña,
A Coruña, Spain
{francisco.bellas,alma.mallo,martin.naya,dsouto,
adeibe,abprieto,richard}@udc.es

Abstract. Teaching robotics in secondary school is common nowadays, although with heterogeneous approaches in different countries, mainly in the specific technology that is used. Even with such lack of standardization, there are many common elements in these early stages of educational robotics, focused on basic assembling, simple sensing, and locomotion, which are a consequence of the simplicity of the robots that are used. In this paper, we aim to go one step ahead, and propose an approach of how the next stage in educational robotics should be. To this end, we analyze the current situation of this subject in reference to the robotics market and society, and then we propose a structure for the curriculum development. This approach is based on the use of the Robobo robot, as a clear example of the type of device that allows to perform such update of the subject, and on the use of a STEAM methodology, where a global view of how to face a robotics problem is presented.

Keywords: Educational robotics · Autonomous robotics · STEAM education · Smartphone · Computer vision · Project-based learning

1 Introduction

It is evident that teaching robotics requires introducing robots in schools as a new teaching element. Not only the robot as a hardware platform, but also including the software to program it, and a set of teaching materials to learn this new discipline. In this sense, an educational robot can be seen as a text book, which must be the same for all the students in the classroom, to make teaching homogeneous. But whereas in other subjects the contents are quite standardized in different countries, and text book contents are similar, in the case of robotics they are not. The main reason behind this is that teaching robotics requires a specific robot to implement the lessons on it. Using many robots during the course, although it would be ideal, is not generally affordable for schools. So, each educational center or institution chooses the robotic platform they consider to be the most adequate, and the curriculum is created for that platform. As a consequence, the number of different educational robots we can find nowadays in the market offering a complete teaching set is large. From simple devices for primary

© Springer Nature Switzerland AG 2020
M. Merdan et al. (Eds.): RiE 2019, AISC 1023, pp. 77–89, 2020.
https://doi.org/10.1007/978-3-030-26945-6_8

schools like Bee Bot [1], Dash&Dot [2], Cubetto [3] or Ozobot [4], to more advanced platforms for secondary schools like the LEGO EV3 [5], Mbot [6] or Thymio [7]. All of them are equipped with a programming environment adapted to the student's age, and with a set of teaching units to start learning robotics.

The degree of formalization of this teaching material is very heterogeneous but, in general, it is more advanced in secondary school platforms than in the primary school ones. Some of them are provided by the manufacturer, and they are usually in the form of tutorials with a low level of formalization, which make them useful for extra-curricular activities but not for official ones. Others, on the contrary, are created with the guidance of teachers and roboticians, and they are more adapted to an official program like [8] or [9]. What is common to all of these educational materials for robotics is that they are focused on introductory concepts. As this is a new subject in secondary education, the initial approaches have been oriented to softly introduce the main concepts underpinning this subject: mechanics, electronics and programming.

Regarding the first of the three areas, there are many robots requiring some kind of mechanical assembly, like the Mbot or Lego EV3, which allow modifying the basic morphology. In terms of electronics, the robots based on Arduino boards are very popular (mBot, Mio or Arduino Robot), and using them in classes implies teaching students the basic aspects of connections, ports, sensors and motors. But, although they are not based on open-hardware, infrared, touch, color, ultrasonic sensors and microphones are present in most of the educational platforms. Regarding motors, their basic properties and controlling parameters are also commonly included in the teaching material. Finally, programming environments are more heterogeneous, but all of them include some kind of block-based programming, like Scratch, Blockly or similar. For secondary and high school, one can find educational robots supporting text-based programming languages like Python or C, although in these cases the teaching material becomes more informal.

The question that arises now is, after these initial approaches to robotics teaching for primary and secondary schools, how should it be in the future? In our view, the teaching community should improve the robotics curriculum in two main aspects: on the one hand, a long-term vision of the curriculum should be designed, starting from primary school and the first grades of secondary school, where all students are involved; continuing with the final years of secondary and high schools, where students are more specialized in technical subjects; and ending with technical topics at university.

On the other hand, the contents required by students in their future should be updated. The robots they will face in their everyday life will be completely different from those they are seeing nowadays in classes. The main difference relies on the autonomy or "intelligence" of future robots. They will interact with people in real environments, solving real-world problems in a global information network. That is, all these robots will use computer vision for object and person detection, sound analysis for speech recognition, tactile screens for touch interaction, and many other advanced sensors. In addition, they will run artificial intelligence algorithms that will allow them to learn autonomously, based on the great advances in the machine learning field, so their interaction with humans will be much more natural and all the aspects related to human robot interaction will be crucial. Finally, they will be connected to a global network, and autonomous cars, buildings, houses, public transport, and service robots will conform a sort of intelligent cloud of devices around us [10].

The current official robotics teaching is out of this future reality, which is seen as something out of the secondary school scope. What we claim here is: *the educational robotics curriculum of the final grades of secondary school and high school should lead to teaching real autonomous robotics*, because it seems mandatory that students understand the basic operation and features of the automatisms that will surround their life, whether they continue with a technological specialization, or not. Thus, after an initial stage with simple robots and introductory concepts as the current one, students should continue their education focusing on topics like computer vision, human-robot interaction, cloud computing, and machine learning. This is a new subject that requires a new curriculum and new robots and, up to the author's best knowledge, only general analysis have been performed until now [11].

The aim of this paper is to provide a plausible approach to this future robotics curriculum for high school (considering in the rest of the paper that high school corresponds to the official education from 14 to 18 years old). It is a long-term work that started four years ago, based on the authors' experience in intelligent robotics teaching and research at the University of Coruña for years [12]. The first step that was taken was to design and manufacture a new type of robotic platform adapted to the type of contents we aimed to teach, called Robobo [13]. The second step was to create a programming environment which allows using advanced sensors such as cameras, microphones or accelerometers from a simple perspective. The third one was to implement sample teaching units to introduce autonomous robotics topics at this level and test them with students in classes. This practical experience was carried out for two years, providing very useful information, and many improvements were implemented both in the programming environment and in the teaching units. Here, we will show the main conclusions extracted from this experience, which is particularized in a curriculum that will be briefly described here.

The remainder of the paper is structured as follows: Sect. 2 is devoted to the presentation of the Robobo robot, as a representative example of the type of next generation of educational robot required. In this section, the main features of the robot will be described, together with the supported programming environments. Section 3 contains a formal description of the curriculum in terms of pre-requirements, structure and methodology. In Sect. 4, four representative teaching units will be briefly described and, finally, in Sect. 5 the main conclusions of this work will be presented.

Fig. 1. The Robobo educational robot (left) and some 3D printed accessories (right)

2   Robobo: The Next Generation of Educational Robots

The Robobo robot was originally presented in [14], and later updated in [15]. It is an educational robot made up of two elements: a mobile base and a smartphone that is attached to the base, as shown in Fig. 1 left. These two elements are linked by Bluetooth, and from this moment, they make up a single robot that can be programmed from a computer or tablet accessing, transparently for the user, the base and smartphone features. The elements included in the base are displayed in Fig. 2, and they are quite similar to those of many existing robotics platforms. They allow the robot to move (wheel motors) avoiding collisions and falls (infrared sensors). In addition, the robot is equipped with a pan-tilt unit that provides two additional degrees of freedom for the smartphone motion, allowing for a more realistic robot interaction. The base includes a set of holes in its bottom part where 3D printed accessories can be attached, so it can manipulate real-world elements. The right images of Fig. 1 show four examples of accessories: two LEGO piece adapters, a pusher and a pen holder.

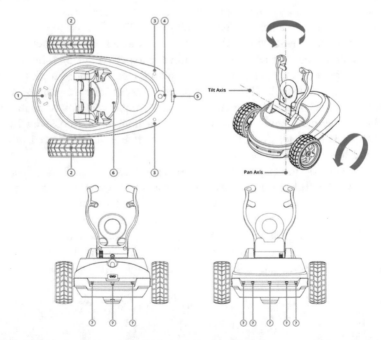

Fig. 2. Robobo base main elements: (1) Front leds (2) DC wheel motors (3) Back leds (4) Power button (5) USB connector (6) Pan-Tilt DC motors (7) Infrared sensors

The smartphone provides Robobo with the following sensors: two high resolution cameras, microphone, 3D gyroscope, 3D accelerometer, 3D magnetometer, tactile screen, light sensor and GPS, among others that could be incorporated in the future. Regarding the actuators, the smartphone has an LCD screen, speakers and flash light; while in terms of connectivity it is endowed with Bluetooth, WI-FI and 4G. Other basic feature of this robot, due to the use of the smartphone, is the CPU speed, which in standard models is currently above 2 GHz. Finally, it must be pointed out that Robobo supports the Android OS, and all the features provided by this system are thus available: speech recognition and production, sound playing and recording, different network services like weather forecast, instant messaging, email, and so on. As we can see, the technological level of this educational robot is much higher than that of the existing ones, and not only regarding sensors or actuators, but also related to connection features that turn this robot into an IoT device with a lot of possibilities for classes.

Fig. 3. Screen capture of the Scratch 3 programming environment

To program Robobo, the computer or tablet and the smartphone must be connected to the same WI-FI. The programming environments have been designed to cover three levels: an introductory one using Scratch, an intermediate one using Python or Java-script, and an advanced one based on ROS. For the lowest level, the current programming environment is displayed in Fig. 3, and it is based on Scratch 3. It runs through a web server and it uses HTML 5, thus it can be executed in any operating system, both in computers or tablets. In those schools where having internet connection available for classes is a problem, the Robobo school pack is equipped with a Raspberry-based hotspot that creates a temporal WI-FI and that runs the Scratch 3 web service. All the Robobo specific blocks can be loaded as an extension, and the a complete list is found in [16]. There are blocks that send commands to the actuators, and others that provide sensing values, both on the base and the smartphone. Many of them are related to the advanced sensors this robot has, like the camera, and they allow detecting colors, faces and QR codes. As this is the first robot that uses computer vision for high school education, the development of simple programming libraries of such complex sensing has been very challenging.

The same set of functionalities available through the blocks are also provided for Javascript and Python, as well as some additional ones [17]. In this case, the user must program the robot using a computer, although there is no limitation in the specific operating system. These two languages were selected due to their global impact in the software development field in recent years [18]. For advanced users, mainly university teaching and research, ROS libraries [17] are available too but, in this case, they can only be used in a computer running the Linux OS, as ROS is only supported by this system.

Table 1. Overall organization of the proposed curriculum

Level	Age range	Robotics	Programming	STEAM
Basic	12–16	- *Locomotion*: wheel motors, IR sensing, encoders - *Tactile interaction*: fling and tap - *Emotions*: speech production, music production and capture, facial expressions, pan-tilt movements - *Colored-objects*: blob position and area - *Environmental:* light and noise sensing	- *Block based* - *Basic algorithms* - *Small programs* • Conditionals • Loops • Operators • Simple variables	- *Integers numbers* - *Decimal numbers* - *Distance* - *Time* - *Cartesian coordinates* - *Angles* - *Basic algebra* - *Basic geometry* - *Curved movement* - *Linear movement* - *Forces* -*Musical scale* - *Technical drawing*
Intermediate	14–18	- *Localization:* gyroscope, accelerometer, compass - *Computer vision:* face detection, QR code *detection* - *Speech recognition* - *Sensor fusion*	- *Text based: Python or Javascript* - *Basic algorithms* - *Medium size programs* • Conditionals • Loops • Operators • Simple variables, arrays and lists • Functions • Events	- *All the previous* - *Accelerated movement* - *Inclined planes* - *Energy* - *Statistics: average, deviation* - *Polynomials* - *Curve fitting* - *Algebra* - *3D printing* - *CAD basics*

3 STEAM Curriculum for Autonomous Robotics

The curricular approach for introducing autonomous robotics in high school has been designed with the following premises:

1. It is oriented to *students that already have some experience in educational robotics*, that is, they should have basic notions about block-based programming, sensors and motor commands.
2. *The smartphone used in Robobo must be the student's phone.* This is a key didactical aspect, because it allows that students realize the potential of their smartphone away from the typical entertainment use. Moreover, it increases the student's motivation due to the personalization of a generic robotic base.
3. The *target age ranges from 14 to 18 years old*, although the lower limit could be decreased down to 12. This is a consequence of the smartphone use, which could be problematic in younger students, and due to the requirement imposed previously that students have already finished an introductory robotics course.

The curriculum is composed of teaching units (TUs) each of them requiring several independent one hour sessions. The TUs have been structured, in this initial approach, into two levels, basic and intermediate, as shown in Table 1. The basic level is focused on introducing the basic features of the Robobo robot, with a higher emphasis in the new sensors and actuators. In this level, all the TUs must be carried out using Scratch. The intermediate level is focused on the combination of different basic features to face more challenging problems, as will be explained later. In this level, the TUs must be developed in Javascript or Python, as chosen by the teacher. An important aspect in this curriculum is that the initial TUs in the intermediate level are focused on the transition from block-based to text-based programming, in order to smooth out such change. In what follows, the main methodological aspects of the curriculum will be described:

Autonomous Robotics: the main objective of this curriculum is to introduce students in the next generation of educational robotics, which is characterized by the following properties. Firstly, the usage of advanced sensing that allows the robot to detect real-world elements without resorting to the creation of unrealistic environments (like putting a line on the floor or constructing a labyrinth with several walls). This type of sensing involves, mainly, computer vision through a camera, speech recognition through a microphone, and tactile interaction through a screen or other type of touch sensor. Although the algorithms required to use such sensing are too advanced for this educational level, the main concepts behind them should be included in the curriculum, as it is important that students understand their basic operation principles. The programming environment must be adapted to the student's level, providing high level libraries. Secondly, the utilization of more interactive actuators like speech production using speakers, LCD screens to show faces or emotions, and the possibility of modifying the robot morphology to adapt it to the task. Again, all these features must be adapted to the age range we are dealing with. Thirdly, the program must be carried out considering the robot autonomy, that is, it must be robust against environment variations and it must operate without any human supervision. In this sense, some intermediate programs can be developed based on fixed conditions to test the solutions, but the final one should be completely autonomous. Finally, although more superficially due to the temporal restrictions, the robot must be seen as an IoT device, which is continuously connected to a global network and it can interact with other devices, whether robots or not. In this sense, students must obtain a more global vision of what an autonomous robot will be in the future, a device that is fully-connected to many others [10]. As a summary, it should be highlighted that this curriculum aims to focus teaching on topics like computer vision, human-robot interaction or multi-robot coordination more than on the classical ones of locomotion, collision avoidance or "robotic fights".

Project-Based Learning (PBL): all the TUs that make up the curriculum are based on a robotics challenge that must be solved by students using Robobo (learning by doing). This challenge must be faced as an engineering project, following all the typical phases of this teaching methodology [19]: group setting, problem identification, objective definition, planning, information retrieval, synthesis and application, solution presentation and feedback. To help the teacher with this methodology, each TU is divided into small robotic activities that lead to its completion in a progressive way. This division is important in our methodology because it is crucial that students understand how to face a real-world robotics problem in a hierarchical fashion.

A key aspect of PBL is that the students are organized into groups. Different group sizes are possible, of course, but we recommend creating groups of four students, contemplating the following roles:

- *Programmer*: responsible for programming the robot
- *Robotician*: responsible of the setup and maintenance of Robobo (both the base and the smartphone)
- *Designer*: responsible for the construction of the environment where the challenge must be solved, and all the Robobo customization accessories
- *Organizer*: responsible for managing the group activity, controlling the time used on each activity and interacting with the teacher in case of questions or comments.

Obviously, each student can help others in a different role in case of necessity, with the aim of all of them being active during the whole class. The roles must be interchanged during the different classes required to solve the challenge.

A very important feature of the proposed approach is that *students must deal with all the aspects of solving a real-world robotics problem*. First, they must build an environment where to solve the challenge, a scaled model of the real problem. As the challenge takes inspiration from a realistic problem, it is a key property that could be solved in the real world. Second, they must customize the Robobo base with the accessories they consider, so the challenge to be solved becomes easier. It is a well-known issue in autonomous robotics that the robot morphology is crucial to simplify the programming, and facing different tasks with a general set of actuators is highly inefficient. Third, students must program the robot to solve the challenge in a reliable way, that is, the task must be solved consistently in different environmental conditions, which is mandatory in autonomous robotics.

STEAM: robotics is a widely established subject in the STEM methodology [20], where an integrated view of technical and scientific contents is conceived. Mathematics, physics, logics, electronics or mechanics are topics covered when solving robotics problems in a global fashion, as proposed here. There are several educational robotics approaches that use the robot as a tool for teaching programming, but this is not the line followed in this work. For the basic level TUs, a previous experience in block-based programming is required. For the intermediate ones, a previous or concurrent formation in text-based programming is recommended too. This is why the TUs that make up the curriculum here have been structured into a hierarchical fashion *with the autonomous robotics contents as the educational core*, and where the required knowledge on programming, mathematics and physics has been introduced considering that it has been previously seen in other subjects of the official curriculum. An overall view of the organization of programming and STEAM topics in each level is displayed in Table 1.

The STEAM approach extends the original one including arts, which can be introduced in many ways. In the specific case of the curriculum proposed here, it is included in the design and construction of the environment for the challenge, and in the proposal of artistic challenges to solve with Robobo in some TUs, like painting, dancing or emotion production.

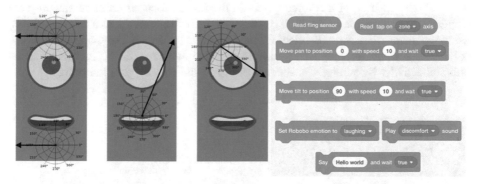

Fig. 4. Scheme provided to explain the fling sensor response (left) and Robobo blocks implied in the solution of this challenge (right)

4 Teaching Unit Examples

In this section, some of the TUs that make up the autonomous robotics curriculum for high school are briefly described to clarify the proposed approach. We present two TUs of the basic level and two of the intermediate one:

86   F. Bellas et al.

Tactile Interaction (Basic Level): in this TU, the challenge was developing a program that allows Robobo to react to the tactile interaction on the smartphone's screen by emitting sounds, speech, facial expressions and movements of the pan and tilt motors. To solve it, the Robobo blocks displayed in Fig. 4 were required. This TU was organized into 4 small activities. The first one was devoted to learning to interact with Robobo through tap gestures (Cartesian coordinates on the screen). The second was focused on learning the response of the tilt motor, and how it can be used to increase the robot expressivity by associating it to "head" movements. In the third one, the fling gesture was introduced together with the pan motor, creating an angular movement as a response to the finger displacement on the screen (left images of Fig. 4 show the type of diagram that is included in the TU to explain such concepts). Finally, the last activity dealt with solving the global challenge and including more expressivity in the robot through the facial expression, sound production and speech production blocks. In this TU, no specific environment was required.

Colored Objects (Basic Level): the challenge was to develop a program that allows Robobo to turn on itself until it finds a green ball in the environment. At that time, it should orient towards the ball and approach it until reached. The environment to test the solution was proposed to be similar to that of Fig. 5 left, so students had to build it. The specific Robobo blocks implied in the solution are those displayed in the right part of Fig. 5. This TU was structured into 4 small activities again. First, learning to measure distances to a colored object, which implied explaining the basics of camera calibration, light conditions, pixel representation, and blob detection. Second, learning to handle the position of a colored object in the image, focused on understanding the Cartesian representation of the blob center of mass. Third, moving to a colored object, using the motor wheels to self-orient the base towards the object. Four, approaching the colored object and stopping at a threshold distance, which could be improved, as an optional activity, by adding a pusher accessory to the base to grab the object.

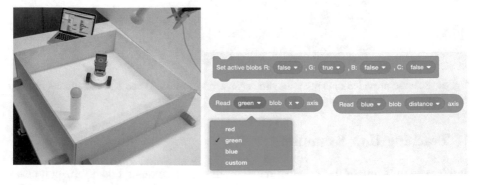

Fig. 5. Proposed environment construction (left) and blocks implied in the solution (right)

Move straight in non-blocking mode and stops when an obstacle is reached

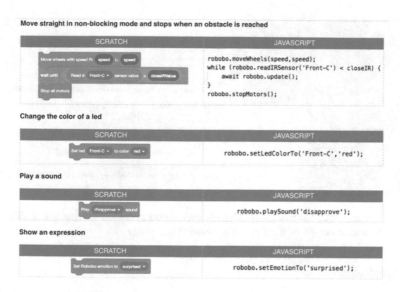

SCRATCH	JAVASCRIPT

Move wheels with speed R: speed L: speed
wait until Read ir Front-C ▾ sensor value > closeIRValue
Stop all motors
	```robobo.moveWheels(speed,speed);```
	```while (robobo.readIRSensor('Front-C') < closeIR) {```
	``` await robobo.update();```
	```}```
	```robobo.stopMotors();```

Change the color of a led

SCRATCH	JAVASCRIPT
Set led Front-C ▾ to color red ▾	```robobo.setLedColorTo('Front-C','red');```

Play a sound

SCRATCH	JAVASCRIPT
Play disapprove ▾ sound	```robobo.playSound('disapprove');```

Show an expression

SCRATCH	JAVASCRIPT
Set Robobo emotion to surprised ▾	```robobo.setEmotionTo('surprised');```

Fig. 6. Example of a piece of code in Scratch (left) and its corresponding Javascript code (right)

Avoiding an Obstacle (Intermediate Level): this is a special TU focused on the transition between block-based and text-based programming, so at the moment of starting it, students already have experience with Robobo and its main capabilities. There are two equivalent TUs, one for Javascript and the other for Python, that cover the same contents. The challenge was to develop a program that allows Robobo to avoid an obstacle placed in front of it at a variable distance, and continue moving straight until a green ball is detected. This problem was selected because it implies using most of the sensors and actuators used in the basic level TUs: motors, IR sensors, emotion blocks and color detection. The TU has been organized into two main parts. The first one is devoted to the challenge solution using Scratch blocks. It is carried out by the teacher and implemented by all the groups in the same way. The second one is focused on the transition, so each part of the original block-based program is translated to Javascript in functional blocks, as shown in Fig. 6. As it can be observed in the figure, there is a high similarity between the two languages, so this methodology is very easy for the students.

Autonomous Parking (Intermediate Level): the challenge in this case was to develop a program that allows Robobo to park between two beacons, labelled using a QR code, similar to those shown in Fig. 7. If the space between the beacons is not enough, the robot must show discomfort; if it fits, it must park and show happiness; and finally, if it is larger than the robot, it must park and stay close to one of the beacons, showing its satisfaction. The TU was designed to be solved using Javascript or Python. Figure 7 shows a small part of the Python code that is provided to the teacher. As it can be assumed, solving a realistic autonomous parking challenge like this is highly motivating for the students, that easily realize the practical utility of the concepts they are learning.

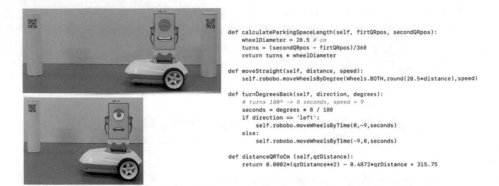

```
def calculateParkingSpaceLength(self, firtQRpos, secondQRpos):
    wheelDiameter = 20.5 # cm
    turns = (secondQRpos - firtQRpos)/360
    return turns * wheelDiameter

def moveStraight(self, distance, speed):
    self.robobo.moveWheelsByDegree(Wheels.BOTH,round(20.5*distance),speed)

def turnDegreesBack(self, direction, degrees):
    # turns 180° -> 8 seconds, speed = 9
    seconds = degrees * 8 / 180
    if direction == 'left':
        self.robobo.moveWheelsByTime(0,-9,seconds)
    else:
        self.robobo.moveWheelsByTime(-9,0,seconds)

def distanceQRToCm (self,qrDistance):
    return 0.0002*(qrDistance**2) - 0.4872*qrDistance + 315.75
```

Fig. 7. Robobo parking in a large space (top left), showing discomfort because it does not fit (bottom left) and a sample of the Python required code (right)

5   Conclusions

The current paper proposes a specific curriculum for introducing the next generation of educational robotics in high school. Students that are already receiving robotics education require a continuation in this subject, which does not exist at the moment in an official way. As described in the paper, the curriculum follows a STEAM methodology where students learn by doing, and where the robotic project is considered globally, from environment construction to robot programming. The device that has been used as the key element to create the TUs is the Robobo educational robot, a smartphone-based platform with the sensors, actuators, computing power and connectivity required for a proper teaching of autonomous robotics. Although many improvements will be carried out in the future, the curriculum presented here is already in use in more than 10 different countries with very successful results.

Acknowledgements.   This work has been funded by the EU's H2020 research programme under grant No 640891 (DREAM) and by the Xunta de Galicia under grant ED431C 2017/12.

References

1. Bee Bot USA. https://www.bee-bot.us
2. Dash&Dot USA. https://www.makewonder.com/robots/dash/
3. Cubbeto robot. https://www.primotoys.com
4. Ozobot web. https://ozobot.com
5. Lego Mindstorms USA. https://www.lego.com/en-us/mindstorms
6. Makeblock Mbot web. https://www.makeblock.com/steam-kits/mbot
7. Thymio II web. https://www.thymio.org/home-en:home
8. Cubetto lesson plans. https://www.primotoys.com/education/lesson-plans/
9. LEGO EV3 Carnegie Mellon curriculum. https://www.cmu.edu/roboticsacademy/robotics curriculum/Lego%20Curriculum/

10. Kehoe, B., Patil, S., Abbeel, P., Goldberg, K.: A survey of research on cloud robotics and automation. IEEE Trans. Autom. Sci. Eng. **12**(2), 398–409 (2015)
11. Murphy, R.F.: Artificial Intelligence Applications to Support K–12 Teachers and Teaching (2019)
12. Integrated Group for Engineering Research. http://www.gii.udc.es/?id=3
13. The Robobo Project. https://theroboboproject.com/en/
14. Bellas, F., Naya, M., Varela, G., Llamas, L., Prieto, A., Becerra, J.C., Bautista, M., Faiña, A., Duro, R.J.: The Robobo project: bringing educational robotics closer to real-world applications. In: Proceedings RIE 2017, pp. 226–237 (2017)
15. Bellas, F., et al.: Robobo: the next generation of educational robot. In: ROBOT 2017: Third Iberian Robotics Conference. Advances in Intelligent Systems and Computing, vol. 694, pp. 359–369 (2018)
16. Robobo programming manual. http://education.theroboboproject.com/en/programming-manual
17. Robobo programming wiki. https://bitbucket.org/mytechia/robobo-programming/wiki/Home
18. Ravisankar, V.: HackerRank Developers Skill report, Online publication (2019). https://research.hackerrank.com/developer-skills/2019
19. Oliveira, A.M.C.A., dos Santos, S.C., Garcia, V.C.: PBL in teaching computing: an overview of the last 15 years. In: 2013 IEEE Frontiers in Education Conference, pp. 267–272 (2013)
20. Goh, H., Ali, M.: Robotics as a tool to STEM learning. Int. J. Innov. Educ. Res. **2**(10), 66–78 (2014)

Design and Analysis of a Robotics Day Event to Encourage the Uptake of a Career in STEM Fields to Pre-GCSE Students

Shane Trimble[✉], Daniel Brice[✉], Chè Cameron[✉], and Michael Cregan[✉]

Queen's University Belfast, Belfast BT71NN, UK
{strimble08,dbrice01,ccameron01,m.cregan}@qub.ac.uk

Abstract. This paper investigates a single day event intended to encourage school students to take up STEM subjects from an early age, in order to access STEM fields later in their educational cycle and thus careers. The event was hosted at Queen's University Belfast in conjunction with the QUB iAMS group and the IEEE. Teaching theories such as Bloom's Taxonomy and constructivism, including social constructivism and constructionism, were used to optimize student learning. The students were given a number of tasks based on these theories and asked to rate them at the end. There was an overwhelmingly positive response from most students, both observed and in ratings. For each activity, the more learning theories applied, the better it was rated. Based on the results, the day was branded a success and will no doubt have a positive effect on encouraging pre-GCSE students to take up STEM subjects and fields.

Keywords: Robotics · Education · STEM · Constructivism · Constructionism · Bloom's Taxonomy

1 Introduction

Due to societies increasing demand for technology in our day to day lives there is a growing requirement for engineers with skills in electronic and software engineering. According to the 2018 "Synopsis and Recommendations" report from UK Engineering there is an anticipated annual shortfall of up to 110,000 engineers with Level 3+ skills in the UK [1].

Robotics is an interdisciplinary field which blends science, technology, engineering and mathematics (STEM) in order to create autonomous systems which can achieve a tangible objective. Such machines are used in a vast array of applications, from manufacturing to medical and bomb disposal to satellite maintenance. Robots are not only relevant in their useful application, they have also captured the minds of humans through literature and film [10]. Young people may also visualize a robot as a toy, [13], with adaptive behaviour according to

© Springer Nature Switzerland AG 2020
M. Merdan et al. (Eds.): RiE 2019, AISC 1023, pp. 90–100, 2020.
https://doi.org/10.1007/978-3-030-26945-6_9

the scenario providing a highly motivational effect [9]. It is for these reasons that the area of robotics was chosen for a day event, with the aim of attracting young students to a career in STEM.

In order to demonstrate the community spirit of engineering and to expose the students to the professional nature of such a career, the event was ran in collaboration with The Institute of Electrical and Electronics Engineers (IEEE) and The Centre for Intelligent Autonomous Manufacturing Systems (iAMS).

The IEEE is the largest professional organization focused on the benefit of humanity through technology. One of their missions is to increase awareness of electrical and electronic engineering through sponsoring awards and events, such as the IEEEXtreme Programming Competition. IEEE members, including the keynote speaker, were present at the event.

iAMS is a pioneer research program started out of Queen's University Belfast with the aim of utilizing a multidisciplinary approach to overcome the challenges of today's advanced manufacturing. All of the instructors were members of this group.

1.1 Constructivism and Constructionism

According to [11], without engagement, learning hardly occurs. The authors of [6] define engagement as participation in the behavioural, cognitive and emotional domains. In order to make the day as effective as possible, several learning theories were applied to the activities, with the idea of maximizing engagement.

Constructivism is an active learning theory that has been used in science pedagogy since the mid 1900s which can be deeply associated with the use of robotics as a learning tool [5]. The theory looks at individuals construction of their own knowledge and understanding of their surroundings through experience and reflection, [16]. The students knowledge base is built up by comparing new experiences to old experiences, allowing them to explore further and ask questions, [12].

Social learning experiences and construction of knowledge in groups can also be a very useful tool. It is referred to as social constructivism [15] and suggests that individuals benefit by using collaborative elaboration to share their perspectives [18], thus learning more effectively together than alone [8]. The authors of [19] emphasize that a cooperative learning environment consisting of small groups maximizes each others learning when working on robotics projects specifically.

The constructionism learning theory further builds on the constructivism learning theory and states that meaningful learning takes place when individuals physically construct an object that is useful or relevant to them or others, and can be shared or demonstrated [15].

When applying constructivism and constructionism theory it is important for the instructors to become facilitators as opposed to solely teaching the subject matter, [13]. An instructor behaving as a facilitator ensures that the student can establish their own understanding of the content, whereas a teacher distributes instructions and information. This shifts the learning from passive to active, by switching the emphasis towards the student [7].

1.2   Bloom's Taxonomy

Bloom's taxonomy is a commonly used and well accepted set of hierarchical models that classifies levels of learning, first published in 1956 [4]. The original version used nouns to label the categories, but the well accepted 2001 revised version, [3], uses verbs to underscore the dynamism. This revision overcomes weaknesses and shortcomings in assumptions of the original, giving teachers and learners a more accurate and intuitive taxonomy to utilize. The taxonomy can also be broken further down into three different domains, addressing different learning areas. Within each domain there are different levels, each requiring a higher level of abstraction from learners. The cognitive domain is knowledge-based, the affective domain is emotion-based and the psychomotor domain is action-based [2]. A summary of how each activity addresses each domain level will be seen in table in the results section, but for the purpose of this paper, the levels will be discussed throughout from the comprehensive categorization of the 2001 taxonomy, seen in Fig. 1.

Fig. 1. The 2001 revised version of Bloom's taxonomy, categorizing the levels in verb form [3]

A brief description of each level is given as follows, from bottom to top: *Remembering* - The recollection of facts and concepts at a basic level. *Understanding* - A simple explanation of these concepts and ideas can be given. *Applying* - Information from the lower levels can be used in new situations. *Analyzing* - Connections can be drawn between ideas. *Evaluating* - Justification can be made of choices, decisions can be defended. *Creating* - Original, novel work can be produced.

1.3 Novelty of Paper

The novelty factor of this paper is our approach to motivating and inspiring students to take up relevant subjects with the aim of possibly pursuing a career in STEM, through the learning theory mentioned in the previous section. Similar learning theory was used to create a full year-long curriculum for students in [14]. The authors of [11] developed a robotic platform for education as we did. We built on this work by combining the two in our own manor over the course of a day, aimed specifically at pre-GCSE students. We also provided the exciting opportunity for students to come to university to learn high level concepts out of their usual environment. What they have learned will enable them to enhance their understanding of how robotic technology exists in the world around them. The students will also have a better idea of how such technology is relevant and applicable to their own lives and the rare chance to get exposure to professional bodies such as IEEE and iAMS.

2 Method

The purpose of the event was to inspire students to take up STEM subjects, through the teaching of robotics fundamentals. To ensure that the day was as effective as possible the assortment of activities were designed to utilize many of the aforementioned active teaching methods. This section will briefly describe the activities in terms of what the student did, and what was the reasoning with regards to what active learning theory was applied. The day was divided into three parts, A, B and C.

Part A consisted of several distinct activities, giving students some fundamental technical knowledge and allowing them to practice some of what they learned. Part B was a team activity, in which students completed a group project. For part C, the students were given a talk from an expert in the field of robotics.

2.1 Part A - Introduction to Robotics

The introduction part of the day was designed primarily based off Bloom's Taxonomy, but elements of contructionism came into play due to the nature of some of the activities.

Interactive Questions and Answers. For this initial activity, the students faced a presenter who asked very high level questions on general knowledge and applications of robots. The purpose of this was to break the ice and engage the students, targeting the bottom two levels of the taxonomy, *remembering* and *understanding*.

Robotics Videos. Students were shown a wide array of videos on the various applications of robotics. The reasoning behind this was to engage the students visually, with the students *remembering* certain applications. This mildly carries into constructivism, with the students building their understanding of robotics by seeing their uses in the world.

Tennis Ball Activity. For this activity, all students stood in one large group. After a brief explanation, the students were directed by the instructor to pass a tennis ball between each other in certain directions based on logic derived from a large playing card held by the instructor. This addresses the bottom three levels of the taxonomy, with students using the bottom two layers when *applying* the third. Social constructivism also is a large factor in this activity.

Maze Solving Presentation. Here a video and presentation was provided, explaining and demonstrating some typical maze solving techniques applied in robotics. This was informative and did not require student response at this point, the student is *remembering* certain facts, and the building of knowledge through the demonstration of the maze solving in the real world again touches on constructivism.

Maze Solving Worksheets. The students were handed a worksheet containing the details of the maze solving algorithms from the previous activity. The sheet contained blank mazes for them to practice what they had learned. Similarly to the tennis ball activity, the bottom three layers of the taxonomy were applied.

Maze Challenge. Pairs were formed and each competed against one another to solve the maze fastest. The students were allowed to pick whichever algorithm that they thought best. This builds on the previous activity, again the bottom three layers are applied, but now the students demonstrate deeper learning by *analyzing* and some even *evaluating* their decisions. Not only is this the bottom five layers of Bloom's Taxonomy targeted here, but social constructivism comes into this activity as they are working in pairs and thus getting the best results together.

For the first part of the day, the instructors observed very high participation rates, with most students fully engaged in all activities.

2.2   Part B - Group Project

For the group project section of the day, the students were tasked with building a fully functioning robotic arm, in order to complete a pick and place task. Students were divided into teams of three to five, in their school groups. Each group was seated at a table along with their accompanying teacher. A kit was distributed, designed in house, consisting of a wooden frame, electric motors, sensors, switches, micro-controller and wires. An instruction manual and a glue-gun were also distributed to each. The entire group was then addressed by an instructor, giving a high level description of the task. After this, the students began the construction project. Two instructors travelled around to aid the students in troubleshooting, facilitating the solution process by asking questions. This ensured that the students learning was active, not passive, a fundamental principle of constructionism [7,13]. The students were given one hour to complete the task. When the robot was completed, the students demonstrated it's functions to the instructors (Fig. 2).

Fig. 2. The completed robotic arm that the students were tasked with building

The main reasoning behind this part of the day was to appeal to the students through constructivism. A cooperative learning environment was necessary to complete the task, thus employing social constructivism, as in [8] and [19]. Constructionism is another obvious form of constructivism that was greatly utilized. The students built a relatively complex system from a large kit, following instructions. This meant that the students were 'learning through building', the basis of this learning theory.

Throughout this part of the day, the students were observed closely by the instructors, in order to gauge group dynamics and individual engagement. Each group worked very well, with most completing the task. Nearly all individuals were fully engaged, with only one student noted who did not want to participate.

2.3 Part C - Keynote Speaker

The day was concluded with a 20 min talk. A highly distinguished special guest speaker was invited to the event to represent the IEEE. The speaker has completed research in many of the sub-fields of robotics and is highly published. The purpose of this part of the day was to demonstrate some real world robotic applications, as in sections of part A, but to show the cutting edge. This touched on the bottom layer, of Bloom's Taxonomy, **remembering**, and the section was more used as an inspirational tool. The instructors noted that many of the students had lost engagement at this point.

3   Results and Discussion

The students were given a brief survey sheet to fill out at the end, consisting of five questions. From the surveys, students responded to the day as follows.

Firstly, the students were asked: "Did you find the taster day useful?". From 87 pupils in total, 71 (86.6%) found the day useful, 3 (3.7%) not useful, and 9 (9.8%) were unsure.

The students were then asked to rate the sessions, from one to ten. The results are given in Figs. 3, 4 and 5.

The introduction to robotics, Part A, received a mean rating of 7.2 and a standard deviation of 1.9. This sits between the two activities, and perhaps indicates the combination of listening and learning by doing. The ratings can be seen in Fig. 3.

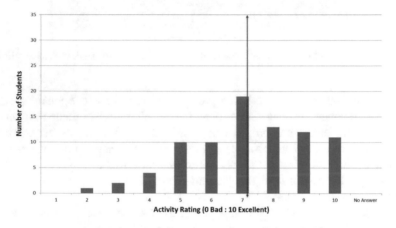

Fig. 3. Rating of part A, introduction to robotics, with an mean of 7.2 and a standard deviation of 1.9, sitting between the other two activities.

Part B, the group construction project, was given the highest rating with a mean of 8.6. A standards deviation of 1.7 indicates that a large quantity of the students enjoyed this part, with minimum variation of the three activities. Results can be seen in Fig. 4. A high rate of participation and enthusiasm among students was noted by the instructors, which is reflected in the rating.

The keynote speaker, Part C, received the lowest mean rating of the activities with 6.5, seen in Fig. 5. It also had the largest standard deviation of 2.3. This wider dispersion indicates that students opinion varied more on this part, some finding it more interesting than others. Instructors noted a drop off in attention at this stage in the event.

The students were also asked about the effect of the day on their future choices. First, they were asked: "Following the taster day, would you be more likely to consider studying STEM related subjects from GCSE?". 31 (37.8%) pupils said yes, 18 (21.9%) said no, and 33 (40.2%) said that they are unsure.

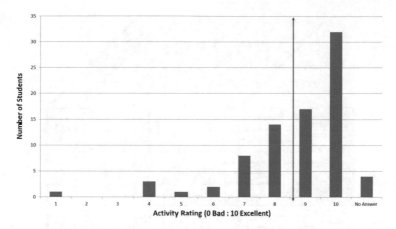

Fig. 4. The group project achieved the highest mean of 8.6 and a standard deviation value of 1.7.

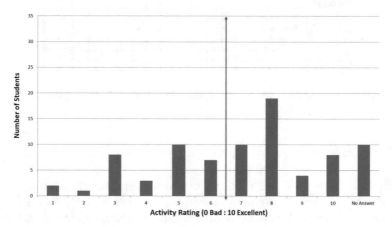

Fig. 5. The lowest rated part of the day was part 3, the keynote speaker. It received a mean rating of 6.5 and a standard deviation of 2.3

Finally, they were asked: "Would you be more likely to consider STEM related subjects at higher level?". 29 (35.3%) said yes, 17 (20.7%) said no, and 36 (43.9%) said unsure.

As briefly touched on in the introduction, Bloom's taxonomy can be applied in the cognitive, affective and psycomotor domains. Figure 6 displays the various levels of each domain that each activity achieves.

It can be noted from the table that most of the activities targeted levels in the first two domains, and nearly nothing in the third.

	Part A						Part B					Part C	
	Interactive Q&A	Robotics application videos	Tennis ball exercise	Maze solving exercise	Algorithm presentation	Maze challenge worksheets	Instruction booklet following	Frame construction	Component placement	System testing & troubleshooting	Wiring	Speaker's account of robotics	Robotics application videos
Cognitive domain													
Knowledge	✓	✓	✓	✓	✓	✓	✓	✓	✓	✓	✓	✓	✓
Comprehension	✓		✓		✓	✓	✓	✓	✓	✓	✓		
Application			✓		✓	✓		✓	✓	✓	✓		
Analysis					✓	✓	✓		✓		✓		
Synthesis					✓	✓				✓	✓		
Evaluation													
Affective domain													
Receiving	✓	✓	✓	✓	✓	✓	✓	✓	✓	✓	✓	✓	✓
Responding	✓		✓		✓	✓	✓	✓	✓	✓	✓		
Valuing	✓	✓						✓			✓	✓	✓
Organizing											✓		
Characterizing													
Psychomotor domain													
Perception				✓									
Set				✓									
Guided response													
Mechanism													
Complex overt response													
Adaptation													
Origination													

Fig. 6. Table demonstrating the various levels that each activity attained for the three domains.

4 Conclusion

From the literature, constructivism, more specifically constructionism, was identified as a highly effective method in which a student actively learns [5,12,16]. Adherence to Bloom's Taxonomy is also a widely accepted learning mechanism. The results support these theories.

Part A of the day was primarily based around Bloom's Taxonomy. All of the sections applied the lower level of the taxonomy, *remembering*. As the students were introduced to more concepts, they incremented up the levels of the taxonomy. Some achieved as high as *evaluating*.

Constructionism was also somewhat applied, social constructivism, for example, in the maze challenge component.

The effectiveness of the application of Bloom's taxonomy and a limited amount of constructivism is reflected in the students ratings, placing this event between parts B and C. A high participation rate was noted by the instructors.

Constructionism was utilized in the group project as the students needed to apply and think for themselves in order to build the robot. The project also employed a high degree of social constructivism, by applying the principles of [17] and [8], in small groups, where a cooperative building environment is utilized to optimize learning [19].

This part of the day also managed to apply many levels of Bloom's taxonomy, which as per the literature, is key to creating an effective learning environment [4]. The bottom two levels, *remembering* and *understanding*, were demonstrated through the students listening to the instruction and reading the manual, in order to reach the next level, *applying*. This was necessary for the implementation of what they had learned. Although step-by-step, the instructions purposefully did not repeat detail for all component placement, meaning the students had to use *analyzing* to proceed at certain points in the construction. The student again used *analyzing*, and also the next level, *evaluation*, when testing out and troubleshooting the robot.

This section of the day received the highest rating, showing that the combination of these two teaching theories had the greatest impact on students. The instructors observed a very high rate of participation, supporting the student ratings.

Part C, the keynote speaker, received the lowest rating. This section applied no real constructivism and vaguely adhered to the bottom layers of Bloom's taxonomy. This is reflected in the student ratings and the instructors observations.

To summarize, the day was seen as a success, in terms of influencing young people to take up STEM subjects from GCSE level. This is both subjective, from the instructors observations, and objective, if correlation is to be drawn from the ratings.

4.1 Future Work

Although the questionnaire supplied both qualitative and quantitative data, some improvements could be made to raise the utility value of the results. It would be useful to increase the number and depth of the questions asked, in order to increase the qualitative value of the data. To improve the quantitative data, the total number of students should be increased. These improvements will be implemented by revising the questionnaire and rerunning the day multiple times, respectively.

Improvement on the activities would mainly revolve around part C, the keynote speaker. For such a group it would be beneficial to target more levels of Bloom's Taxonomy, through higher engagement.

It would also be beneficial to conduct a longitudinal study in order to analyze the actual uptake and attrition rate in GCSE STEM subjects. When this is done, statistical analysis will mean more robust and accurate results, allowing for deep reflection on how to improve the day, thus hopefully attracting more young people to STEM subjects.

References

1. Engineering UK 2018, Synopsis and Recommendations. EngineeringUK (2018)
2. Amer, A.: Reflections on Bloom's revised taxonomy. Electron. J. Res. Educ. Psychol. **4**(1), 213–230 (2006)
3. Anderson, L.W., Krathwohl, D.R., Airasian, P.W., Cruikshank, K.A., Mayor, R.E., Pintrich, P.R., Raths, J., Wittrock, M.C.: A Taxonomy for Learning, Teaching, and Assessing: A Revision of Bloom's Taxonomy of Educational Objectives, Abridged Edition. Longman, White Plains (2001)
4. Bloom, B.S., et al.: Taxonomy of Educational Objectives, Cognitive Domain, vol. 1, pp. 20–24. McKay, New York (1956)
5. Fosnot, C.T., Perry, R.S.: Constructivism: a psychological theory of learning. Constr. Theory Perspect. Pract. **2**, 8–33 (1996)
6. Fredricks, J.A., Blumenfeld, P.C., Paris, A.H.: School engagement: potential of the concept, state of the evidence. Rev. Educ. Res. **74**(1), 59–109 (2004)
7. Gamoran, A., Secada, W.G., Marrett, C.B.: The organizational context of teaching and learning. In: Handbook of the Sociology of Education, pp. 37–63. Springer, Boston (2000)
8. Greeno, J.G., Collins, A.M., Resnick, L.B., et al.: Cognition and learning. In: Handbook of Educational Psychology, vol. 77, pp. 15–46 (1996)
9. Kabátová, M., Pekárová, P.J.: Lessons learned with lego mindstorms: from beginner to teaching robotics. Researchgate **74**, 216–222 (2018)
10. Khan, Z.: Attitudes Towards Intelligent Service Robots, vol. 17. NADA KTH, Stockholm (1998)
11. Kim, C., Kim, D., Yuan, J., Hill, R.B., Doshi, P., Thai, C.N.: Robotics to promote elementary education pre-service teachers' stem engagement, learning, and teaching. Comput. Educ. **91**, 14–31 (2015)
12. Mascolo, M.F., Fischer, K.W.: Constructivist theories. In: Cambridge Encyclopedia of Child Development, pp. 49–63 (2005)
13. Mauch, E.: Using technological innovation to improve the problem-solving skills of middle school students: educators' experiences with the lego mindstorms robotic invention system. Clearing House **74**(4), 211–213 (2001)
14. McKay, M.M., Lowes, S., Tirhali, D., Camins, A.H.: Student learning of stem concepts using a challenge-based robotics curriculum. In: American Society for Engineering Education Annual Conference and Exposition, pp. 1–25 (2015)
15. Papert, S.: Mindstorms: Children, Computers, and Powerful Ideas. Basic Books, Inc. (1980)
16. Piaget, J.: The Moral Judgment of the Child. Routledge (1965)
17. Van Meter, P., Stevens, R.J.: The role of theory in the study of peer collaboration. J. Exp. Educ. **69**(1), 113–127 (2000)
18. Vygotsky, L.: Interaction between learning and development. Read. Dev. Child. **23**(3), 34–41 (1978)
19. Lau, K.W., Tan, H.K., Erwin, B.T., Petrovič, P.: Creative learning in school with LEGO(R) programmable robotics products, vol. 2, pp. 26–31 (1999)

Robotics Education To and Through College

Brian R. Page[1], Saeedeh Ziaeefard[2], Lauren Knop[3], Mo Rastgaar[1],
and Nina Mahmoudian[1(✉)]

[1] Purdue University, West Lafayete, IN 47907, USA
{page82,rastgaar,ninam}@purdue.edu
[2] Ohio State University, Columbus, OH 43210, USA
Ziaeefard.1@osu.edu
[3] Michigan Technological University, Houghton, MI 49931, USA
lknop@mtu.edu

Abstract. Robotics education has made great strides to enable the next generation of engineers and workers with early education and outreach. This early education effort is able to engage students and promote interest, however an integrated pathway to and through college is needed. This pathway needs to build upon early experiences with opportunities to advance across age groups. This paper presents the authors experience developing robotics curriculum across age groups. Middle and high school education has been implemented in a summer camp environment utilizing two co-robotic platforms, a water sensing robot called GUPPIE and an assistive robot named Neu-pulator, engaging 201 total students between Summer 2014–2017. The university course is a senior level technical elective introducing autonomous systems through a mobile robotic platform, a smart car, with 72 total students in Spring 2017 and 2018. In this work, the survey results gathered from Summer 2017 pre-college and Spring 2018 college level activities are presented. Overall observations and lessons learned across age groups are also discussed to better create a pathway from young learners to practicing engineers. The key to success of robotics programs at any age are hands-on, exciting activities with sufficient expert support so that students are able to learn in a frustration free environment.

Keywords: Project-based learning and robotics · Robotics curricula · Robotics education · Marine robots · Assistive robots · Mobile robots

1 Introduction

The field of robotics is rapidly expanding into most aspects of everyday life. To encourage technological competency of the future workforce, we need to be able

This work is partially supported by the National Science Foundation under grant numbers 1350154, 1453886, 1921060, 1921046, and 1923760. Results presented in this paper stem from work on Summer Youth Programs and Autonomous Systems courses while at Michigan Technological University. The authors recently moved to Purdue University.

© Springer Nature Switzerland AG 2020
M. Merdan et al. (Eds.): RiE 2019, AISC 1023, pp. 101–113, 2020.
https://doi.org/10.1007/978-3-030-26945-6_10

to introduce fundamental concepts of robotics at an early age. Significant effort has been put forth in engaging young students particularly in the middle and high school age groups [1]. These efforts have shown a large degree of success, however, the progression from early robotics education to university level education is not clear. This paper presents the authors experience teaching robotics programs to students from middle school through graduate school. For middle and high school students, the authors have organized a one week Summer Youth Program (SYP) over recent years introducing collaborative robotics. 201 students have attended this camp which introduces the GUPPIE and Neu-pulator. These two co-robots are based on the theme of robots helping people. At the university level, 72 students have completed a 3-credit senior level Mechanical Engineering undergraduate course on autonomous systems. In this course, students build and program a small autonomous vehicle to navigate through a model town. Both groups follow a similar hands-on, project-based approach. This paper extends our previous work [2–9] by presenting the most recent survey results from the 2017 summer youth programs and 2018 autonomous systems courses along with the generalized lessons learned across all age groups.

All our educational efforts have followed a project-based approach, promoting interdisciplinary learning opportunities in order to generate a more meaningful learning experience. The choice of platform is one of the determining factors in crafting these interdisciplinary learning opportunities. Table 1 shows a brief comparison between platform options. For the middle and high school group, the interdisciplinary learning theme is manifested in our co-robotics approach (Neu-pulator and GUPPIE) that pairs robotic applications with helping human life to teach STEM concepts. At the university level, the smart car is able to help students connect robotics to the broader applications of engineering. Our approach divides robotics into five fundamental disciplines: (1) engineering modeling and design, (2) electronics and circuitry, (3) programming, (4) assembly and pro-

Table 1. Comparable robotic platforms

Platform	Domain	Target audience	Cost (USD)
GUPPIE [2–7]	Marine (AUV)	Middle school	27
Neu-pulator [3,8]	Manipulator	Middle school	121
Elegoo Car + Pixy [9]	Ground	University	140
KUKA youBot [10]	Ground/Manip.	University	30000
AERobot [11]	Ground	Middle school	10
Duckiebot [12]	Ground	University	150
SeaPerch [13]	Marine (ROV)	K-12	179
OpenROV [14]	Marine (ROV)	University	899
DENA [15]	Marine (ROV)	University	Unknown
Adventure-I [16]	Marine (AUV)	University	Unknown
Service-Arm Type CS-113 [17]	Manipulator	University	Unknown
LEGO NXT Arm [18]	Manipulator	Middle/high	350

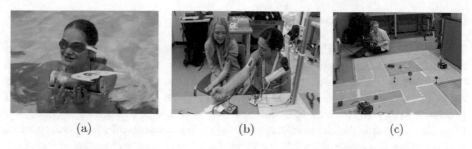

(a)   (b)   (c)

Fig. 1. (a) The GUPPIE during pool testing. (b) The Neu-pulator during experiment and demonstration. (c) The smart car while going through model town.

duction, and (5) testing and troubleshooting. Students are able to practice the engineering design process through each hands-on activity.

The Neu-pulator (Neurally Controlled Manipulator) and GUPPIE (Glider for Underwater Problem-solving and Promotion of Interest in Engineering) are used to introduce middle and high school students to assembling, programming, and testing of robots. This paper builds upon previous iterations of the GUPPIE and Neu-pulator project over recent years [2–6,8] with the most recent results and minor design updates. Over the course of a week-long camp, students learn robotics concepts in a hands-on, project-based manner. This is guided by our custom workbook that is distributed to students during the week. The students are introduced to engineering design process, mechanical engineering, electrical engineering, biomedical engineering, and environmental engineering. Utilizing Neu-pulator and GUPPIE, the overall theme of the curriculum is helping people. Every task is framed in such a way that the students are able to understand how robotics and STEM can be incorporated to improve peoples lives. Students work in pairs or in small groups to encourage team building skills and reduce the stress associated with solo development.

The smart car used at the university level also helps students learn robotics concepts in a hands-on, project-based manner. The car enables students to apply skills learned in other courses such as linear control and circuits to the real-world. The entire course is built around how autonomous mobile platforms function with students learning all the key concepts of autonomous cars. Due to the structure of the course, students work on one key concept per week, culminating in autonomous operation in a model town. While each student has to develop their own system, the course is set up so that the weekly lab sessions are highly collaborative to accelerate learning.

The remainder of this paper presents the platform designs in Sect. 2, the curriculums in Sect. 3, recent survey results for SYP and university level in Sect. 4, and a generalized observations & lessons learned in Sect. 5.

2 Robotic Platform

The Neu-pulator, GUPPIE, and smart car are all easy to assemble robots with common core functionality and components. The Neu-pulator is an assistive robot resembling a prosthetic, the GUPPIE is an exploratory, water sensing robot, and the smart car is a toy car. The robots use low-cost, off the shelf products to enable students to continue creating new systems after they graduate from the program. All the platforms have significant overlap in hardware choice, which unifies the curriculum between the three platforms and across age groups.

The **GUPPIE**, Fig. 1a, is an extremely low-cost underwater glider. It's design has been simplified down to the critical components for basic operation to allow students to focus on the main requirements and functionality of underwater gliders. The GUPPIE is built inside of a sealed tube and contains a buoyancy drive, control system, and energy storage. Mounted on the outside of the hull is a wing and trim weights to enable flight. The total cost for the GUPPIE is 27 USD. The current revision is an iterative improvement over previous GUPPIEs to focus on reducing the cost and increasing the accessibility of the platform [2–6].

The **Neu-pulator** robot [3,8], Fig. 1b, is designed to resemble the characteristics of a human arm and introduces students to basic concepts in robotic manipulation; including types of joints, degrees of freedom, end-effectors, and how these all tie into the forward kinematics of a robot. The Neu-pulator is composed of two revolute joints that are actuated by two low-cost servo motors. These joints resemble the motion produced by the elbow and wrist joints of a human arm. In between each joint are wooden linkages, approximately 15 cm in length, which act as the upper and forearm of the robot. Using these 2 degrees of freedom (DOF), the robot's end effector can reach to many different positions within its joint space. The Neu-pulator's total cost is 121 USD. Due to this, the platforms are shared between small groups of students rather than in pairs.

The **smart car**, Fig. 1c, combines two off the shelf hobby level robotics solutions. The Elegoo Smart Car kit provides the mobile platform including everything required for basic navigation at a low cost. With the addition of a Pixy camera, the car is able to navigate based on visual information. The smart car is a simple four wheel, skid-steerable mobile platform controlled by Arduino. The full kit used in the course is 140 USD enabling each student to have their own car. By ensuring a 1:1 robot:student ratio, every student is required to show competency in all aspects of robotics, as expected in a university level course.

3 Curriculum

The developed curriculum across age groups follows the engineering design process. The core idea is that all robotic systems operate in a 'see-think-act' cycle [19]. The robotic system starts by sensing the environment before thinking about where it is and what it needs to do. Once a decision is made, the robotic system performs an action. Each age groups curriculum builds on this idea with age and skill appropriate content.

In the SYP program, each of the concepts is covered throughout the week [3]. The program is a five day residential summer camp held at Michigan Tech including 28 h of instruction following our custom workbook. The program is kept at a 4 to 1 student:teacher ratio. Instructors for the program are both graduate and undergraduate students. To inspire young learners, the overall theme of the curriculum is based on helping human life. Every task is framed in such a way that the students are able to understand how robotics and STEM can be incorporated to improve peoples lives.

The SYP curriculum focuses on hands-on learning. Students are introduced to the engineering design process from concept generation through prototype testing in an accelerated environment. This is guided by a custom workbook that features picture-based instructions. As an example, during one of the early programming projects the workbook lists all materials needed, has a cartoon circuit diagram, and then guides the students through setting up the code with description of each step. Projects progress through the engineering design process, mechanical/electrical engineering, controls, and programming.

At the university level, the curriculum focuses more on introducing higher level robotics specific concepts with a focus on controls and programming. Similar to the middle school curriculum, the goal of the university level course isn't to create technical experts. It is to create engineers who are able to intelligently understand and discuss autonomy. Additionally, the Autonomous Systems course forms a strong, hands-on foundation for further work on becoming a technical expert.

The university curriculum is broken into a traditional lecture section based on [19] and a custom developed hands-on laboratory section. In the weekly, 2-hour labs over the 14-week semester students proceed through 9 labs from basic programming fundamentals to vision based navigation. Every lab has a 1–2 page guide which lays out lab requirements. The early labs have more thorough explanations of steps to take, possible pitfalls, and other hints. As the course progresses, the guides become sparser with the final project guide having just the project requirements. The semester culminates with a 4-week long intensive final project where students must program their vehicles to autonomously navigate through a model town. The lab sections are kept at approximately 10:1 student:teacher ratio to provide students sufficient assistance while developing their skills.

The remainder of this section covers key concepts introduced in the youth program using GUPPIE and Neu-pulator as well as highlights of the university level course. Only changes and critical curriculum components are mentioned with more in depth coverage of the curriculum in our other work.

3.1 Engineering Design Process

The youth program introduces students to the engineering design process. This is accomplished initially through a hands-on activity constructing micro-gliders [7]. Micro-gliders are simple wood stick and paper clip assemblies that are deployed into a fish tank. Micro-gliders operate similar to the GUPPIE on a downward trajectory. The goal of the micro-glider is to travel as far as possible on the downward trajectory in the tank. This activity teaches students about force interactions of buoyancy, gravity, drag, and lift. In order to succeed at the micro-glider challenge, students quickly learn that they need to iterate through the design process. For example, changing the relative location of paper clips and sticks. Additionally, it serves as the first team building event of the week to help get the students comfortable with speaking up. This is particularly necessary to increase engagement from under-represented groups. The activity helps students figure out how to balance the glider, and in effect, how to control the GUPPIE. This experience aids students understanding of how to control pitch by shifting the center of gravity relative to the center of buoyancy.

3.2   Mechanical/Electrical Engineering

Through using a Computer Aided Design (CAD) software, key concepts of mechanical design as well as the assembly process of both the GUPPIE and Neu-pulator are introduced. In the CAD segment, students design and model the Neu-pulator components that they will later use. This includes modelling of the links and joints to examine the range of motion of the design. During this section, we also introduce the concept of additive manufacturing as it pertains to the manufacturing process for GUPPIE components.

Prototyping and 3D printing helps to make the connection from the CAD sessions early in the week to the real parts during fabrication. The GUPPIE relies on the use of 3D printed components, unfinished printed circuit boards (PCBs), and a pre-fabricated hull. The hull includes a polycarbonate tube and 3D printed end cap that are epoxied together. To assemble the GUPPIE electrically, the students are taught how to solder components onto a PCB such as LEDs, resistors, and wires. Due to the low cost of the GUPPIE, every pair of students is able to have a custom GUPPIE. Similarly, the assembly process for Neu-pulator involves building of the mechanical and electrical systems. Both the Neu-pulator and the GUPPIE use similar components which simplifies assembly by the students due to increased familiarity.

3.3   Controls

Controls topics are not directly introduced to the students, however key concepts of controls are introduced in every portion of the program. Through hands-on activities, students are able to gain experience with different concepts in controls. For example, the micro-glider introduces balancing forces, the Arduino starter kit introduces automatic control, and the Neu-pulator prototype arm introduces closed-loop control. This approach means that students are not overwhelmed with abstract concepts and are instead able to learn controls gradually through practical implementation. More explicit controls topics are covered during the programming and testing of the robotic platforms. The goal of the controls education section is for students to gain an applied understanding of how control systems work. For the middle school age group this means knowing how micro-controllers interact with the world and learning the concepts of feedback and closed-loop control. At the university level, more controls topics are introduced culminating in creating a state based linear control system using visual feedback to drive through a model town.

Control of the GUPPIE is accomplished using a bang-bang feedforward control system with feedback from a single limit switch, Fig. 2a. The buoyancy engine in the GUPPIE is built around a continuous servo with a power screw to convert the rotational energy from the servo into linear motion of a plunger. The plunger is weighted to increase the mass shift associated with buoyancy changes and cause pitching motion. To dive, the plunger is pulled forward in the vehicle which also pulls water into the vehicle. The plunger is pulled in until a limit switch is depressed. This switch triggers the controller to advance to the next state where it waits a pre-determined amount of time for the vehicle to dive. Once the vehicle has completed the dive it expels water, pitches up, and glides upward for a timed maneuver until the process is repeated. This control system operates in an open loop, timing based fashion.

Control of the Neu-pulator is completed using a proportional feedback control system, Fig. 2b. This control system uses a proportional value of the students EMG values as the reference input into the servos. The servos themselves have feedback control systems that maintain the desired angles. This basic closed-loop control encourages students to experiment with the code, changing the response of the motors to the input. This often leads students to start thinking about more complex control scenarios for the robotic arm such as using inverse kinematics to solve the joint angles for a target end effector location.

Control of the smart car is more advanced than the middle and high school platforms. Due to the extended schedule and additional background knowledge available for the university course, the control system grows to involve multiple feedback loops and multiple discrete control states, Fig. 2c. The primary feedback sensor is the Pixy camera which outputs an array of information related to what it sees. The controller processes this information to localize and decide where it is in the model town. The vehicle state machine then determines which control method is implemented and sent to the h-bridge for motor control.

3.4   Programming

Introducing students to programming is the major component of curriculum consisting of 12 h out of the total 28 h program. Beginning on the first day, we introduce Arduino through several small projects from the Arduino starter kit. In particular, the students learn the structure of Arduino, how to write and upload codes, and how to search for help. The starter kit also introduces students to electrical engineering concepts such as circuits by having students build a circuit on a breadboard consisting of resistors, LEDs, and buttons. Concepts required for control of the GUPPIE and Neu-pulator are then introduced during the remaining days including conditional decision making, timing, *for* and *while* loops, reading sensors, and controlling servos. Once these concepts are

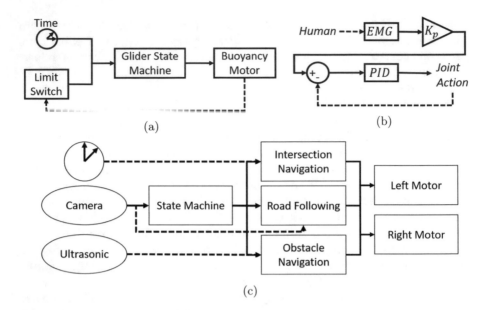

Fig. 2. (a) The control system in the GUPPIE uses a timing based, feedforward controller aided by a limit switch to control the buoyancy system. (b) The Neu-pulator control system is a proportional controller based on EMG signal that is fed into the servos onboard PID controller. (c) The smart car control system is more complex than the other platforms. It includes multiple sensors and a state machine deciding between different control modes depending on the environmental conditions.

presented, students assemble and program a prototype Neu-pulator that they control using potentiometers. To completely program the GUPPIE and Neu-pulator, students need to demonstrate a firm understanding of how the vehicle works, how to instantiate and manage variables, conditional decision making, and timing. The developed codes are tested multiple times on prototype platforms prior to deployment on the actual robots. This enables students to learn through experience with a low-stress test environment.

At the university level, the course initially focuses on development of baseline programming ability. The majority of students in the class have either no experience or limited experience with programming. The university students start with the same tutorial sessions as the middle school students, then rapidly progress to more advanced topics.

4   Survey Results

Using the described curricula and robotic platforms, the authors have engaged a total of 201 middle and high school students over four years through the SYP program including 106 girls. Additionally, 72 university students have completed the autonomous systems course. This paper presents the results from the most recent round of curriculum revision with 51 SYP students during Summer 2017 and 36 university students in Spring 2018. These results are from discrete groups of students so the long term impact of consistent robotics education cannot be concluded. This section presents specific survey results for each group, while overall observations and lessons learned across age groups is presented in Sect. 5.

4.1   Summer Youth Programs 2017 Results

In the Summer Youth Programs a pre-survey, daily surveys, and post-survey were used to evaluate the performance of the curriculum and robotic platforms.

When asked about how their perceived ability to do robotics changed? Students overwhelmingly stated that their ability increased. In fact, 36/51 of the students stated that their coding skills improved. This matches the goals and teaching distribution of the course as the majority of time was spent on teaching students how to program.

When asked what the students favorite segments were, the majority of students (36/51) stated that they enjoyed building or programming. While the mechanical assembly of the actual robotic platforms was not the focus of the course, it was what students had the most experience with prior to the camp and where they were most comfortable. The activities helped the majority of students (85%) to learn about Arduino programming and building circuits. The students stated that they liked that they learned how things work, how programs interface with parts, and how to do programming from scratch and troubleshooting.

Other questions on the survey included career choices, personal history, challenges encountered, and favorite robot. Of particular note in the survey results are some of the individual student responses when asked what are robots useful for. Students gave answers ranging from prosthetics to "military action". The majority of students gave a response focusing on helping human life such as "I think that robots can be useful for assisting elderly and help perform minor medical diagnoses."

When asked about teachers performance, overall, the students liked that the teaching team was knowledgeable and capable of explaining problems when they asked for help, yet the students were also allowed to work independently and solve problems on their own. Further, the students liked that the teaching team was friendly and approachable and each instructor had different strengths and ways of explaining solutions, "I liked how since they were many of them, if you didn't understand one way it was explained you could get a second opinion and they could word it in an understandable way." Although most students appreciated the teaching efforts, some students became frustrated especially when they had to wait for assistance. "I really would only change the number of staff because sometimes you had to raise your hand for a while." These results indicate that the selection of instructors is very important for youth robotics. In particular, having multiple instructors with different ways of explaining abstract concepts helps students from different backgrounds to understand.

When asked about how they would improve the course? Students responded that they would increase the quantity of coding and assembly by possibly building more robots and a longer camp. Several student mentioned that they wanted to customize the platform more, "time and parts to let us tinker." A consistent theme in the survey was that students would prefer to have a workbook with no errors and to have more robust components.

4.2   Spring 2018 Autonomous Systems Results

At the university level a more extensive pre-survey and post-survey were used to evaluate performance of the curriculum. With the extensive survey, we were able to identify several key questions that are able to indicate program effectiveness.

At the beginning of the course we asked students if they had any programming experience. All students said that they had some experience with programming in MATLAB as it is a required component in other courses, however, 16/36 students said that they had some experience with the language used in the course, Arduino. This result combined with our experience teaching Autonomous Systems in the Spring 2017 led us to expand the programming tutorial sections at the beginning of the semester to establish a sufficient baseline knowledge.

When asked what students wanted to do after graduation students gave a wide range of responses from "I want to work in the aerospace industry! I'd love to work at NASA and help further our knowledge of the solar system." to "After receiving my undergrad degree, I will continue my schooling and pursue a graduate degree in mechanical engineering. After that I want to work in the powersports industry as an NVH (Noise, Vibration, and Harshness) engineer." In total, 12/36 wanted to immediately apply the knowledge gained in autonomous systems to a manufacturing related career. This is partly due to the university's long term relationship with the automotive industry.

When comparing the pre- and post-surveys, the confidence in computer programming was identified as the most important question. At the pre-survey, students indicated that they were moderately confident with a mean score of 3.92 on a scale from 1 to 7 versus 4.76 on the post-survey. Additional post-survey results indicate that students thought the work in autonomous systems was more interesting than their other courses (6.37) and they learned more than in other courses (5.57). When combined, these results demonstrate that students appreciated the hands-on nature of the course and the way the material was presented.

Two post-survey results stand out in addition to the mentioned Likert scale questions. When asked if the course has helped them to receive an offer of employment, 20/36 students indicated that it had. This result indicates that the market is in need of new engineers with hands-on robotics experience. Additionally, when asked if they were interested in an advanced level autonomous systems course, 30/36 students answered yes and 36/36 said they would recommend the course to a friend.

5   Observations and Lessons Learned

While the survey results are focused on the 2017 summer youth program and the 2018 autonomous systems course, the iterative nature of the improvements to the curriculum means that we can generate overall observations and lessons learned based on the 201 total (106 female) middle and high school students introduced to GUPPIE and Neu-pulator as well as the 72 university students introduced to the smart car.

Common across all age groups and iterations of the program is the need for high quality support to aid learning. For the youth groups we maintain a 5:1 student:teacher ratio or better, while at the university level 12:1 was sufficient. With this amount of support, students were able to succeed. Teaching assistants in the youth programs need to have a functional understanding of the specific robotic platform but do not need to be experts in robotics and programming. At the university level though, the teaching assistants need to have a deep understanding of mobile robotics and be good at troubleshooting. This is because of the customization allowed during a full semester, every student will develop a very different project as the course progress resulting in unique challenges.

Additionally, students of all age groups appreciate the hands-on nature of the projects. Younger students appreciate having something interesting to show off and experiment with, particularly as it relates to helping human life. University level students are able to see hands-on applications of what they have learned in other courses. Several of our university level students have leveraged the project-based experience in Autonomous Systems to gain employment opportunities.

As the ubiquity of robotics increases over coming years, students of today will require interdisciplinary knowledge to live in the coming smart society where autonomous systems are integrated into everyday life. This interdisciplinary knowledge requires time and practice to develop, and by starting early, students are given the best chance of succeeding in the future. However, just providing early engagement opportunities is not enough. Continued engagement through university level is necessary and can be built around a common learning experience. This common experience follows the same 'see-think-act' regardless of the specific project and technical competency. Using these lessons learned, a fully integrated continuous learning program from middle school through graduate school can be created.

Overall, an integrated pathway from middle school through university must maintain a motivating learning context in order to engage the students. For example, we have tried to focus on helping human life in pre-college and autonomous cars at the university level. Additionally, we have found that younger students are surprisingly adept at technical challenges, however, managing frustration is key to learning. We have had good success with integrating play and learn together. As the students get older, they become less agile with new concepts but are better able to manage stress resulting in more consistent learning.

References

1. Benitti, F.B.V.: Exploring the educational potential of robotics in schools: a systematic review. Comput. Educ. **58**(3), 978–988 (2012). https://doi.org/10.1016/j.compedu.2011.10.006. http://www.sciencedirect.com/science/article/pii/S0360131511002508

2. Ziaeefard, S., Ribeiro, G.A., Mahmoudian, N.: GUPPIE, underwater 3D printed robot a game changer in control design education. In: 2015 American Control Conference (ACC), pp. 2789–2794 (2015). https://doi.org/10.1109/ACC.2015.7171157

3. Ziaeefard, S., Miller, M.H., Rastgaar, M., Mahmoudian, N.: Co-robotics hands-on activities: a gateway to engineering design and stem learning. Robot. Auton. Syst. **97**, 40–50 (2017). https://doi.org/10.1016/j.robot.2017.07.013. http://www.sciencedirect.com/science/article/pii/S0921889017301240

4. Ziaeefard, S., Page, B.R., Knop, L., Ribeiro, G.A., Miller, M., Rastgaar, M., Mahmoudian, N.: GUPPIE program—a hands-on stem learning experience for middle school students. In: 2017 IEEE Frontiers in Education Conference (FIE), pp. 1–8 (2017). https://doi.org/10.1109/FIE.2017.8190546

5. Ziaeefard, S., Mahmoudian, N.: Marine robotics: an effective interdisciplinary approach to promote stem education. In: Lepuschitz, W., Merdan, M., Koppensteiner, G., Balogh, R., Obdržálek, D. (eds.) Robotics in Education, vol. 630, pp. 154–165. Springer, Cham (2018)

6. Mitchell, B., Wilkening, E., Mahmoudian, N.: Developing an underwater glider for educational purposes. In: 2013 IEEE International Conference on Robotics and Automation, pp. 3423–3428 (2013). https://doi.org/10.1109/ICRA.2013.6631055

7. Ziaeefard, S., Mahmoudian, N.: Building micro underwater gliders: lesson plan for exploring engineering design and understanding forces and interaction. Mich. Teach. Assoc. **61**, 48–56 (2016)

8. Knop, L., Ziaeefard, S., Ribeiro, G.A., Page, B.R., Ficanha, E., Miller, M.H., Rastgaar, M., Mahmoudian, N.: A human-interactive robotic program for middle school stem education. In: 2017 IEEE Frontiers in Education Conference (FIE), pp. 1–7 (2017). https://doi.org/10.1109/FIE.2017.8190575

9. Page, B.R., Ziaeefard, S., Moridian, B., Mahmoudian, N.: Learning autonomous systems—an interdisciplinary project-based experience. In: 2017 IEEE Frontiers in Education Conference (FIE), pp. 1–7 (2017). https://doi.org/10.1109/FIE.2017.8190555

10. Bischoff, R., Huggenberger, U., Prassler, E.: Kuka youbot - a mobile manipulator for research and education. In: 2011 IEEE International Conference on Robotics and Automation, pp. 1–4 (2011). https://doi.org/10.1109/ICRA.2011.5980575

11. Rubenstein, M., Cimino, B., Nagpal, R., Werfel, J.: Aerobot: An affordable one-robot-per-student system for early robotics education. In: 2015 IEEE International Conference on Robotics and Automation (ICRA), pp. 6107–6113 (2015). https://doi.org/10.1109/ICRA.2015.7140056

12. Paull, L., Tani, J., Ahn, H., Alonso-Mora, J., Carlone, L., Cap, M., Chen, Y.F., Choi, C., Dusek, J., Fang, Y., Hoehener, D., Liu, S., Novitzky, M., Okuyama, I.F., Pazis, J., Rosman, G., Varricchio, V., Wang, H., Yershov, D., Zhao, H., Benjamin, M., Carr, C., Zuber, M., Karaman, S., Frazzoli, E., Vecchio, D.D., Rus, D., How, J., Leonard, J., Censi, A.: Duckietown: an open, inexpensive and flexible platform for autonomy education and research. In: 2017 IEEE International Conference on Robotics and Automation (ICRA), pp. 1497–1504 (2017). https://doi.org/10.1109/ICRA.2017.7989179

13. Nelson, S.G., Cooper, K.B., Djapic, V.: SeaPerch: how a start-up hands-on robotics activity grew into a national program. In: OCEANS 2015 - Genova, pp. 1–3 (2015). https://doi.org/10.1109/OCEANS-Genova.2015.7271419
14. OpenROV: DIY kit—openrov underwater drones (2018). https://www.openrov.com/products/openrov28/
15. Tehrani, N.H., Heidari, M., Zakeri, Y., Ghaisari, J.: Development, depth control and stability analysis of an underwater remotely operated vehicle (ROV). In: IEEE ICCA 2010, pp. 814–819 (2010). https://doi.org/10.1109/ICCA.2010.5524051
16. Liu, H., Wang, B., Xiang, X.: Adventure-I: A mini-AUV prototype for education and research. In: 2016 IEEE/OES Autonomous Underwater Vehicles (AUV), pp. 349–354 (2016). https://doi.org/10.1109/AUV.2016.7778695
17. Cabré, T.P., Cairol, M.T., Calafell, D.F., Ribes, M.T., Roca, J.P.: Project-based learning example: controlling an educational robotic arm with computer vision. IEEE Revista Iberoamericana de Tecnologias del Aprendizaje 8(3), 135–142 (2013). https://doi.org/10.1109/RITA.2013.2273114
18. Serrano, V., Thompson, M., Tsakalis, K.: Learning multivariable controller design: a hands-on approach with a lego robotic arm. In: Chang, I., Baca, J., Moreno, H.A., Carrera, I.G., Cardona, M.N. (eds.) Advances in Automation and Robotics Research in Latin America, pp. 271–278. Springer, Cham (2017)
19. Siegwart, R., Nourbakhsh, I., Scaramuzza, D.: Introduction to Autonomous Mobile Robots. Intelligent Robotics and Autonomous Agents. MIT Press, Cambridge (2011). https://books.google.com/books?id=4of6AQAAQBAJ

Technological Literacy Through Outreach with Educational Robotics

Georg Jäggle$^{(\boxtimes)}$, Lara Lammer, Hannah Hieber, and Markus Vincze

ACIN Institute of Automation and Control, Vienna University of Technology,
Vienna, Austria
gjaeggle@acin.tuwien.ac.at

Abstract. Educational robotics has gained increased importance and attention worldwide as an excellent teaching tool for STEM (Science, Technology, Engineering, Mathematics). However, catching the enthusiasm of young learners who are not already interested in STEM remains challenging. In this paper, we describe our outreach concept where young people from Vienna and surroundings visit the technical university with their teachers for a three-hour program. Our focus is on technological literacy, the understanding of what a robot is and how it works and may look like, as well as different robotic application areas. Theory and hands-on are combined in an age-appropriate concept based on constructivism. In order to evaluate our approach, we have developed a short post-questionnaire. Our results with 255 young people ages 7 to 17 show that after the visit 84% are more interested in technology and 80% are more interested in robotics. 85% of the young learners find that robots are complex machines after the visit. Despite that fact, 85% of those who find robots complex are more interested in robotics, 85% want to come back to learn more about robotics, and 91% will tell their families about these activities.

Keywords: Educational robotics · Technological literacy · Interest in STEM · Perception of robots · STEM career

1 Introduction

Many educational robotics activities evolve around the motivation to evoke interest in STEM fields [1, 2]. In some cases, there are more specific motivations, like teaching robotics [3], mathematics [4], physics [5], computational thinking [6], engineering [7], programming design and project management [8] or mechanics [9]. Consequently, the focus of most approaches is on teaching STEM and robotics [10] as well as hands-on problem solving, teamwork, and innovation [11]. However, activities being short-term, high-intensity, technology-focused, and competition-driven may be limiting participant diversity [12]. Alimisis [13] similarly argues that the way robotics is currently introduced in educational settings is narrow. If young people with a wider range of interests are to be addressed, broader perspective projects and new and innovative ways to increase the attractiveness and learning profits of robotics are needed. Kandlhofer and Steinbauer [14] confirm this by arguing that the "concept of educational robotics should not only focus on separate, isolated topics but [..] also be applied as an

© Springer Nature Switzerland AG 2020
M. Merdan et al. (Eds.): RiE 2019, AISC 1023, pp. 114–125, 2020.
https://doi.org/10.1007/978-3-030-26945-6_11

integrated approach, fostering a holistic understanding and acceptance of different areas and fields".

In this paper, we are going to describe our approach that elucidates technology and robots for all children in order to promote technological literacy and interest in technology and robotics. We developed an outreach concept and a fitted POST-questionnaire, which was evaluated scientifically through a pilot-phase and a subsequent evaluation process. The paper presents findings from our evaluation with 255 children over a time-period of nine months, from May 2018 to January 2019.

2 Related Work

Motivation and emotions play an important role in learning. Adapting learning activities to children's lives and interests [15] and empowering children to learn through play [16] will motivate them. These ideas build the base of the theoretical framework of constructivism. Furthermore, according to Bruner's [17] constructivist learning theories, the act of learning involves three almost synchronous processes: acquisition (gaining new information), transformation (changing old information into new information), and evaluation (judging if the information change makes sense). Social support enhances the learning experience; knowledge and strategies are shared and developed through social interaction with other people. Language, tools and artefacts are important media to externalize ideas, which in turn is key for communication with others [18].

One study published several reasons for the loss of interest in STEM: There is too much content in several curriculums. The teacher uses the wrong teaching methods. The reputation of the discipline is not so popular in peer groups. The learning process dissuades students from pursuing the field, as most of the students receive bad marks in STEM subjects and think STEM disciplines are too difficult. The wrong teaching methods are reading instructional texts and are too theoretical. [19]. One factor to influence on pursuing a STEM career positively is to increase the interest in STEM with learning through hands-on exercises during the lesson [1]. In this way, the opportunities for experiential learning could be increased, resulting in the pupils broadening their horizons through 'learning by doing' like hands-on activities [20]. Another factor that positively influences the pursuit of a STEM career is through increasing the self-efficacy [21], which is linked to a positive STEM task performance [22]. More practical hands-on activities increase the students' self-efficacy and influence the positive attitude in STEM are [23, 24]. Also, it is necessary to participate in out-of-school activities to increase the interest in STEM [25–27].

For the design of our activities, we use a powerful concept that considers all above mentioned factors. For example, young learners come to the technical university to become researchers and investigate robots and robot behaviours. Alternatively, we talk about technology and nature together and different robotic application areas, especially healthcare, to involve those kids who are not necessarily interested in robots or technology but social topics. We start with the lecture and repeat key elements throughout the whole visit to enhance the learning. Our tutors from different backgrounds (students of engineering, architecture, psychology or literature science, or electrical engineer in retirement) act as role models. During the activities, we underline the design process and combine black-box with white-box to demystify robots and empower the students to think about these as useful tools to make human lives better.

3 Outreach with Educational Robotics

In our program 'Outreach with Educational Robotics' our focus is on technological literacy, the understanding of what a robot is and how it works and may look like, as well as different robotic application areas. Theory and hands-on are combined in an age-appropriate concept based on constructivism. School classes of elementary and secondary schools from Vienna and surroundings visit the technical university with their teachers for a three-hour program. We start with the activity block "lecture about robots" for the whole class and then divide them into groups. Each group visits each of the hands-on activity stations "explore a robot", "innovation lab", and "interaction with a humanoid robot". Important elements, like the children being scientists or sensors being an important part of a robot, are repeated through all activity blocks. When the children are young (under 9 years old), the "innovation lab" is skipped, and the visit shortened to 2,5 h. In the following we describe each of the activity blocks.

3.1   Activity Block: Lecture About Robots

The main objective of the lecture is to elucidate technology and robots to the visiting class (and the accompanying adults although this is not our focus). Following topics are addressed and also repeated through all activity blocks:

- What is technology? What is nature?
- What is a robot? What is the difference to other machines?
- What application areas are there for robots? (Focus on social topics and healthcare)
- How do robots look like? Do they really have to look and behave like humans?
- Which parts does a robot have? How do sensors work?

The visit starts in a lecture hall where the class is greeted and welcomed to the university. The children are pointed to the fact that they are going to do the same things as students at the university do, hear a lecture, work hands-on with robots and be scientists. The lecture is adapted to the age level of the children and lasts between 45 and 60 min. In an interactive session, the children first discover the definition of technology as human-made and useful artefacts, as opposed to nature which could exist or grow without human interference. The knowledge and process of making artefacts and using them are also included in this technology definition. Then, the young learners are introduced to our definition of a robot as an autonomous self-driven technology with a physical embodiment that senses its environment, reacts to changes in it and eventually changes it. The lecture then continues with different application areas and robot appearances. We put a focus on showing the students that robots do not have to look like humans or imitate human behaviour. Finally, we finish the lecture with robot parts, especially sensors and microcontrollers. Depending on the age level we go in depth and show practical uses of STEM subjects like mathematics or physics. Figure 1 shows one example of different robot appearances categorized in machine-like, animal-like, cartoon-like, human-like after Fong and Dautenhahn [28].

Fig. 1. An overview of different robot appearances

3.2   Activity Block: Explore a Robot

The aim of this 30 min activity is to foster explorative learning and a positive attitude towards robots. The hands-on use of Thymio robot allows the participants to make their first observations about the different sensors and their functions autonomously, and thus become scientists at the technical university. Thymio is a small robot that is programmed via its "Push-the-button" function. It has six programmes (colours) which use different sensors (touch and infrared sensors, microphone) and show different behaviours (follow an object, avoiding obstacles, run away from an object, follow a trail, follow button command, handclap reaction).

In teams of two or three, the first task is: "We found this thing here and want to find out what it is. Please help us find out what it is and what it can do." After the participants have a first idea of the Thymio robot and its programs, there is an in-depth exploration of these programs. Participants are asked to find out which behaviour the Thymio robot shows in each program and which sensors are active in each case and note these findings in a distributed worksheet. The last part consists of problem-solving tasks. The participants are asked to choose the right programs so that all Thymios can go in a row one behind each other. Depending on the level of the participants they are asked to put a pen in the pen holder of Thymio robots and create geometric forms on the paper (Figs. 2 and 3).

Fig. 2. Generation robots (https://www. generationrobots.com/de/401213-mobiler-roboter-thymio-2.html)

Fig. 3. Explore a robot

3.3   Activity Block: Innovation Lab

The aim of this activity is to make participants understand that a robot is built of different components (sensor, processor, actuator) and that building robots is a long process of having ideas, prototyping and refining. Different functions of sensors (distance sensors, touch sensors, muscle sensors) are introduced. Participants are involved actively by exploring the sensors and brainstorming about functions and possible application fields (Fig. 4).

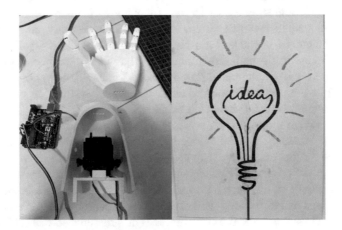

Fig. 4. Innovation lab

3.4   Activity Block: Interaction with a Humanoid Robot

Participants are expected to develop an understanding of the joint work of sensors, processors and actuators and their importance for the interaction between humans and humanoid robots. Furthermore, participants are required to be aware of the differences between a human and a humanoid robot. They are introduced to different application fields of humanoid robots and should be able to assess the necessity of humanoid robots for certain activities critically. The participants are involved actively to explore

Pepper robots functions by interacting with it, and also see the code "that makes the robot talk and move" on a big screen. With different interactive games, they discover the joint work of sensors and actuators for the human-robot interaction. Differences between the robots Romeo, Nao and Pepper are discussed (Fig. 5).

Fig. 5. Interaction with a humanoid robot

4 Evaluation Design

The evaluation design is divided into two parts. A quantitative part with questionnaires and the qualitative part with image interpretation. The combination of both research methods gives a better understanding of the perception from young people at the robotic field. The students bring their drawings of robots before the outreach robotics workshop and fill out the questionnaires after the activity blocks. The evaluation process is accompanied by the following main questions:

1. How is the perception about robots by students?
2. How did robotic activities influence the interest of students in STEM?
3. How did students like the robotic activities?

4.1 Quantitative Method

In order to find out to what extent the participation in the workshops influences the attitude towards STEM, robotics and the interest in a STEM career, we constructed a 4-scale- questionnaire that the children fill out after the participation. The evaluation process started with a pilot-phase from April to Mai with 70 participants. After the pilot

Table 1. Questionnaire

Shortcut	Statements
Q1	This is my first visit at a university
Q2	I have visited a robotics workshop like this before
Q3	After this robotics workshop, I am more interested in technology
Q4	After this robotics workshop, I am more interested in robotics
Q5	I would like to come back and learn more about robotics
Q6	I will tell my family about today
Q7	Robots are complex machines
Stars	How many stars would you give the workshop?

phase, we developed a POST-Questionnaire and evaluated the outreach robotic workshops from May 2018 to January 2019 with 255 young students ages 7 to 17 (shown in Table 1).

The statement Q1 gives more information about our participant and the impact of out-of-school activities at the technical university. We want to reach young people without experience at the university. The statement Q2 gives an answer about the young people and their experiences in robotics workshops. It could be interesting in the long term if young people will participate more in robotics workshops than at the moment and if there is a difference in gender, age and migration background. The statements Q3 and Q4 give an answer about the influence in pursuing a STEM career. We hypothesize that the outreach robotics workshop influences the interest in STEM and the path to the university. We included the statement Q6 because the family plays an important role in the pursuit of a career of young people and it is important that they share their learning experiences at the technical university in their social group. The statement Q7 in combination with Q5 shows us the resilience and personal self-efficacy to come back and learn more about robotics although they think that robots are complex machines. The stars give us results for the 3rd main question, how did the young people like the outreach robotics workshop.

4.2   Findings

The questionnaires after the outreach robotics workshop filled out from 255 young people, which were 54% girls, 46% boys and 51% had not German as their first language. 189 (75%) of the 255 participants visited a robot workshop for the first time, 64 (25%) have visited a robotics workshop before. 73% of the participants visited a university for the first time and had at the statement Q3 and Q4 a mean of 4 compared to the other group with a mean of 3. The participants, who have a robotics activity at the university for the first time are more interested in technology and robotics after the visit. This result shows the impact of the out-of-school activities at the university (Table 2).

Table 2. Results of the questionnaire

	Q1	Q2	Q3	Q4	Q5	Q6	Q7	Stars
Percentile 25	2	1	3	3	3	3	3	4
Minimum	1	1	1	1	1	1	1	2
Median	4	1	4	4	4	4	4	4
Maximum	4	4	4	4	4	4	4	4
Percentile 75	4	3	4	4	4	4	4	4

The table shows that most of the students are more interested in technology and robotics after the robotics workshop. Also that most of the students would like to come back and learn more about robotics and that most of the students will tell their family about this day. The students gave in mean four stars for the activities. The lowest score was two stars.

The gender comparison of Q3 and Q4 show different results. The mean was three for girls and four for boys. The boys are more interested in technology and robotics after the workshop than girls. However, in the gender comparison of the primary school children, there is no difference between Q3 and Q4. This result is interpreted that the outreach robotics workshop is an multiple-entry point for all young children (elementary school level). Additionally, the children from primary schools are more interested in technology and robotics after the workshop than the students from the secondary school. Students without German as their first language are also more interested in technology and robotics after the workshop than students with German as the first language (Table 3).

Table 3. The statement Q7 compared with Q5 and Q6

		I would come again to learn more about robots						I will tell these activities my parents			
		Yes		No		Total		Yes		No	
		Number	Percent	Number	Percent	Number	Percent	Number	Percent	Number	Percent
Robots are complex	Yes	182	85,4%	31	14,6%	213	100,0%	194	90,7%	20	9,3%
	No	31	83,8%	6	16,2%	37	100,0%	33	89,2%	4	10,8%
	Total	213	85,2%	37	14,8%	250	100,0%	227	90,4%	24	9,6%

213 participants answered the statement "Robots are complex" with strongly agree or agree. 182 (85,4%) of those participants answered the statement "I would come again to learn more about robots" with strongly agree or agree. 227 (90,7%) of those answered the statement "I will tell these activities my parents" with strongly agree or agree. This result shows that the students will come again although robots in their perception are complex machines and that most of the students will share their experience with their community.

4.3   Qualitative Method

With our Draw-A-Robot working sheet, we wanted to investigate children's perceptions of robots and secondly document how stereotypic images change over the years. Therefore we developed a combination of the Draw-A-Scientist Test (DAST) and the 5 step plan of Lammer [29]. The Draw-A-Scientist Test (DAST) is an open-ended test designed to investigate primary school children's perceptions of a scientist [9]. The 5-step plan is designed for children in primary and secondary school and based on design methods that empower children by giving them a child-appropriate structure for their creative process. The plan can be integrated into different teaching or research contexts, and adapted to different age groups or even to adults who are not familiar with robotics.

The result of the combination of both methods mentioned above was the worksheet Draw-A-Robot. First, participants are asked to draw a robot as they imagine it and get a list of the following points to support them:

Step 1 – Robot Users (Who is meant to use your robot?)
Step 2 – Robot Tasks (What abilities has your robot and which tasks does it take care of?)
Step 3 – Robot Communication (How can you communicate with your robot?)
Step 4 – Robot Design (How does your robot look like?)
Step 5 – Robot Behavior (Which behaviour does your robot show?)
Step 6 – Robot Parts (Which materials and components do you need for your robot?)

Second, participants are asked to annotate their drawings with words and sentences in order to clarify abstract ideas.

4.4   Evaluation Process

We integrated the Draw-A-Robot worksheet in the preparations for two of our Outreach workshops. 41 secondary school children at the age of 11 to 12 filled out the worksheet in their handicraft lesson at school before coming to the workshop. Lammer [29] suggests that children design robots in order to address actual problems in their lives (including their family), e.g. being alone at home, needing support with homework or wanting somebody to play with. In our analysis, we, therefore, concentrated on the standard indicators which derived from Lammer's [29] findings. Additionally, we were interested in the shape the participants gave their robot and if they draw the robot being inside their house or not.

4.5   Results

Table 4 shows which feature and tasks the participants gave their robots. Figure 2 shows four examples of children's drawing ideas.

Table 4. Features and tasks

Features	Total occurrence
In the house	12
Outside of the house	6
Humanoid	34
Not Humanoid	8
Tasks derived from Lammer (2017)	
Play or entertain	4
Do or help with homework	7
Help or serve or both	3
Help in household	4
Cook or serve food	13
Bring or carry or lift objects	6
Talk or make conversation	15
Be a friend (comforting)	3
Protect	2
Play music or sing	12

In most of the drawings, it was not recognisable where the robot is located. It was placed in the house more often (12 occ.) than outside (6 occ.). The most preferred morphology was anthropomorphic (34 occ.), followed by zoomorphic (3 occ.) and machine-like (3 occ.). In total, 13 children explicitly stated that their robot should be nice or friendly. These findings affirm Lammers identification of social robots designed as assistants, companions or pets. Figure 6 shows a robot that is not anthropomorphic and is designed to address the needs of humans (carry and lift things). Figure 7 shows a strong perception of the robot as a living being. It is drawn with several human attributes (name, gender, size) and is given human anatomy.

 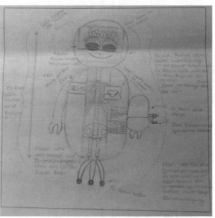

Fig. 6. Draw a robot 1   **Fig. 7.** Draw a robot 2

5 Conclusion and Outlook

In our work, we always place children and their needs first. We design our activities with our knowledge of pedagogical design and focus our scientific approach to understanding the short-term impact – intended or not intended – of our educational activities. The combination of hands-on activities with out-of-school-activities in an educational robotic context offers more interest in STEM and particularly influences self-efficacy. Especially, the novel situations and out-of-school workshops force an adaptation or adjustment of students at the new activity space and foster their technology literacy towards the topics of robotics. The approach of constructivism learning in out-of-school workshops increases the interest in STEM of young students. The outreach concept is a multiple-entry point for all students irrelevant which gender, age or culture. The evaluation shows that most of the students have more interest in STEM and will share their experience with others. The next steps are to find out the impact of role models with a developed questionnaire, which will be extended with pictures and symbols to reach all young people with high reading and low reading skills. It will also be interesting to have a long-term study of Draw-a-Robot to investigate the changing perceptions of robots and the interest in STEM and technological literacy sustainably.

We will also develop a new evaluation tool to evaluate the gained technological literacy during the outreach robotics workshop.

Acknowledgements. The Outreach Program is sponsored by the Faculty of Electrical Engineering and Information Technology at the TU Wien. We would like to thank Habibe Idiskut, Markus Ortner, Martin Piehslinger, and Jan-Ove Wiesner for their help in carrying out the workshops and collecting the data.

References

1. Mead, R.A., Thomas, S.L., Weinberg, J.B.: From grade school to grad school: an integrated STEM pipeline model through robotics. In: Robots in K-12 Education: A New Technology for Learning, pp. 302–325. IGI Global (2012)
2. Bredenfeld, A., Leimbach, T.: The roberta initiative. In: Workshop Proceedings of International Conference on Simulation, Modeling and Programming for Autonomous Robots, (SIMPAR 2010), pp. 558–567 (2010)
3. Yudin, A., Salmina, M., Sukhotskiy, V., Dessimoz, J.-D.: Mechatronics practice in education step by step, workshop on mobile robotics. In: Proceedings of 47th International Symposium on Robotics, ISR 2016, pp. 1–8 (2016)
4. Norton, S.J.: Using Lego construction to develop ratio understanding. Math. Educ. Third Millenn. Towards **2010**, 414–421 (2004)
5. Williams, D.C., Ma, Y., Prejean, L., Ford, M.J., Lai, G.: Acquisition of physics content knowledge and scientific inquiry skills in a robotics summer camp. J. Res. Technol. Educ. **40**(2), 201–216 (2007)
6. Catlin, D., Woollard, J.: Educational robots and computational thinking. In: Proceedings of 4th International Workshop Teaching Robotics, Teaching with Robotics & 5th International Conference Robotics in Education, pp. 144–151 (2014)
7. Rogers, C.: Engineering in kindergarten: how schools are changing. J. STEM Educ. Innov. Res. **13**(4), 4 (2012)
8. Wolz, U.: Teaching design and project management with Lego RCX robots. ACM SIGCSE Bull. **33**(1), 95–99 (2001)
9. Chambers, J.M., Carbonaro, M., Murray, H.: Developing conceptual understanding of mechanical advantage through the use of Lego robotic technology. Australas. J. Educ. Technol. **24**(4) (2008)
10. Benitti, F.B.V.: Exploring the educational potential of robotics in schools: a systematic review. Comput. Educ. **58**(3), 978–988 (2012)
11. Feil-Seifer, D., Matarić, M.J.: Human robot human–robot interaction (HRI). In: Encyclopedia of Complexity and Systems Science, pp. 4643–4659. Springer, Heidelberg (2009)
12. Hamner, E., Lauwers, T., Bernstein, D., Stubbs, K., Crowley, K., Nourbakhsh, I.: Robot diaries interim project report: development of a technology program for middle school girls (2008)
13. Alimisis, D.: Educational robotics: open questions and new challenges. Themes Sci. Technol. Educ. **6**(1), 63–71 (2013)
14. Kandlhofer, M., Steinbauer, G.: Evaluating the impact of educational robotics on pupils' technical-and social-skills and science related attitudes. Robot. Auton. Syst. **75**, 679–685 (2016)
15. Piaget, J., Inhelder, B.: The Psychology of the Child London, Henley Roudedge Kegan Paul (1969)

16. Montessori, M.: The Montessori Method, Rome 1912 (1964)
17. Bruner, J.S.: The Process of Education. Harvard University Press, Cambridge (2009)
18. Vygotsky, L.S.: Mind in Society: The Development of Higher Psychological Processes. Harvard University Press, Cambridge (1980)
19. Thomas, M., Weigend, M.: Informatik und Natur: 6. Münsteraner Workshop zur Schulinformatik, 1. Aufl. Books on Demand (2014)
20. Honebein, P.C., Duffy, T.M., Fishman, B.J.: Constructivism and the design of learning environments: context and authentic activities for learning. In: Designing Environments for Constructive Learning, pp. 87–108. Springer, Heidelberg (1993)
21. Kramer-Bottiglio, R.: Intersecting Self-Efficacy and Interest: Exploring the Impact of Soft Robot Design Experiences on Engineering Perceptions (2018)
22. Britner, S.L., Pajares, F.: Sources of science self-efficacy beliefs of middle school students. J. Res. Sci. Teach. Off. J. Natl. Assoc. Res. Sci. Teach. 43(5), 485–499 (2006)
23. Ornstein, A.: The frequency of hands-on experimentation and student attitudes toward science: a statistically significant relation (2005-51-Ornstein). J. Sci. Educ. Technol. 15(3–4), 285–297 (2006)
24. Foley, B.J., McPhee, C.: Students' attitudes towards science in classes using hands-on or textbook based curriculum. Am. Educ. Res. Assoc. (2008)
25. Dabney, K.P., et al.: Out-of-school time science activities and their association with career interest in STEM. Int. J. Sci. Educ. Part B 2(1), 63–79 (2012)
26. Holstermann, N., Grube, D., Bögeholz, S.: Hands-on activities and their influence on students' interest. Res. Sci. Educ. 40(5), 743–757 (2010)
27. Jäggle, G., Lepuschitz, W., Girvan, C., Schuster, L., Ayatollahi, I., Vincze, M.: Overview and evaluation of a workshop series for fostering the interest in entrepreneurship and STEM. In: 2018 IEEE 10th International Conference on Engineering Education (ICEED), pp. 89–94 (2018)
28. Fong, T., Nourbakhsh, I., Dautenhahn, K.: A survey of socially interactive robots. Robot. Auton. Syst. 42(3–4), 143–166 (2003)
29. Lammer, L., Weiss, A., Vincze, M.: The 5-step plan: a holistic approach to investigate children's ideas on future robotic products. In: International Conference on Human-Robot Interaction Extended Abstracts, Bamberg (2015)

Robotic Theater: An Architecture for Competency Based Learning

Enrique González[1], Andrés De La Pena[1], Felipe Cortés[1], Diego Molano[1],
Benjamín Baron[1], Nicolas Gualteros[1], John Páez[1,2(✉)], and Carlos Parra[1]

[1] Pontificia Universidad Javeriana, Bogotá, Colombia
egonzal@javeriana.edu.co
[2] Universidad Distrital Francisco José de Caldas, Bogotá, Colombia
jjpaezr@udistrital.edu.co
http://www.javeriana.edu.co
http://www.udistrital.edu.co

Abstract. In this paper, an educational model for competency based learning is proposed; this model is applied in some small cities in Colombia. For this purpose, a platform for robotic theater is developed to represent social problems. Furthermore, the Quemes robot, the VED EDR kit robotic kit and some kits of sensors and actuators are used. All of the components and robot used are Arduino-compatible. The model was successfully applied with the participation of about 200 participants. At the end of the learning process, the young participants are able to propose and create innovative solutions applicable to their regions.

Keywords: Education · Robotics · Robotic theatre ·
Constructionism · Arduino

1 Introduction

During the last few years, technology has been rapidly changing and evolving, producing important impacts on society. In Colombia, ICT contributes with a wide range of applications that help to develop high influence projects in the country and its regions. However, not all social groups have been able to have technology access and knowledge, despite the effort that the government of Colombia has done. In order to be a positive factor that mitigates this situation, the Smart Town Project[1] has integrated an educational model and a robot-based technological platform in order to create an experience for boys and girls in small cities and towns to develop competencies related with: citizenship and appropriation of territory, eco-environmental sensitivity, innovation and entrepreneurship, science and technological skills. At the end of the learning process, the young participants are able to propose and create innovative solutions applicable to their regions. This process is organized in three educational stages: awareness,

---

[1] https://sophia.javeriana.edu.co/smarttown/index.html.

© Springer Nature Switzerland AG 2020
M. Merdan et al. (Eds.): RiE 2019, AISC 1023, pp. 126–137, 2020.
https://doi.org/10.1007/978-3-030-26945-6_12

appropriation and projection. In the first stage, the learners get familiar with robotics and are able to understand the potential of this technology. In the second moment, they learn and apply the engineering design methodology and its phases by using and appropriating a technological platform for robotic theater. In the third moment, they propose and develop an applied automation project that responds to a problem or take advantage of an opportunity of their community or region. In order to carry out this educational experience, a series of educational primer, a software platform and electronic boards were developed. Moreover, two kinds of robots were used during the implementation of the project: Quemes, a line follower robot, and the VEX EDR robotic kit. The robotic theater platform controls the interaction between robot actors. The internal architecture of the robot is based on a BDI emotional model. These tools not only allow to developed technical skills, but also soft skills as required; moreover, they encourage the participation of the learners and represent an amusing tool that motivates the learning process.

2 Robotics and Education

Robotics has been acknowledged as a learning tool that motivates students, fosters the development of logical thinking, supports the learning process and promotes the development of complex thinking, which are very useful skills in the modern times. At the end of the twentieth century, Seymour Papert proposed the Constructionism [1] as strategy to support the learning process through the use of robots [2]. The Quemes robot is an example of the application and design of a learning tool using the Constructivism pedagogical model. The platform was developed not only to learn technological subjects, but also includes a strategy to develop cooperative work and citizenship skills [3], [4]. There are many other similar examples, for instance using LEGO robots to learn physics concepts [5], to acknowledge new ways of thinking [6], to foster the communicative skills in areas which have been traditionally away from technology. However, robot platforms are not the only critical factor, the implementation of robotics in scholar environments requires also to take into account issues as design of adequate learning spaces, teacher's training to use robots correctly [7], and new vision of teacher's role as mediator [8]. In conclusion, the robots fosters four group of competences: dynamic *(motivation)*, strategic *(metacognition)*, cognitive *(learning tools)* and specifics *(technical)*, [9], [10]. During the 90s in Colombia and other countries, the use of robots in educational contexts was promoted by two factors: the teacher's curiosity to know about the robotics' benefits and the governmental policies that recognize the technology's area as a need in the curriculum. Currently, the use of robots is present in two scenarios: formal education in schools [11] and non-formal education which has a greater visibility in technology parks and science museums. Most of the successful cases are related to the STEAM areas, *(Science, Technology, Engineering, Art and Mathematics)*. This diverse use has led to the maturation of the educational robotics field and it has increased the proposals to use robots in academic projects and as even in entrepreneurship regional proposals as *Robots y enseñanza en Colombia, 2013*. The Smart Town project is positioned within the constructionism paradigm and uses robotics as a motivation and effective tool, not only to acquire technological skills but also to develop competences related to citizenship, innovation, environment, and even communication and reflection skills.

3   Educational Model

The *AC4* learning model was created by an interdisciplinary team that includes educators, psychologists, and engineers while the development and the research project named *Smart Town, Talent and Innovation Applied to the Territory*. This model arises from a pedagogical emerging perspective, in which knowledge is assumed as a collective construction and has meaning according to the specific contexts where it is produced [12]. In this sense, this model finds epistemological support in the socio-constructivism paradigm which understands learning as inter-learning. This approach is feasible when there are connections among cultural and historical elements from each of the learning participants. In the project, tutors, learners, researchers and research assistants receive and exchange information about the environment to adapt it structurally and foster the development of the cognitive structures. The AC4 model, *(Cooperation, Construction, Community and Creativity)*, was implemented in learning environments named co-laboratories. Human development procedures are assumed as a series of interdependent, simultaneous and individual and group learning processes that participants generate through the exchanges and interactions in specific socio-historical contexts which are crossed by cultural, political and economic traditions, [12]. Thus, the AC4 proposal recognizes that development is simultaneous and interdependent to the process in which the participants adapt to the culture through the inclusion of social and symbolic models. It means that when people learn, there are cognitive dynamic and social actions which allow each one learn from the other to acquire the competences required to transform and perpetuate culture and knowledge. The AC4 model assumes creativity as a strategy carried out by the participants to transform their lives according to their needs and the community needs. From this perspective, adolescents are recognized as key authors in the transformation of social structures that favor the production and reproduction of their adversities. For this project, the adversities are a challenge to the curiosity. The adversity related to a problem favored the development of solutions and so the change and the transformation of the local and regional environments: "the characteristic of curiosity would be joined deeply to the scientific and technological level, to the critical-reflexive disposition and people's access to the continuous training. In fact, the man progress in the nature control is the history of a long process, in which each overtaken obstacle has been an unfavorable condition which might have become in a trouble; and so it causes searches and explorations" [13] (p. 70). Here, creativity is the condition to build other stories and proposals, or as [14] claimed "it supposes a fact of permanent overcoming, reconstruction and independence" (p. 280). Under these circumstances, the proposed learning model promotes the construction of solutions having in mind the particularities given for the region where the participants live, their age, genre and cultural resources. Thus, these elements are fundamental to give sense to the learning experiences. In this model, the appropriateness of its contents and methodologies are linked to the territories where the learning practices are carried out: in the case of the project the small size cities Zipaquirá, Soacha, Girardot located in the Cundinamarca region. These towns are located in the center area of Colombia with 125.000, 350.000 y 144.000 residents respectively. The AC4 learning model allows integrating permanently the learning environments with the participants' comprehensions, their contexts and needs. For example, the apprentice is assumed as a learner who has an attitude to learn permanently, pointed to the acquisition of results and participation in specific processes. Thus, it is able to lead a perspective centered in the cause-effect connection and retake the principle of indivisibility among learning contents and life.

4   Educational Model Application

Some activities were proposed at each educational stage in order to apply the AC4 educational model; such activities were set out through amusing stories in a learning book. For this purpose, some hardware and software tools were designed and implemented. In this educational stage, students get familiar with robots in order to get aware of their potential. By identifying different kinds existing robots, learners not only learn the fundamentals technical concepts, but also can appreciate their utility for society. During this process, they get encouraged to study this technology in depth, and also get confidence in their capacity to do this. The main tool for this moment is a line follower robot, called Quemes.

4.1   First Educational Stage: Awareness

The first activity that is proposed to the learners is to look up for information about different kinds of robots that exist, their features and functions, and also their applications. With this information they compare and analyze the advantages and disadvantages of the different alternatives to build robots. Finally, they make an oral presentation to their classmates. These activities are performed by teams of four students. Then, the second activity is hands on task using the *Quemes* robot in the context of a challenge. The proposed problem is based on the following hypothetic situation: a very well-known public person is coming to visit their city, thus it is required to have host that guide the visitor to the main touristic places. For this purpose, each team plans a tour for the visitor and prepares a simulated scenario of the city; the routes are simulated with black lines that can be followed by the robot. Finally, the robot is programmed to follow the desired planed path.

4.2   Second Educational Stage: Appropriation

In this moment, the learners know and apply the engineering design methodology and its stages by means of practical activities. These activities are embedded in the creation of a dramatization related to important issues of their community. The main purpose is that they get familiar with the methodology to solve a problem, but also they acquire the basic technological skills involved in the construction of a robot. The development of these methodological and technical competences in a restricted and controlled problem make easier the learning process and it's the fundamental basis for the more general and applied projects that they carry out at the third educational stage. The technological tool to support this activity is the robotic theater platform created in the Smart Town project; a detailed description of this tool is presented in the next sections. The first activity is to choose a social problem of their city; the theater play will be created around this problem trying not only to recreate the situation and issues, but also to propose solutions and actions that can be done by the people to mitigate the problem. Thus, they must search relevant information concerning the selected problem to prepare their dramatization. It is necessary that they be conscious of the available technological components in order to establish the potentials and constraints. The platform allows to create new embedded commands in the robot that can be used when creating the drama script. In the context of the engineering design methodology this first activity corresponds to the analysis phase. The next stage, the design phase, consists in conceiving and building the actor robots that would play in the performance. For this activity, the *Quemes* robot and the VEX EDR robotics kit are used. The scenario for the show is also created and constructed at this time. This problem solving approach, related to real problems of the communities, is used to develop technological skills related to coding, electronics and mechanics.

4.3   Third Educational Stage: Impact

In this moment, the aim was that the students propose a solution to a problem in their region and implement a functional prototype. The selected problem should be solved using basic automation tools. Notice that the purpose is not to generate a robotic solution, but to use in a practical fashion the methodology and the technical knowledge learned during the previous appropriation moment. As the project also aims to develop innovation competences oriented to create solutions that give value to the communities, the engineering design methodology is complemented with an innovation methodology. For this purpose, in the project some tools proposed by the Design Thinking methodology [15] are adopted. This methodology consists of five stages: empathize, define, ideate, prototype and evaluate. The empathize stage is centered on getting closer to the associations, government and companies of the region; the learners visit them in order to identify and understand necessities of the different stake holders. The define stage is carried out through an empathy map and a critical reading checklist. The ideation stage is supported by the brainstorm method; all of considered solutions are tackled from the automation and control theory approach. Then, for the prototyping stage, the learners build basic prototypes of the best ideas in the former stage; thus, the final users of their solution can interact and provide feedback. Even, if at this time the prototypes still have some mistakes, they are presented in order to be able to detect errors, obtain new ideas and generate improvements. Finally, in the evaluating stage the verification that all the requirements have been accomplished is performed. The final validation and adjustments includes the participation of the end users. In order to carry out the activities of this final educational stage, the students use the *Quemes* robot, the VEX EDR robotics kit, the parallax sensor kit, the starter kit for Arduino (ARDX) and the dfrobot sensor kit. Using these tools they are able to propose and to develop their projects.

5   Quemes Robot and VEX Robotic Kit

The *Quemes* robot is shown in Fig. 1. It consists of an Arduino microcontroller, a connection board, two wheels, two servomotors, a receiver-transmitter Bluetooth device, a set of batteries, a switch, and six infrared sensors *(four in the wheels' axis and two in the front of the mechanical structure)*. The electronic board is a circuit designed to be used in learning environments that facilitates the connection between the Arduino microcontroller and the other electronic components of the robot. This board was designed to prevent damage caused by connection mistakes such as short circuit and inverted connections. It also allows to easily incorporate additional components to the robot basic platform; for example, led lights, servomotors, sensors, among others can extend the robot functionalities. The basic *Quemes* robot can perform three basic motion functions: follow line until the next crossing point, turn left and turn right. These robot motion commands are controlled by the embedded software in a reactive way using a closed-loop control mechanism. Additionally, there are two commands to open and close a grip if it is mounted on the robot. Thus the robot can move in scenarios that include crossing lines, where each crossing point represents an important landmark in the context of the robot specific task. For instance, in

Fig. 1. Quemes robot

the theatre setup, crossing points can represent the places that characters must visit during the play. When required, other functions can easily be added according to specific needs. These basic functions are programmed by using Ardublock. The STBlocks software allows to use the preconfigured functions of the robot in order to program new routines. Motion and new specific commands are sent to the robot via Bluetooth; once the robot has performed the action, it sends a command acknowledge to indicate that the action has been completed, causing the STBlocks controller to send the next command and so on.

5.1   VEX EDR Robotic Kit

In order to create automated solutions to community problems, a flexible mechatronic platform is required. In this project, the VEX EDR robotic kit has been used to accomplish this objective. This kit offers a set of structural and electronic components. Therefore, the students are able and feel motivated to build different actor robots. This kit has step-by-step instructions to assemble a predefined robot, which helps the students to easily understand and work with the kit. However, what is more interesting is that the VEX EDR robotic kit can be integrated with the *Quemes* robot. In fact, the basic robot can be extended to improve its features and to provide it with new functionalities. In this case, learners can use their creativity to build amazing characters by adding motors, sensors and mechanical structures to the *Quemes* robots. Thus, most of the students take the basic robot and create additional structures allowing to build a variety of characters; however what is most important, from the educational pint of view, is that learners by doing this task acquire basic skills in electronics, mechanics and informatics. This is the practical fashion to foster the creativity and construction abilities, two of the fundamental pillars of the AC4 educational model.

6   ARDUBLOCK Programming Environment

This section introduces Ardublock,[2] one of the principal support software tool used in the Smart Town project. In particular, it explains in more detailed way how this tool is used and how it's extended. Ardublock is a graphical software that generates code to program the Arduino UNO microcontroller and it is distributed as a plugin under an open software license (GNU). Ardublock is the tool selected as the programming learning environment for the students in the Smart Town classroom. It was selected because of its graphical features, as ease of use, naturally transition to the C++ code, which it is interpreted by the Arduino IDE, as well as its ability to extend the graphical tools. The graphical workspace of Ardublock is shown in Fig. 2. This workspace has a left panel that provides access to a large amount of blocks developed for third parties. For instance, some devices bought for the Smart Town project are from DFRobot, which happens to

---

[2] http://blog.ardublock.com.

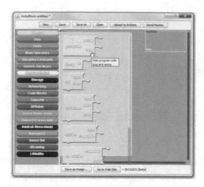

Fig. 2. Ardublock's graphical user interface.

have a menu with blocks for all DFRobot sensors. Additionally, for the robotic theatre development, it was required to have new graphical blocks that allow to control the *Quemes* robots and to support the communication between the robot and the *RoboAct* software. Fortunately, Ardublock allows to create brand new blocks with a simple design and some few code modifications. Its user interface is well decoupled, so the first step in the construction of new blocks is to create its sprite and graphical design. This has been achieved through the modification of the Ardublock.xml descriptor file that defines every graphical component; this file is interpreted a displayed by the Openblocks framework. Once, the graphical block has been created, the translation of the block to C++ code is accomplished by the creation of a single class and the corresponding mapping to the graphical block. As it is illustrated in Fig. 3, this mapping is done throughout a single file named mapping.properties. This file stored the mapping in pairs which indicate the name of the block in *ardublock.xml* file and the name of its corresponding class.

Fig. 3. Ardublock's extension workflow.

7   *RoboAct* Software

The *RoboAct* software was designed and implemented in order to meet the requirements and model presented in the section III; specifically, it is used as a support for the activities of the second educational stage. In this educational

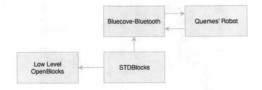

Fig. 4. STBlocks component diagram.

stage, the main objective is to create a theatrical presentation using extended-*Quemes* robots. In order to accomplish this objective, the learners had to take a walkthrough where they learn basic aspects of mechanics, electronics and programming. *RoboAct* is a software tool that facilitates learning algorithmic thinking and provides tools to control multiple robots. In particular, it provides a visual approach that helps the student to plan the coordinated motions and actions when multiple robots are involved. The *RoboAct* software is composed of two main modules, both developed under the Java language. The first module, named STBlocks, was constructed based on the graphic design OpenBlocks component of Ardublock. The second module is the *RoboAct*-UI (user-interface) build upon the BDI-BESA, a multiagent framework previously developed at Javeriana University. This software provides a platform to build multi-agent systems based on the Belief-Desire-Intention approach [16]. STBlocks is similar to Ardublock, which means that visually the STBlocks look the same as the Ardublock plugin for Arduino. However, the code executed in STBlocks is a Java program and the communication with the Arduino microcontroller is done through the Bluetooth communication messaging facilities. Figure 4 shows the components involved in the STBlocks module. The Openblocks component provides the framework that handles the creation of user interface blocks. Bluetooth Bluecove component is a library that provides mechanisms to manage the communication with the *Quemes* robots. In the figure, the *Quemes* Robot component represents the program embedded in the Arduino microcontroller; this is crucial for the correct operation of the STBlocks module because the sent messages must be received and handled by the embedded program. During the work with the learners, the learning of how to create embedded codes through Ardublock is included in the contents at the beginning of the second educational stage. Thus, when they must use *RoboAct* for the play, the embedded code creation is not a concern. On the other hand, *RoboAct*-UI module can be divided into two components: the framework component and the user's interface component. The framework component is an implementation of the architecture and design features for robotic performers. The basic architecture was first proposed and implemented by [17] using the BESA-BDI framework. The user's interface component also uses the STBlocks user's interface in order to communicate with the *Quemes* robots. The visual approach used facilitates the creation and sequencing of logical blocks denominated Activities. These activity blocks are the basic structures of a script play. Using this blocks it is possible to specify the coordination dependencies between the actions performed, even simultaneously, by several robot characters. Finally, it consolidates the group of tasks and communication exchanges that each single robot actor has to accomplish to perform the play.

8   Case of Study: Soacha, Zipaquirá and Girardot

A pilot test experience was carried out in order to implement and evaluate the model and developed tools in a real operational environment. The Smart Town project was tested in three small cities of Cundinamarca-Colombia: Soacha, Zipaquirá and Girardot. This pilot experience in the co-laboratories learning spaces was performed during 16 weeks and its intended audience was young people between the ages of 15 and 18 years from those cities. A three month's preparation phase was required to organize the logistic aspects and to promote the inscription of the participating girls and boys. Finally, more than 200 learners took part of the robotic experience. One of the key points of the AC4 model is that the learning process must be oriented to solve problems of the local communities and to take advantages of the potentials of each specific territory. Then, an analysis of the three towns was made, not only for the specific purposes of the pilot test but also to take into account their common features while designing the general model adopted in the project. A common characteristic of these cities is that, even if they are not very large, they are some of the most populated of Cundinamarca. A brief description of this small cities is the following: Soacha is located in the south edge of the high plateau called Sabana de Bogotá. The major problems that the city has are unemployment, which is estimated in 45%; insecurity, and the lack of routes. Soacha has companies in the industrial, commercial and service sector; nevertheless, several of these companies only have a weak administrative structure, thus they are not conscious of their own needs. Zipaquirá is located 47 km from Bogotá. The major economic activities are mining, tourism, agriculture, ranching and commerce. The major crops in Zipaquirá are peas, carrots and potatoes. The most important tourist attraction in Zipaquirá is its famous salt cathedral. Girardot is located at the south of Cundinamarca. The major economic activities in Girardot are tourism, agriculture and ranching. There are also some manufacturing industries, especially of drinks. Insecurity and the pollution of its rivers are the most important problems of this city. In the first educational stage, the visitor's problem was flexible enough for the specific context of each city. In general, the learners, even if most of them had never had contact with robotics, were able accomplish the challenges. They were introduced to the basic concepts of robotics, but what was more important is that they get conscious of their own capabilities. In the second educational stage, the robotic theater demonstrated to be a perfect motivation frame to develop methodological and technical competences. Most of the learners were able to build their own robotic character and to integrate these robots in a well-coordinated script. The subjects of the plays were related to the social community problems that were identified in the pilot test's cities. Finally, during the third educational stage, the learners could generate small functional prototypes related to problems were automation could be part of the solution. The most outstanding projects that the students developed were: a prototype system for automated irrigation, this system is used to water a crop depending on the moisture of the soil; an automated food dispenser for street dogs, this system allows to give food to street dogs through and automated gate; and a prototype of a bag system classifier, where by light and dark color identification bags can be classified. During the pilot experience the Blackboard LMS platform was used. This platform allowed the communication between the students, tutors, mentors and the research team by means of discussion forums and its chat facilities. All the information and materials regarding the pilot test were available and shared through the platform.

9   Conclusion

The project Smart Town integrates multiple methodologies: the AC4 educational model, the Engineering Design method and the Design Thinking methodology. Based on this approach, several educational and technological tools were created in order to develop competences in young people by the use of robotics and automation to solve needs in their communities. The learning process is organized in three educational stages: awareness, appropriation and impact. At the first moment, the aim is that students have their first contact with robotics; at the second one, it is pretended to coordinate a robotic theatre play to represent social issues; and at the third moment, to propose an automation solution to problematics concerning their municipality. The *Quemes* Robot, the VEX EDR kit of robotics and some kits of sensors and actuators are used as hardware tools during the development of the project. In addition, the *RobotAct* a software platform was developed to coordinate the robotic theater plays. After the project was finished, the student's autonomy level was increased regarding the use of robotics for problem solutions. Likewise, the learners were motivated to explore other technological tools. However, what is more important is that they open their mind to new ways of solving problems and that they recognize their own capabilities to contribute and generate positive impacts on their communities. In the near future, the research team aims to extend and implement the Smart Town model and tools to bring these benefits to girls and boys in more towns of Colombia, and that the model could be also adopted even in other territories.

Acknowledgments. The research project "Smart Town: talent and innovation applied to the territory" was financed with resources of the General System of Royalties - Science, Technology and Innovation Fund, managed by the Gobernación de Cundinamarca.

References

1. Papert, S.: Situating constructionism. constructionism. i. harel and s. papert. norwood (1991)
2. Bers, M.U., Flannery, L., Kazakoff, E.R., Sullivan, A.: Computational thinking and tinkering: exploration of an early childhood robotics curriculum. Comput. Educ. **72**, 145–157 (2014)
3. González Guerrero, E., Páez Rodríguez, J. J., José Roldán, F.: Robots cooperativos, quemes para la educación. Revista Vínculos **10**(2) (2013)
4. Guerrero, E.G., Rodríguez, J.J.P., Roldán, F.J.: Uso de robots cooperativos para el desarrollo de habilidades de trabajo cooperativo en niños. Revista de Investigaciones UNAD **12**(2), 43–56 (2015)
5. Mitnik, R., Nussbaum, M., Soto, A.: An autonomous educational mobile robot mediator. Auton. Robots **25**(4), 367–382 (2008)
6. Lindh, J., Holgersson, T.: Does lego training stimulate pupils' ability to solve logical problems. Comput. Educ. **49**(4), 1097–1111 (2007)
7. Fagin, B.S., Merkle, L.: Quantitative analysis of the effects of robots on introductory computer science education. J. Educ. Resour. Comput. (JERIC) **2**(4), 2 (2002)
8. Catlin, D.: Using peer assessment with educational robots. In: International Conference on Web-Based Learning, pp. 57–65. Springer, Heidelberg (2014)
9. Denis, B., Hubert, S.: Collaborative learning in an educational robotics environment. Comput. Hum. Behav. **17**(5), 465–480 (2001)
10. Ovadiah, Y.H., Samboni, G.M., Rodríguez, J.P.: Cooperative robots used for the learning process in the cooperative work. In: Trends in Practical Applications of Heterogeneous Multi-Agent Systems. The PAAMS Collection, pp. 165–172. Springer, Heidelberg (2014)
11. Mubin, O., Stevens, C.J., Shahid, S., Mahmud, A.A., Dong, J.-J.: A review of the applicability of robots in education. J. Technol. Educ. Learn. **1**(209–0015), 13 (2013)
12. Vygotsky, L.S.: Mind in Society: The Development of Higher Psychological Processes. Harvard University Press, Cambridge (1980)
13. Fernández, L.M.: El funcionamiento institucional. Instit uciones educativas. Dinámicas institucionales en situaciones críticas, pp. 53–72 (1994)
14. Gómez, Á.I.P.: La cultura escolar en la sociedad neoliberal. Ediciones Morata (1998)
15. Plattner, H.: An introduction to design thinking process guide. The Institute of Design at Stanford: Stanford (2010)
16. González, A., Angel, R., González, E.: BDI concurrent architecture orientedto goal managment. In: 2013 8th Computing Colombian Conference (8CCC), pp. 1–6. IEEE (2013)
17. De la Peña Santana, A.A.: Roboact modelo de control autónomo y cooperativo para el teatro robótico. Master tesis degree. Pontificia Universidad Javeriana, Bogotá, Colombia (2014)

First Steps in Teaching Robotics with Drones

Using Flying Robots to Introduce Students to Coding and Problem-Solving

Benedikt Breuch$^{(\boxtimes)}$ and Martin Fislake

Universität Koblenz-Landau, Campus Koblenz, Universitätsstraße 1,
56070 Koblenz, Germany
{bbreuch, fislake}@uni-koblenz.de

Abstract. Teaching with drones will report and reflect about a conducted educational concept using a flying robot to motivate and teach programming to students at different learning levels. It will show how to enable students to experience that programming is a form of interaction between men and machine or in this case a human-robotics-interaction (HRI). They also learn that a code is nothing else than a language which is used for HRI and apply the newly gained knowledge **to** a problem-based programming task.

Keywords: Flying robot · Drones · Teaching coding · Teaching concept

1 Introduction

There already is a variety of approaches to introduce children and young people to coding, computational thinking and problem-solving. The following teaching concept is the trial of an alternative which revealed to score, particularly, with motivation. In this way, it is possible to reach learners that normally shut themselves away off to this topic.

Concepts for teaching problem-based coding to students require the selection of the right medium from a variety of educational robots that the market offers. Especially, against the background of machines doing more work than humans by the year 2025[1], it is important to prepare young learners to interact with robots.

Especially flying drones that deliver goods will have a significant impact on the logistics sector [1]. As part of the digitalization more and more processes become automated. Therefore, it is a logical consequence to not only teach students how to code but also foster their problem-solving competence.

---

[1] https://www.technologyreview.com/the-download/612121/machines-will-do-more-work-than-humans-by-2025-says-the-wef/.

© Springer Nature Switzerland AG 2020
M. Merdan et al. (Eds.): RiE 2019, AISC 1023, pp. 138–144, 2020.
https://doi.org/10.1007/978-3-030-26945-6_13

2 Literature Review

After more than 20 years of an ongoing development of educational robotics there are still new ideas for the classroom. Due to smaller, faster and better integrated circuits, actors and sensors there is a still growing variety of robots, new functions and a wider scope of applications.

Therefore, researchers and colleagues in schools pick up this chance and try to find out new teaching formats like [2] drone challenges based on the typical characteristic of a particular robot to enhance robotics skills in K-12 education.

Others discuss the didactic functions of education robots as tools for teaching computational thinking (CT), computational thinking practice (CTP), and information and communication technologies (ICT) or computer sciences principles (CSP) that are involved in robotics [3–5], for teaching robotics itself [6, 7] and implicitly as tools for differentiation [8–10].

In contrast to these approaches [11] used robots as mediators and as tools for activating abilities through a didactic approach based on action. As a result, they reported that educational robotics leads didactics to the integration of real and artificial, through an approach that is functional both to contents and to methodology, promoting the motivation with a positive effect on learning process.

Nevertheless, it is important first to understand "robotics as a learning tool for educational transformation" as [6] emphasized and to differ between the technical artefact itself and the format of the learning method possibly fostered by the artefact.

3 Educational Specifications

The Airblock® drone has been developed to motivate and foster teaching block-based programming, the principles of aerodynamics and the use of critical thinking skills to children [12].

The programming language Scratch that is used to program the robot, is block-based and therefore easy to learn for children aged 8 years and older. This makes the drone very suitable for teaching programming to students at different levels.

Since this process can be done independently from the app's connection to the drone, students can focus on the programming process and afterwards connect their device to the drone for testing.

The drone not only visualizes the program but also helps to create real-life challenges and experiences which help to motivate students that are easily overstrained by tasks that involve abstract thinking and problem solving.

Furthermore, the Airblock® drone's ability to crash and fall apart without breaking very easily contribute to the motivating factor because when crashing hard enough, the drone simply falls apart and can be reassembled quick and easy by the students. This makes fun, helps to reduce inhibitions that can obstruct creative thinking processes, and enables even students with minor potentials to use their own way to find a solution like, for instance, the trial-and-error-method (Fig. 1).

Fig. 1. Airblock@ drone before and after crash

4 How to Teach Programming with the Airblock® Drone

This example lesson was taught to a 9th grade in the subject Technical Education at the Katholische Hauptschule Bülowstraße in Cologne, Germany. The Students were between 14 and 16 years old. In general, most students showed a lack of motivation during previous units which dealt with engines and woodwork and seemed to have problems to hold their attention for longer periods.

In order to make the process of learning how to code easier, it is advisable to remove distractions from the programming environment for these lessons. Thus, the operation of the makeblock® app and the operating principle of the Airblock® drone should be taught in preceding lessons.

Since coding is usually not part of these students' lifeworld and, therefore, a very abstract learning topic, it is difficult to teach it to students at all learning levels.

Simulation games can bridge the gap between the abstract world of programming and the students' lifeworld. It is also an appropriate method to foster the problem-solving competence [13]. So, the idea of this teaching concept is based on a simulation game where students act as the staff of an imaginary company that wants to cater customers by using drones.

4.1 Lesson and Method

The lesson should begin with a short overview of the course of the lesson and its main goal in order to make it transparent to the students. The goal should refer to a real-life situation in which the lesson is embedded.

The lesson goal in this case is "Today, we learn how to program our pizza drone." Thus, the lesson simulates a real-life challenge which makes the tasks less abstract and fosters the learners' motivation.

The lesson begins with an icebreaker activity called "program your teacher". A student (programmer) gives the teacher verbal commands to program him to leave the room. The teacher should think in advance about appropriate commands for the task and only react to compliant student commands. At the beginning, the students tried simple commands like "walk out the door" or "turn left". The teacher did not react to the first one and started turning left but kept turning until the student said "stop". Now, other students who understood the problem were allowed to help the programmer by naming precise commands until the teacher walked out the door.

In the second phase of the lesson the learners have to work in pairs to go through a learning circle with different stations in a specific order. To secure the learning success, the pairs have to complete a routing slip at every station. The questions on the routing slip should require an answer that contains the main learning goal of each station.

On the first station, the focus is on the realization of the program as a sequence of commands. The learners have to use their own smartphones to program their partner in order to collect a certain item that the teacher has placed beforehand. Of each pair, one student will be the programmer who has to make the other learner (receiver) collect the object. The receivers remain in the classroom while the programmers write the commands. Not all students fulfilled this task but realized the sequential nature of a program either by creating or executing.

The second station is about codes as a form of language which conveys commands. By scanning a prepared QR-code with their smartphones the students receive a document which contains an explanation for the term "code" as signs that convey commands and an example of a key. Now, the learners solve the riddle in which two robots communicate via the use of symbols from this key. The riddle's solution is a command that sends the students to the next station. This learning station gives the learners the opportunity to find out that programming requires a common language for a successful HRI or machine-to-machine communication (Fig. 2).

Fig. 2. Students scanning QR codes to receive their task

All students were able to solve the task and moved to the third station. They secured their learning objective by correctly answering the questions "What is a code?" and "Who communicates through the use of codes?" on their routing slip. Finally, they wrote down the decrypted message.

At the third station, students have use their newly acquired knowledge to perform a problem-oriented programming task with their smartphones and the Airblock® drone: In order to enhance the customer satisfaction of the fictive company, the drone should do a funny trick to thank the customer for his or her order. The goal is to program a one-button operation where the drone starts, ascends, performs a looping, and lands again. The learners need to read the task closely because it contains all commands for which they have to find the equivalents in the makeblock® app. Due to the fact that the drone forgives mistakes by crashing and falling apart, this station particularly requires trial-and-error processes in order to remove mistakes from the program and make it work. Whenever a drone crashes, the programmers withdraw from the station and focus on the debugging process and the next pair can test their program. If a program proves to be correct, the students fill out the routing slip (Fig. 3).

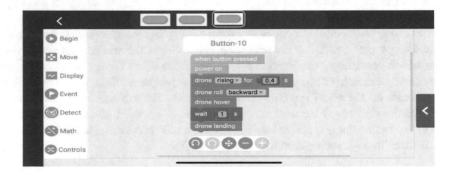

Fig. 3. Coding environment with functioning program for third station

Except for one pair, all learners were able to successfully code the program within the given time. According to the teacher's observations and the information on the routing slips, all students needed two to four tests until success was achieved. After students have finished every station, they report their experiences in a final class discussion. The teacher will ask for the answers to the questions on the routing slip. "Yes" and "No" answers can be counted on a board for visualization.

4.2   Differentiation and Grading

To meet the needs of inhomogeneous courses, differentiation is essential to enable all students to achieve the goal of the lesson. In the learning circle, students work and learn at their individual speed. Additionally, at the second and third stations QR-codes provide students with tips that help them to achieve the learning goal.

The coding itself proved to be challenging for the students. Even though the pairs were able to discern all necessary commands from the text and find the equivalent blocks, all drones crashed during the first test phase. The teacher pointed out the "tip codes" for these stations. Surprisingly, the students refused to use the "tip codes". They had become ambitious and let the teacher know that they wanted to solve the problems on their own.

The students were graded on the one hand on the basis of their achievements of the learning goals which were documented in their routing slips. Also, the teacher's observations during the lesson were taking into considerations. These include the effectiveness of the debugging process as well as the accuracy of the program with regard to the task.

5 Conclusion

The practical application of this teaching concept showed that this approach to teaching coding by using a flying robot in a simulation game, fits even learners who are less motivated. The problem-based coding approach even induced ambition in those young people. At the end of this and the following lessons of this unit, the students gave positive feedback and asked what the programming task for the next lesson might be.

Also, the word has spread about what the students made the drone do which caused students from other classes and grades to ask for the conduction of the unit in their learning group. As a result, the unit of which the presented lesson is a substantial element will become part of the school's internal curriculum for technology lessons.

References

1. Hofmann, E., Mathauer, M.: Wettbewerbskräfte im Logistikmarkt der Zukunft. Internationales Verkehrswesen **2**(70), 37–39 (2018)
2. Bermúdez, A., Casado, R., Fernández, G., Guijarro, M., Olivas, P.: Drone challenge: a platform for promoting programming and robotics skills in K-12 education. Int. J. Adv. Robot. Syst. **16**, 1–19 (2019)
3. Misirli, A., Komis, V.: Robotics and programming concepts in early childhood education a conceptual framework for designing educational scenarios. In: Karagiannidis, C., Politis, P., Karasavvidis, I. (eds.) Research on e-learning and ICT in Education: Technological, Pedagogical and Instructional Perspectives, pp. 99–118. Springer, New York (2014)
4. Shoop, R., Flot, J., Friez, T., Schunn, C.: Can computational thinking practices be taught in robotics classrooms? Presented at the International Technology and Engineering Education Conference. National Harbor, Washington, DC (2016)
5. The Board of Studies: Teaching and Educational Standards NSW (BOSTES). A guide to coding and computational thinking across the curriculum. https://educationstandards.nsw.edu.au/wps/portal/nesa/about/news/news-stories/news-stories-detail/guide-to-coding-and-computational-thinking-across-the-curriculum. Accessed 21 Nov 2016
6. Eguchi, A.: Robotics as a learning tool for educational transformation. In: Alimisis, D., Granosik, G., Moro, M. (eds.) Proceedings of 4th International Workshop Teaching Robotics, Teaching with Robotics & 5th International Conference Robotics in Education, Padova (Italy), pp. 27–34 (2014)
7. Schäffer, K., Mammes, I.: Robotik als Zugang zur informatischen Bildung in der Grundschule. GDSU-J. Heft **4**, 59–72 (2014)
8. Fislake, M.: Robotics in technology education. In: de Vries, M. (ed.) Handbook of Technology Education, pp. 361–384. Springer, New York (2018)

9. Janka, P.: Using a programmable toy at preschool age: why and how? In: Workshop Proceedings of SIMPAR. International Conference on Simulation, Modeling and Programming for Autonomous Robots, Venice (Italy), pp. 112–121 (2008)
10. Friebroon Yesharim, M., Ben-Ari, M.: Teaching robotics concepts to elementary school children. In: Lepuschitz, W., et al. (eds.) Robotics in Education - Latest Results and Developments, pp. 77–88. Springer, New York (2018)
11. Alessandri, G., Paciaroni, M.: Educational robotics between narration and simulation. Procedia – Soc. Behav. Sci. **51**, 104–109 (2012)
12. makeblock Homepage. https://www.makeblock.com/steam-kits/airblock. Accessed 28 Jan 2018
13. Hüttner, A.: Technik unterrichten, 3rd edn. Verlag Europa-Lehrmittel, Haan-Gruiten (2009)

Project-Based Learning Focused
on Cross-Generational Challenges

Georg Jäggle[1], Munir Merdan[2](✉), Gottfried Koppensteiner[3],
Christoph Brein[3], Bernhard Wallisch[4], Peter Marakovits[5],
Markus Brunn[5], Willfried Lepuschitz[2], and Markus Vincze[1]

[1] Institute of Automation and Control, Vienna University of Technology,
Vienna, Austria
{jaeggle,vincze}@acin.tuwien.ac.at
[2] Practical Robotics Institute Austria, Vienna, Austria
{merdan,lepuschitz}@pria.at
[3] HTL Technologische Gewerbemuseum, Vienna, Austria
{gkoppensteiner,cbrein}@tgm.ac.at
[4] HTL Donaustadt, Vienna, Austria
wall@htl-donaustadt.at
[5] HTL Ottakring, Vienna, Austria
{peter.marakovits,markus.brunn}@htl-ottakring.ac.at

Abstract. The combination of project-based learning and educational robotics
to solve real-world challenges can have an impact on the development of students' self-efficacy, communication and collaboration skills. This paper presents
a cross-generational project, which engages high school students in the development of different practical solutions for specific users' problems and needs.
The students take over the role of co-researchers within the frame of different
out-of-school activities and are challenged to identify and solve complex
problems. We present the project implementation process that integrates analysis, development and test phases. Also, the evaluation of the impact of the
project regarding the involved students is reported in this work.

Keywords: Project-based learning · Educational robotics ·
Real-world challenges · Participative research · Technical high school

1 Introduction

In order to deal with the challenges of the 21st century, pupils and students (future
workers) need to be able to think critically and solve problems [1]. Since technology is
developing so fast, we do not know what knowledge and skills will be required from
them in their future careers [2]. Project-based learning (PBL) is considered an approach
for teaching in which students offer the solution for real-world problems or challenges
through an extended inquiry process [3]. PBL engages students to research in solving
authentic problems by creating objects collaboratively. It can be described as a student-
centred teaching method that occurs over a specific time, during which students select,
plan, investigate and produce a product, presentation or performance that answers a
real-world question or responds to an authentic challenge [4]. The characteristics of

© Springer Nature Switzerland AG 2020
M. Merdan et al. (Eds.): RiE 2019, AISC 1023, pp. 145–155, 2020.
https://doi.org/10.1007/978-3-030-26945-6_14

PBL are developing students' thinking skills, allowing them to have creativity, encouraging them to work cooperatively, and leading them to access the information on their own and to demonstrate this information [5]. An effective PBL environment consists of five components: (a) an authentic and engaging driving question, (b) student-generated artefacts, (c) student collaborated research, (d) an audience of community, and (e) the use of technology-based cognitive and communication tools [6]. Content-wise, robotics in education has emerged as a superb tool for practical learning, not only of robotics itself but also in general topics of STEM, offering major new benefits to education. It enables students the flexibility to develop interdisciplinary projects and to discover exciting topics, integrating all the skills needed for designing and constructing machines, software and communications system [7, 8]. Besides the use of educational robotic activities offers the possibility of concretely and contextually approaching different concepts used in classroom practices thus establishing connections between subjects [9].

In the project iBridge[1], we combine project-based learning with the application of robotics aiming to improve student skills and their interest in technology by engaging the students in cross-generational research topics. On the one hand, the project intends to get young people involved in the field of assistive technologies for senior citizens. On the other hand, the students are also concerned with the development of robots for children. In both cases, the students have to face the challenge the respective user groups to technology usage and address the needs of older adults or children respectively for understanding their abilities and desires. In this context, the students have to consider what functionality a robot or assistive technology has to provide and how it should look like, behave and interact in order to be well accepted. Considering that the challenges are broad enough and offer many different ways to respond to, the students have a chance to practice their creativity to develop a possible solution. In previous work, we introduced the project and presented its overall concept [10]. In this paper, we report on the realization of the project and evaluation results after one year.

The paper is structured as follows: The following Sect. 2 briefly introduces the project activities and their related schedule. Section 3 describes the analyzing process and different types of workshops, which are organized within the framework of the project. Section 4 gives a short overview of the developed prototypes. Section 5 presents different types of activities in the testing phase. Finally, the impact of the project is described in Sect. 6, and a conclusion is given in Sect. 7.

2 Project Activities

The iBridge project methodology is based on the PBL teaching technique and encompasses the following four phases: analysis, development, tests and user manual, which refer to a final solution (see Fig. 1). At the very beginning, within the analysis phase, actual scientists introduce the requirements of older adults and children as well

---

[1] https://www.sparklingscience.at/en/projects/show.html?–typo3_neos_nodetypes-page%5Bid%5D= 1263.

as electrical, mechanical, and programming aspects of robotics in general to the students. In this phase, the students also learn about the user needs and requirements through direct interviews with them as well as through observations of the users when they are using specific technologies (PCs, robots). In the development phase, a relevant prototype is developed based on the results of the user survey and students ideas but also considering the technical feasibility. The prototype is tested and evaluated in the test phase by the students and users in order to identify necessary improvements for the hardware and software. The gained insights from the tests are used in the last phase for technical improvements as well as for the creation of a user manual.

Fig. 1. Project methodology [10]

3 Analyzing Process

At the beginning of the analyzing process, all students participated in a workshop about participative research. They learned about qualitative methods such as interviews and diaries but also trained their skills through peer interviews. The analyzing process continued with user workshops, which were led by students. On the one hand, workshops with seniors denoted as ICT Workshops were held aiming that students gain knowledge about the seniors' needs. On the other hand, workshops called Innovation-Lab were organized with children from primary schools in order to learn about their interests and skills.

3.1 ICT Workshop

Each ICT Workshop started with a PC tutorial, where students acting as tutors explained to the elderly people the basic functions of computers as well as how to use the internet (see Fig. 2a). This was very important since several studies found that there are benefits to be derived from intergenerational interactions associated with the use of technology, such as knowledge sharing and increased understanding in interactions across generations [11]. These activities tend to bridge the generation gap and establish trust between seniors and students. Students reported that several seniors talked about their ICT problems in everyday life during the tutorial. Students helped the seniors to solve these problems but also notified the observations. In the further course of the workshops, students interviewed elderly people about their knowledge, experiences and views on robots (see Fig. 2b). At the beginning, the students asked them about their general abilities (how well they see, hear, move, etc.), but also collected information about their preferences for technology usage (how often they use computers or the internet) as well as for which purposes they are doing that (skype, news, emails, etc.). In the further stage of the interview, they asked the seniors generally about robotics: (i) what do you imagine under a robot, (ii) have you ever seen a robot in real life or on TV, etc., and (iii) if so, what kind of robot was that and which skills did it have? Finally, the students asked them about the skills that a robot should have to support them meaningfully as well as which pros and cons can they imagine in using a robot to help in everyday life. Finally, the students introduced some ideas about features that a robot assistant should have (emergency button, medicine and appointment reminder, etc.) and asked the seniors about their opinion. Regarding their requirements on robots, the seniors answered that robots should cook, clean the floor and iron clothes. Others suggested that the robot could take the role of a dog. Two seniors proposed that robots should clean windows, mow the lawn and clean dishes, but also be a teacher for different topics and dance with them. Regarding the question related to the function of a robotic assistant, seniors answered that it should have functions such as playing music, playing and talking with them, loading the mobile battery of a cell phone battery, and having an alarm clock. They liked a lot the students' ideas about the assistant that reminds them to take medicines and has an emergency signal for the doctor and family.

(a)   **(b)**

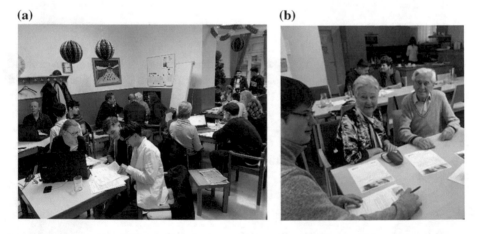

Fig. 2. (a) PC tutorial and (b) students interviewing seniors

3.2   Innovation-Lab Workshop

During the Innovation-Lab workshop, students built simple robots[2] with children from primary schools in order to explore their interests, skills and how they interact with

---

[2] https://www.wunderwuzzi.at/.

robots. At the beginning, the children were sceptical and reserved since they did not believe that they would be able to build a robot on their own. However, during the workshop they were more and more interested and enthusiastic to accomplish the task, asking the students about motors, movement, performances as well as how to program a robot. After the workshop with the children, the students discussed their observations with experts and teachers. They brain-stormed the guiding questions what the robot for children should look like, what it should be able to do and how the children should work with it. They proposed ideas such as a continuous track robot that can avoid obstacles easily (see Fig. 3a), three toothbrush robots connected to each other in a V-shape and controlled by an Arduino (see Fig. 3b), a baby robot, or a cube robot able to move around (see Fig. 3c), or even a robot animal. They finally concluded that it should look like a toy, be freely configurable and remotely manageable by an app, be easily programmable and have wheels.

(a)   **(b)**   **(c)**

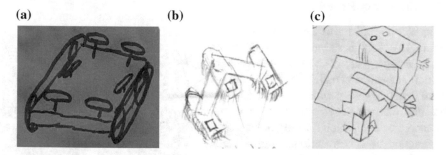

Fig. 3. (a) Continuous track robot, (b) toothbrush robot, and (c) cube robot

4   Developing-Process

In the developing phase, students from three technical high schools (denoted as HTL in Austria) focused within the frame of their prescientific school graduation thesis on the development of prototypes related to a selected user group (seniors or children) based on their ideas and scope. Each prototype was developed based on the results of the workshops as well as on a technical feasibility study. Two students groups (see Sects. 4.2 and 4.3) decided to focus more on assistive technologies instead of developing a robot prototype.

4.1   Cuddly Toy Robot for Seniors

The aim of the cuddly toy project was to develop a sensitive cuddly toy "Paul" [12] that provides technical support for elderly

Fig. 4. Cuddly toy "Paul"

people. Functions such as a medicine alarm clock, sending an emergency signal, reading from audio books, warning of low battery status by vibration, playing a game (e.g. "Simon says"), as well as a heart rate monitor and a reminder for appointments, are integrated into a robotic prototype (see Fig. 4). With the idea to make the toy affordable for the general public in the future, the cheapest possible (but reliable) components and open source software are used. A Hedgehog controller [7] was used as a control unit of the system. This controller can be programmed in Python or C programming language. In order to manage the cuddly toy, an age-appropriate webpage was created to enable the administration of some functions like book database, appointment and medicine reminder, phone numbers for an emergency call, etc. The website was programmed using HTML, CSS and JavaScript. Moreover, in order to enable an easier setting of the webpage as well as to learn about robot functions, the students prepared a detailed manual.

4.2   Emergency Bracelet

Being able to call faster for help means that seniors gain a higher sense of safety in the case of an emergency. The focus of this school graduation thesis was mainly on the support of old or handicapped people in everyday life. The idea was the development of an inconspicuous bracelet, which on the one hand has the function of a conventional digital clock and on the other hand of an emergency button. In the case of emergency after double pressing the built-in button a responsible person or helper is alerted. The emergency bracelet (see Fig. 5) consists of an ESP8266 microcontroller,

Fig. 5. Emergency bracelet

which has an OLED display and a WIFI module. The signal is sent after the key combination via WLAN to any receiver.

4.3   "Unconscious Recognition" System

The idea behind the development of an "unconscious recognition" system is to make the seniors safer at home. The "unconscious recognition" system (see Fig. 6) should recognize inanimate bodies on the ground and send a notification to a coordination office, which then retrieves the image and calls the ambulance. As soon as the system detects a possible "inanimate body" case, it triggers an alarm via a small speaker and a small warning light in the room to alert the possibly unconscious person. To avoid false alarms, a

Fig. 6. "Unconscious   recognition" system

button needs to be pressed. To prevent unauthorized access to the camera, retrieving the image is impossible without a specific setting on the built-in Raspberry Pi. So if someone generally does not want to provide images, the system can be set up to protect privacy. The communication works with several PiCams and one host.

4.4 Yeet-Bot for Children

The student group of the technical high school in HTL Donaustadt focused on the development of a robot for children, which is named Yeet Bot (see Fig. 7). The Yeet Bot should become a training kit that can be individually assembled from various modules. The target group is represented by children from the 3rd to fifth school level, who should be introduced to electronics and robotics in a playful way. The main module provides the power supply for all other modules. The switcher matrix is used to enter motion commands that the robot should perform. Light sensors measure the distance from the robot to the nearest obstacle. Using a potentiometer module, the children can steplessly regulate the resistance value and influence the signals transmitted in the circuit in such a way that, e.g. the speed of the vehicle can be adjusted. The display is designed as an LED matrix, which is running simple animations that depict the behaviour of the electricity and thus to create a better understanding of electricity for the children. These modules can be combined with each other in various ways to introduce the function of each component to young people. To support the work with the robot, didactic methods and explanations are presented in a manual, which serves as assistance for workshops with the robot kit.

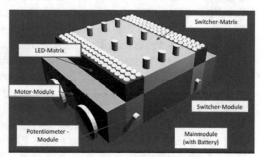

Fig. 7. Yeet-Bot for children

5 Testing Process

During the test phase, the completed prototypes were tested by the final users to identify possible improvements to the hardware and software. This phase was based on the four major steps: (1) development of a questionnaire for the interviews, (2) student-student workshops, (3) interviews with user groups and tests of the user group with the prototype for giving feedback in the form of interviews, and (4) evaluation of results.

- **Step 1**: In the first step students prepared the questionnaire that should enable users to rate the developed technology. The questions encompassed all available functions and features as well as general reactions on the prototype. Among the other, the questionnaire included questions if the users would use prototype (why and why not), which functions could especially help, what was particularly difficult to use and what they would change.

- **Step 2**: During this second phase, the students that developed prototypes held workshops with younger students that were going to present prototypes and perform interviews with user groups. The older high school students (fifth year, 13th school grade) showed their experiences during the development of the prototype to younger students (third year, 11th school grade), but also taught them the prototype functions (see Fig. 8a) and presented the questionnaire. This workshop was held within the frame of the regular lessons "Media Technology", where the students generally learn about usability engineering. They also explained, based on their own experience, how to lead the interview with the user group.

- **Step 3**: Within the third step, the users tested the developed prototype and were interviewed about it. The workshop again started with the PC and internet course to establish trust between the final users and "experts", which were young students this time. After that, the younger students demonstrated the prototype features to the user group and led the interviews (see Fig. 8b). Most of the interviewed seniors were from 60 to 88 years old, and most of them have expertise with using a computer. The idea behind student-student workshops was that younger students as neutral "experts" learn how to present technology and how users usually test and value the developed products. Moreover, they should learn how to perform interviews with users and what the user value in a specific product, but also what is not quite appreciated and why.

- **Step 4**: Finally, the interviews were evaluated. The results of the interview showed that the seniors preferred mostly the functions of the appointment reminder as well as the audio books. The features emergency button and the medicine reminder were also well appreciated. As the biggest problem was registered that the speaker was not loud enough. Moreover, the users also had some difficulties in using the webpage and in setting the book database and appointment as well as medicine reminders.

(a)   **(b)**

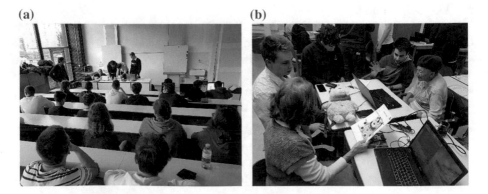

Fig. 8. (a) Student-student workshop, and (b) user workshop

6 Impact of the Project

At the end of the test process, the students (fifth grade and third grade) were interviewed about their impressions about the project activities as well as the level of their
skills before the project and now. Being asked about interest in more out-of-school
activities, 65% answered positively. Furthermore, 68% stated that this kind of activities
enables them to learn more about the state-of-the-art technologies. Besides, 69%
confirmed that they are now more interested in study one of the STEM fields. 76% also
said that they are now more interested in engineering. The influence of out-of-school
learning on communication skills (see Fig. 9) is evaluated through the following
statements:

- S1: I improved my communication.
- S2: I improved my skill to connect better with other people.
- S3: I start a conversation with other people more easily.
- S4: I can be tuned better into the conversation partner.
- S5: Out-of-school activities improved my communication skills.

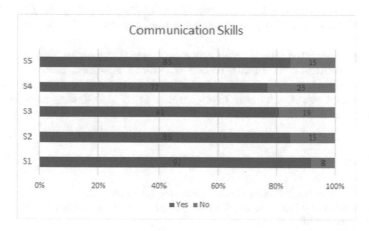

Fig. 9. Influence of out-of-school learning on communication skills

Generally, the project fostered more than just content knowledge and engaged
young people with technology in different ways. Young people took an expert role in
peer to peer interviews with users but also in conversation with teachers and scientists.
The role of the teacher and participating researchers in the project was to give main
directions and support students by the organization of workshops or specification of
questions for interviews. Nevertheless, the students were motivated to show initiative
and pursue the solution of problems on their own. Often, students were confronted with
unexpected problems that they had to solve in order to create functional prototypes.
The identification of the problem and its solution required creativity as well as per-
sistency. Moreover, the students had to apply different areas taught in school considering the interdisciplinarity of the aimed solution making the development process
more dynamic.

7 Conclusion

This paper presents the implementation of the project iBridge that aims at introducing students from high schools into a project-based learning process of developing a prototype, which meets the actual requirements and needs of specific users. Through many different activities (analysis of the application field, interviews with users, development of the prototype and tests with users), the participating students were given the opportunity to engage with technology in different ways. The students took over an expert role in the entire prototype development process. A project-based approach enabled students to explore different technologies and fields but also prove their creativity. Through participation in the project, the students increased their interest in STEM and improved communication skills. We conclude that the combination of project-based learning and educational robotics is a promising approach to improve learning effects while benefitting the community through the focus on real-life challenges.

The future objective is to improve all prototypes and hold more user workshops to test them. We also plan to enhance the prototypes based on the users' suggestions and comments. Besides, a further comprehensive evaluation of the impact of the project on the students' skills will also be accomplished until the end of the project.

Acknowledgements. The authors acknowledge the financial support by the Sparkling Science program, an initiative of the Austrian Federal Ministry of Education, Science and Research, under grant agreement no. SPA 06/294. We would also like to thank the students Paul Bernhard Mazzolini, Simon Appel, Christoph Kern, Yuming Miao and Vincent Schwartz for their exceptional engagement in the project.

References

1. Dunn, K.: Learning Robotics Online: Teaching a blended robotics course for secondary school students. Master thesis, Department of Education, University of Canterbury, Christchurch (2014)
2. Bellanca, J.A., Brandt, R.S.: 21st Century Skills: Rethinking How Students Learn, vol. 20. Solution Tree Press, Bloomington (2010)
3. Lattimer, H., Riordan, R.: Project-based learning engages students in meaningful work. Middle Sch. J. **43**(2), 18–23 (2011)
4. Holm, M.: Project-based instruction: a review of the literature on effectiveness in prekindergarten through 12th grade classrooms. Insight: Rivier Acad. J. **7**(2), 1–13 (2011)
5. Chiang, C.L., Lee, H.: The effect of project-based learning on learning motivation and problem-solving ability of vocational high school students. Int. J. Inf. Educ. Technol. **6**(9), 709–712 (2016)
6. Cervantes, B., Hemmer, L., Kouzekanani, K.: The impact of project-based learning on minority student achievement: implications for school redesign. NCPEA Educ. Leadersh. Rev. Doctoral Res. **2**(2), 1–50 (2015)
7. Krofitsch, C., Hinger, C., Merdan, M., Koppensteiner, G.: Smartphone driven control of robots for education and research. In: IEEE 2013 International Conference on Robotics, Biomimetics, Intelligent Computational Systems, Yogyakarta, Indonesien (2013)

8. Lepuschitz, W., Koppensteiner, G., Merdan, M.: Offering multiple entry-points into STEM for young people. In: Robotics in Education - Research and Practices for Robotics in STEM Education. Advances in Intelligent Systems and Computing. Springer, Heidelberg (2016)
9. Silva, V.F., Jucá, S.C.S., Moura, V.V., Pereira, R.I.S., Silva, S.A.: Robotics education in public schools using recycled materials and principles of project-based learning. Int. J. Innov. Educ. Res. **6**(8), 145–152 (2018)
10. Jäggle, G., Vincze, M., Weiss, A., Koppensteiner, G., Lepuschitz, W., Merdan, M.: iBridge - participative cross-generational approach with educational robotics. In: Robotics in Education. Advances in Intelligent Systems and Computing, vol. 457, pp. 263–274 (2018)
11. Bailey, A., Ngwenyama, O.: Bridging the generation gap in ICT use: interrogating identity, technology and interactions in community telecenters. Inf. Technol. Dev. **16**(1), 62–82 (2010)
12. Paul (2019). https://www.kuscheltier-paul.at/. Accessed 30 Jan 2019

Technologies for Educational Robotics

A Generalized Matlab/ROS/Robotic Platform Framework for Teaching Robotics

Nuria Rosillo[1(✉)], Nicolás Montés[1], João Pedro Alves[2,3,4], and Nuno Miguel Fonseca Ferreira[2,4]

[1] Universidad CEu Cardenal Herrera, Valencia, Spain
{nrosillo,nicolas.montes}@uchceu.es
[2] Institute of Engineering of Coimbra, Polytechnic Institute of Coimbra, Coimbra, Portugal
jpalves.unico@gmail.com, nunomig@isec.pt
[3] University of Trás-os-Montes and Alto Douro, Vila Real, Portugal
[4] INESC TEC-INESC Technology and Science, Porto, Portugal

Abstract. In previous works, an educational platform based on Matlab/Simulink/Lego EV3 has been developed that allows interacting in real time with the robot. However, this platform is limited to the capacity of Simulink to adapt to different robotic devices, limiting its use to LEGO and Arduino or Rasberry Pi robots.

In this paper we present a generalization of this platform to any type of robot, based on Matlab/ROS/Robot.

Robot Operating System (ROS) is a robotic middleware, that is, a collection of frameworks for software development of robots [4]. Despite not being an operating system, ROS provides standard services such as hardware abstraction, control of low-level devices, etc.

MATLAB® support for ROS is a library of functions that allows you to exchange data with ROS-enabled physical robots.

Matlab/ROS/Robot interaction is tested in a generic robot with Arduino Mega and Raspberry Pi 3, demonstrating the viability of the presented educational robotic platform.

Keywords: Robots in education · Low cost platforms · MATLAB® · ROS

1 Introduction

Over the last few years, robotic platforms have been used to improve the teaching experience in several subjects, [1], which is why the use of low-cost platforms has become widespread [2]. Low cost platforms have been used to improve the teaching experience in several subjects, in particular those related to Robotics, but also for developing other kinds of applications focused on data processing, advanced measuring and communication, amongst other possibilities. Furthermore, these low-cost platforms have such high capabilities that they can even be employed in some research developments. Most useful low-cost platform are [2]: Arduino, the Raspberry Pi, Kinect, etc. In the case of mobile robots, in the last years it has been possible to acquire low-cost

© Springer Nature Switzerland AG 2020
M. Merdan et al. (Eds.): RiE 2019, AISC 1023, pp. 159–169, 2020.
https://doi.org/10.1007/978-3-030-26945-6_15

mobile robotic platforms as Adept [3], Moway [4], epuck [5] and LEGO Mindstorms [6]. These low-cost platforms do not offer the same accuracy as industrial robots, but they are valid for educational activities and even research and experimental tests.

On the other hand, there are platforms used to teach coding by means of robotic low-cost platforms. There are open source platforms but there are another stablished and relevant platforms like MATLAB® that are not open source. Although it is not free, it is extensively used in Universities thanks to the student licenses. MathWorks have developed a free package for Matlab-Simulink to connect with LEGO and Arduino robots. It allows to construct Simulink blocks as a Simulink code and send to the robot.

In the last years, Robot Operating System (ROS) is stablished as a robotic middleware, that is, a collection of frameworks for software development of robots [7]. Despite not being an operating system, ROS provides standard services such as hardware abstraction, control of low-level devices, etc. Besides, ROS is free software under BSD license terms and has been established as the standard OS for robots, widely used in educational Robotics, where most manufacturers offer libraries to work with their robots for free, [8, 9].

1.1 Goal of the Paper

In previous works, an educational platform based on Matlab/Simulink/Lego EV3 [10] has been developed that allows interacting in real-time with the robot. However, this platform is limited to the capacity of Simulink to adapt to different robotic devices, limiting its use to LEGO and Arduino or Rasberry Pi robots.

In this paper we present a generalization of this platform to any type of robot, based on Matlab/ROS/Robot. This paper is organized as follows. Section 2 shows our previous work, an educational platform based on Matlab/Simulink/Lego EV3. Section 3 shows how to replace Simulink part from ROS part to generalize the platform. Section 4 shows some experimental test on how to use it. Section 5 shows the conclusion and future works.

2   Previous Works. An Educational Platform Based on Matlab/Simulink/Lego EV3

In MATLAB® and Simulink® environments, free libraries can be downloaded for programming and communicating with Lego NXT and EV3 robots. This library is very easy to use, as it works as common Simulink blocks. When the code is finished, Simulink translated to C++ and transfers it to the Lego Mindstorms EV3 through Wi-fi connection. Other Simulink blocks can be also employed in the Lego robot, with a similar functioning.

Regarding the code execution in the LEGO EV3 robot, some options are available to the programmer. One of them, used in our previous work, [10], is the Simulink® External Mode. In real-time the External Mode, the Simulink® Coder™ dynamically links the generated algorithm code with the I/O driver code generated from the I/O blocks. The resulting executable file runs in the operating system kernel mode on the host computer and exchanges parameter data with Simulink® via a shared memory interface.

Parameters changed in the Simulink® block diagram are automatically updated in the real-time application thanks to the Simulink® external mode. The external mode executable is fully synchronized with the real-time clock. In the proposed Matlab-Simulink-EV3 framework, a Simulink® template was created for modelling the system, see [10].

In the MATLAB environment, "tunable" parameters were defined to control the robot. Each time a parameter value is changed in Matlab, the value must be reloaded in Simulink. A schema of the proposed parameter update system is depicted in Fig. 1.

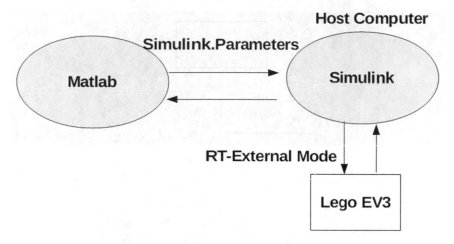

Fig. 1. Linking Matlab-Simulink-EV3.

The framework proposed in our previous work, [10], was tested in a 2×2 m2 white square table located on the floor and a web camera located on the ceiling, as shown in Fig. 2 (left). The camera is connected via USB to the PC.

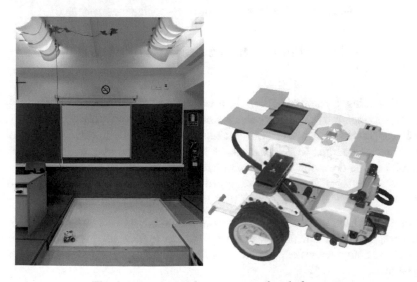

Fig. 2. Experimental setup to test the platform

The Lego Robot was equipped with three blue points that locate its position and orientation via images captured by the camera, Fig. 2 (right). The goal point is a red disk.

Figure 3 shows four snapshots about the results. As we can see, the user can interact with the framework in real-time, in this case changing the goal, see [10].

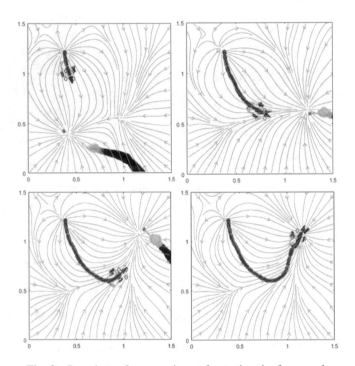

Fig. 3. Snapshots of an experiment for testing the framework.

3   A Generalized Platform Based on Matlab/ROS/Robotic Platform

Robot Operating System (or ROS) is a commonly used framework for designing complex robotic systems. It is popular for building distributed robot software systems, as well as for its integration with packages for simulation, visualization, robotics algorithms, and more. ROS has become increasingly popular in industry as well, especially in the development of autonomous vehicles.

Robotics System Toolbox has equipped MATLAB and Simulink with an official interface to ROS, Release R2015a. This interface lets you:

- Connect to ROS from any operating system supported by MATLAB and Simulink.
- Leverage built-in functionality in MathWorks toolboxes – for example, control systems, computer vision, machine learning, signal processing, and state machine design.

- Automatically generate standalone C++ based ROS nodes from algorithms designed in MATLAB and Simulink.

3.1   Linking Matlab-ROS

When communicating ROS with Matlab, ROS uses a series of environment variables to establish communication between different nodes that are connected to a network. These variables have information about the IP address of the equipment, as well as its identifier (name) as can be seen in the following code.

```
export ROS_WORKSPACE=~/catkin_ws
#export ROS_IP=192.168.43.254
export ROS_MASTER_URI=http://pi-Robot3.local:11311
export ROS_HOSTNAME=pi-Robot3.local
```

ROS is already operating so it only remains to join it to Windows, for it Matlab has the rosinit command which must be executed with the IP of the virtual machine in the following way:

```
rosinit('http://192.168.0.25:11311')
```

After executing this command, ROS is already connected to Matlab.

The functions included in the Robotic System toolbox allow you to subscribe to the topics, publish in them, create and read messages or know the topics and nodes existing in the network. Here there are the most important:

- rossubscriber: Allows you to subscribe to the messages of a topic, that is, to read them.
- rospublisher: Allows you to post messages in a topic.
- rosmessage: Create ROS messages.
- receive: Receive new messages from ROS.
- rosnode: Shows information about the nodes of the ROS network.
- rosservice: Shows information about the services of the ROS network.
- rostopic: It shows information about the ROS network's topics.
- rosinit: Connects the Matlab node with the ROS network.
- rosshutdown: Close the Matlab node in the ROS network.

With these main functions and others of less frequent use, you can interact with the rest of the nodes and topics as if you were in ROS.

3.2   Linking Matlab-ROS-Robot

The starter of the robot is started at the command line, the complete ROS must be installed to cv_bridge which binds the ROS to the OpenCV for Computational View, the bridge to arduino, the multimaster-fkie to work in a multi-agent environment.

The robot starts up by ssh at the command line

```
#!/bin/bash <- shbang the script
roscore &    <- starter ROS master

sleep 8
rosrun master_discovery_fkie master discovery
_mcast_group:=224.0.0.251 & <- multimaster enviorement

sleep 8
rosrun master_sync_fkie master_sync & <-multimaster tim-
ing

roslaunch video_stream_opencv camera.launch & <- Video

sleep 8

roslaunch robotcraft_bot driver.launch & <-launch /
cmd_vel and the remaining robot controls as led sensors
and odometry
```

Therefore, the schema depicted in Fig. 1 is updated with ROS to the depicted in Fig. 4.

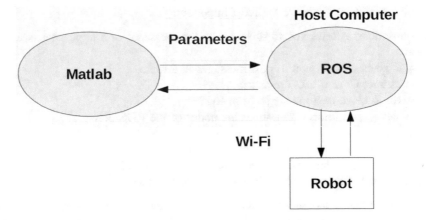

Fig. 4. Linking Matlab-ROS-Robot platform.

Finally, the robotic platform is tested with Matlab-ROS connection in the laboratory in order to test the proposed platform in Fig. 5.

Fig. 5. Matlab-ROS-Robot in the laboratory

4   Testing the Platform with an Educational Application

In order to test the benefits of the proposed platform a tracking application for a red ball has been designed, a robotic system depicted in Fig. 6 is selected.

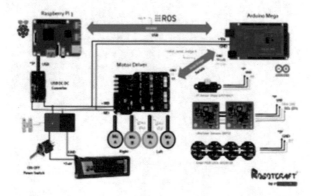

Fig. 6. Main hardware part of the robotic system.

The sensors chosen and the development platform are relatively flexible and fully featured, so that the same basic robotic hardware could be used on other type of robots.

The use of the Raspberry Pi and Arduino over other choices, was to give a fully featured Linux operating system. The sensors adopted were the SRF02 Ultrasonic range finder i2c Sensor 15 cm–250 cm and Sharp analog distance sensor 10–80 cm. For a robotic chassis and traction, a micro metal gear motor HP with extended motor shaft was used. The low-level programming using Arduino Mega, which will be used mainly for navigation, will be followed by the high-level programming using Raspberry Pi 3. The robot chassis chosen is shown in Fig. 7.

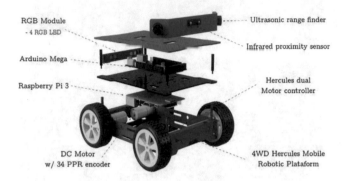

Fig. 7. Base of the assembled robots.

In addition, a front camera has been added to provide it with artificial vision, so finally the real robot has been developed in Fig. 8.

Fig. 8. Real robot in the laboratory with front camera

At the time of working with the robot camera, a preprocessing of the image from the Rasperry has been performed using ROS or Matlab.

Then in Matlab a subscription of the robot's camera is made in jpeg format if we made the preprocessing using ROS or directly via Matlab. For instance, if the preprocessing is performed using ROS we must subscribe the Robot camera.

```
s.imsub = rossubscriber('/robot2/image_raw/compressed')
```

In the following figures the robot is shown with the red ball and then the preprocessing done in the image of the ball for its recognition (Figs. 9 and 10).

Fig. 9. Robot with the red ball

Fig. 10. Image of preprocessed red ball

Finally, in Matlab a subscription of the robot's sensors is made, and the code can implemented:

```
s.sensor = rossubscriber('/robot2/distance_sensors');
s.ts = rospublisher('/robot2/cmd_vel') <- connection to /
cmd_vel
s.velmsg = rosmessage(s.ts) <-activation of the messaging
system for / cmd_vol
```

To stay in constant communication with ROS a timer to trigger an event in matlab is created.

```
t = timer('StartDelay',0);
set(t,'ExecutionMode','fixedSpacing','Period',0.04)
set(t,'TimerFcn',{@evento1,s},'StopFcn','')
start(t)
```

An event function is created to work as a laboratory for students where they can modify matlab code and see how their code has an effect on the robot.

5 Conclusions

In the present paper the Robotics System toolbox has been researched and developed, which allows to join the ROS network and to manage a robot with ROS from Matlab that is a generalization of the platform presented in our previous work, matlab/simulink/lego ev3, to any type of robot, based on matlab/ros/robot.

ROS was born facing the difficulty of dealing with the problems that were generated when trying to combine environments and tools for the development of applications for robots. A community-supported and easily accessible structure was necessary to make this possible. ROS brings all these facilities and users have seen it. It has been developed with the collaboration of several institutions and universities of recognized prestige in robotics. Currently supports many types of robots and a variety of hardware, with great popularity and a very active community.

Robot System Toolbox is a toolbox of Matlab 2015 that provides algorithms and connectivity with hardware for the development of autonomous mobile robotics applications. It allows to represent maps, plan trajectories and follow them efficiently with autonomous robots using Matlab or Simulink and integrate them with the algorithms of Robotics System Toolbox.

This toolbox provides an interface between Matlab/Simulink and ROS that allows you to test and verify the applications of the robots included in ROS and use their simulators. It is compatible with the generation of C++ code, which allows generating a ROS node from Matlab and introducing it into the ROS network, working with the ROS messages, publishing and subscribing in topics, accessing the ROS parameter servers or tree of transformations. In short, to be able to handle ROS from Matlab.

The presented platform can be used for whatever kind of robot, as we can see in the performed application, ball tracking, that is supported by a strong toolbox that allows, above all by code, to carry out almost all the actions that can be carried out in ROS. From reading and writing of topics and nodes to the use of highly useful algorithms for robotics, the students also can work with artificial vision, using MATLAB or ROS. However, the main disadvantage, compared with matlab/simulink/lego ev3, is that the date refreshment, frequency period, is slow. Future works are focused to overcome this problem.

References

1. Ozuron, N., Bicen, H.: Does the inclusion of robots affect engineering students achievement in computer programming courses? J. Math. Sci. Technol. Educ. 4779–4787 (2017)
2. Irigoyen, E., Larzabal, E., Priego, R.: Low-cost platforms used in control education: an educational case study. In: 10th IFAC Symposium Advances in Control Education, pp. 256–261 (2013)
3. Epuck. e-puck education robot (2014). http://www.e-puck.org. Accessed 26 Oct 2017
4. ADEPT. Adept Mobile Robots (2014). http://activrobots.com. Accessed 26 Oct 2017
5. Moway. Moway Robots (2017). http://moway-robot.com. Accessed 26 Oct 2017
6. Mindstorm LEGO. LEGO Mindstorms (2017). http://mindstorms.lego.com. Accessed 26 Oct 2017
7. Quigley, M., Gerkey, B., Conley, K., Faust, J., Foote, T., Leibs, J., Berger, E., Wheeler, R., Ng, A.Y.: ROS: an open-source robot operating system. In: Proceedings of Open-Source Software Workshop of the International Conference on Robotics and Automation, Kobe, Japan, May 2009
8. Araujo, A., Portugal, D., Couceiro, M.S., Rocha, R.P.: Integrating Arduino-based educational mobile robots in ROS. In: 2013 13th International Conference on Autonomous Robot Systems (Robotica), pp. 1–6. IEEE, April 2013
9. Fonseca Ferreira, N.M., Araújo, A., Couceiro, M.S., Portugal, D.: Intensive summer course in robotics – robotcraft. Appl. Comput. Inform. (2018). https://doi.org/10.1016/j.aci.2018.04.005
10. Montés, N., Rosillo, N., Hilario, L., Mora, M.: Real-time matlab-simulink-LEGO EV3 for teaching robotics subjects. In: Robotics in Education, Rie 2018. Advances in Intelligent Systems and Computing, vol. 829, pp. 230–240. Springer, Heidelberg (2018). ISBN 978-3-319-97084-4
11. Old-Geoffroy, Y., Gardner, M.A., Gagné, C., Latulippe, M., Giguere, P.: ros4mat: a Matlab programming interface for remote operations of ROS-based robotic devices in an educational context. In: 2013 International Conference on Computer and Robot Vision (CRV), pp. 242–248. IEEE (2013)

Turtlebot 3 as a Robotics Education Platform

Robin Amsters$^{(\boxtimes)}$ and Peter Slaets

Katholieke Universiteit Leuven, 3000 Leuven, Belgium
{robin.amsters,peter.slaets}@kuleuven.be

Abstract. Teaching robotics to engineering students can be a challenging endeavor. In order to provide hands-on experiences, physical robot platforms are required. Previously, obtaining these platforms could be expensive, and required a lot of technical expertise from teaching staff. However, more recent models address these issues, therefore providing more opportunities for hands-on sessions. In this paper, we describe how we used the Turtlebot 3 mobile robot in master courses at KU Leuven. We provide an overview of the main functionalities, and suggest a number of improvements to further lower the learning curve for students. Additionally, we elaborate on the curriculum and learning outcomes of two courses that utilized Turtlebots in practically oriented sessions.

Keywords: Turtlebot · Robotic Operating System ·
Educational robotics · Mobile robotics · Project-based learning

1 Introduction

When designing the curriculum for a particular course, one can include a variety of different learning activities. Lectures are often used to convey theoretical concepts to large groups of students, while practically oriented lab sessions or projects can encourage students to apply their knowledge to real-world problems. Standardized test scores were found to be correlated to the frequency of hands-on learning [1]. In science education specifically, these practically oriented learning activities can take the form of simulations, remote experiments or hands-on experiments. An extensive literature survey by Ma and Nickerson [2] concluded that there are proponents and opponents to all types. However, some form of experimental work may offer benefits over a pure simulation approach. Sauter et al. [3] found that students who conducted remote labs were more engaged in the experiment they were performing than those who used computer simulations. Additionally, remote lab users wrote higher-quality research questions and were more open to performing repeated experiments. In robotics education, a physical robot is required for hands-on or remote experiments. However, selection of an educational robotics platform is not straightforward. In the past, one often had to choose between purchasing expensive models, or building a custom platform from scratch. Table 1 provides an overview of several wheeled robotics platforms,

© Springer Nature Switzerland AG 2020
M. Merdan et al. (Eds.): RiE 2019, AISC 1023, pp. 170–181, 2020.
https://doi.org/10.1007/978-3-030-26945-6_16

Table 1. Overview of wheeled robot platforms

Robot	Year of release	Cost [€]	Target audience	Sensors
Pioneer 3DX [4]	1995	3233.00	Higher education	Sonar/encoder
Khepera [5]	1999	2999.00	Higher education	Sonar/IR
Amigobot [6]	2001	3150.00	Higher education	Sonar
Scribbler 3 [7]	2010	220.86	Secondary education	IR/ambient light/encoder
Turtlebot 2 [8]	2012	1673.00	Higher education	RGB-D/gyroscope/encoder
Edison [9]	2014	51.58	Secondary education	IR/sound/ambient light
Cozmo [10]	2016	199.99	Secondary education	Camera/IMU/IR
Turtlebot 3 (burger) [11]	2017	585.95	Higher education	LIDAR/IMU/encoder
Turtlebot 3 (waffle) [11]	2017	1399	Higher education	LIDAR/RGB/IMU/encoder
Rosbot 2.0 [12]	2018	1661.00	Higher education	RGB-D/LIDAR/encoder/IR

with release dates ranging from 1995 up to 2018. This is list is not meant to be exhaustive, rather, it is intended to show a trend. One can observe that while robotics education has become significantly more affordable, platforms with sensors that enable functionalities such as mapping and autonomous navigation are still relatively expensive.

When selecting a platform for introductory robotics courses at KU Leuven Campus group T, we set forward the following requirements:

- **Wheeled robot:** Most of the envisioned practical sessions focus on wheeled robotics, rather than humanoids or robot arms.
- **Low cost:** Due to the large number of students, the cost per robot needs to be limited.
- **Low learning curve:** The introductory mobile robotics course that is to make use of the platforms, only has three lab sessions (see Sect. 4). The time required to get familiar with the platforms should therefore be minimized, in order to maximize the available learning time.
- **Advanced perception:** While simpler robots are capable of tasks like line-following, this is not sufficient for the envisioned courses and projects. In master level courses, more advanced perception is required. Some form of depth perception should at least be available (e.g., in the form of an RGB-D camera, or a laser scanner).
- **Standard software framework:** It is imperative that the skills students acquire are as broadly applicable as possible. By using a standard software framework, students are more likely to use the same software again in their later career. Additionally, such a standard framework usually has more third party software available, broadening the possibilities of the platform.

Based on our requirements, the Turtlebot 3 burger stands out from the others, as it offers advanced sensors for a significantly lower price. Moreover, it makes use of the "Robotic Operating System" (ROS), which is an open source framework. In the rest of this work, we will discuss our experiences in using this platform for robotics education. The paper is structured as follows; Sects. 2 and 3 provide an overview of the hardware and software, and the additions that were made. We made use of these platforms in several educational projects, two case studies are described in Sects. 4 and 5. Section 6 discusses the learning outcomes, before and after the introduction of new robot platforms. Finally, a conclusion is presented in Sect. 7.

2   Hardware Overview

Turtlebots are sold in 2 main configurations, referred to as the "burger" and "waffle" models (see Fig. 1). The waffle model is larger, has an additional camera sensor and is significantly more expensive than the burger model. As we had no need for a larger robot or the camera, we opted for the burger model. The main hardware components of the platform are:

- **LIDAR:** a 360° laser scanner is mounted on top of the robot. This sensor transmits a rotating laser, which reflects of nearby obstacles. By using the reflection time, distances to nearby objects can be obtained, which can be used for mapping or obstacle avoidance.
- **Raspberry Pi:** one level below the LIDAR is a Raspberry Pi 3 (RPi) single-board computer. Since its introduction a few years ago, the Raspberry Pi has gained incredible popularity among hobbyists and researchers due to its small size, low cost and power use. It is most useful in applications that require a computer, but not a large amount of processing power. On the Turtlebot, it reads data from the sensors and communicates with a 'master' pc, which does the actual computations.
- **OpenCR:** one level below the RPi, the OpenCR hardware control board is located, which is based around an Arduino Uno. The board contains an Inertial Measurement Unit (IMU) with a three axis accelerometer, gyroscope and magnetometer. Its main function is to connect the Raspberry Pi to the motors and sensors, such that data can be read and commands can be sent. Additionally, the board provides power connections for all components.
- Finally, on the lowest level, we find the **motors** and the 11.1 V Lithium Polymer (LiPo) **battery**. Both the burger and the waffle models have a differential drive configuration, which means that they have 2 independently driven motors and 1 caster wheel for stability. Dynamixel XL430-W250 servo motors are used to drive the wheels.

3   Software Overview

The operating system on the Raspberry Pi is Ubuntu MATE 16.04, a lightweight version of the Linux distribution Ubuntu. This version is better suited for single-board computers due to the lower hardware requirements compared to the complete desktop version [14]. A number of modifications were made to the default version to make it more suited to our needs:

Fig. 1. Overview of the Turtlebot 3 models [13]

– The Raspberry Pi functions as a wireless access point (AP). Previously, we utilized a central AP, to which all robots and control computers were connected. However, this infrastructure resulted in unstable connections due to the large number of devices connected to the router. If the robots function as their own AP, less devices need to be connected to each network. Therefore, the requirements of such a network are lower. We found that the individual networks did not interfere with each other significantly. Even when 10 robots were used in the same room, connections remained stable. Another advantage of this approach is that students do not need to stay in the same room to test their programs. They can temporarily move to a more open space if needed. Turning a Raspberry Pi into a wireless AP was achieved with the help of a few online tutorials [15, 16]. When connected to the robot WiFi, a Secure Shell (SSH) connection can be established on a fixed IP address. The first connected device will also be assigned a fixed IP address.
– Superfluous software such as LibreOffice, Thunderbird, etc. was removed. Additionally, the Pi was set to boot without a graphical interface.
– A number of Bash aliases were defined, which replace frequently used lengthy commands with shorter versions. For example, the complete command to initialize sensor communication on the robot is:

```
roslaunch turtlebot3_bringup turtlebot3_robot.launch
```

which we replaced by:

```
bringup
```

Besides a Linux distribution, the Robotic Operating System was also installed on the SD card of the RPi. ROS is an open-source, meta-operating system used in robotics research and industry [17]. It provides functionality like hardware abstraction, localization, navigation, visualization, etc. in a standard communication structure. ROS works via a publisher-subscriber interface. Programs are broken up into smaller units called 'nodes'. Nodes communicate with each other trough 'topics'. A node can make information available by 'publishing' it on a topic, in the form of a 'message' (ROS data formats). Another node can then 'subscribe' to receive this information. This approach enables modularly built programs and abstraction. While ROS simplifies getting started in robotics education and research, the learning curve can still be quite steep. Therefore, we also make use of the Robotics System Toolbox, which enables publishing and subscribing to ROS topics in MATLAB [18]. At our campus, students already complete several projects with MATLAB before starting their Masters degree. Therefore, it is preferred over other ROS interfaces such as C++ and Python. A custom toolbox was developed as a final abstraction level for the mobile robotics exercise sessions (Sect. 4) and the control systems project (Sect. 5). The focus of these courses is not to learn ROS, therefore the internal software processes of the robot are abstracted away. The custom code, the documentation and the exercises can be found on our GitHub page.[1]

The resulting software architecture is shown in Fig. 2. The robot acts as a slave to a master PC on which the program is executed. A 'roscore' (central ROS process that connects nodes to each other) needs to be started manually by executing a command on the RPi over an SSH connection. The Robotics System Toolbox starts a similar process in the background on the master PC. Following initialization, the two processes share data, which can be used on either the PC or the RPi. The use of the MATLAB toolboxes is optional, as the robot can also be directly controlled via a roscore on the master PC. This is the case in, for example, master theses that make use of these robot platforms. These students will work on their project much longer and therefore have more time to gain a better understanding of ROS.

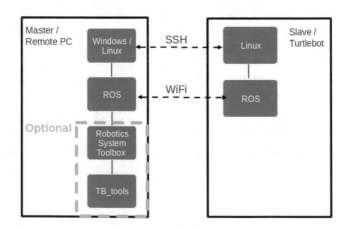

Fig. 2. Software architecture

---

4   Example Use Case: Mobile Robotics Exercise Sessions

As a first use case, we will discuss the autonomous vehicles course offered at KU Leuven Campus Group T. This course consists of an introduction to robotics, and an introduction to mobile robotics. In the introduction to robotics chapter, concepts such as workspace analysis, forward and reverse kinematics are discussed, while the mobile robotics course focuses on autonomous navigation. Course material is based on the book "Robotics, Vision and Control" by Corke [19]. The associated MATLAB toolbox is also used [20].

To enable autonomous navigation, a robot needs to know where it is. Therefore, localization is the first chapter of the course and is discussed in depth. Dead-reckoning is introduced first as it is arguably the simplest the simplest form of localization. By integrating displacement measurements, one is able to acquire a position estimate relative to the initial position. However, every measurement has a slight error, therefore the overall error only grows over time. Nonetheless, it is a useful source of information, as it is generally always available due to the use of proprioceptive sensors (such as IMU's and encoders). To combat drift, additional sources of information are needed. The Kalman filter is discussed as a sensor fusion algorithm. Additionally, the non-linear Extended Kalman Filter is introduced. Besides localization, an autonomous robot needs to be able to plan a trajectory between the starting location and the destination. Therefore, several of the most used path planning algorithms (A*, D*, distance transform, etc.) are covered in the lectures and the exercise sessions. The special case of a robot that does not need a specific plan, is also discussed. These robots make use of so called 'reactive navigation', and react directly to sensor input (such as driving towards a light, or following a line) [19]. Finally, an autonomous robot needs to be able to execute its plan. Path following falls into the category of control theory, and is therefore not a part of this course.

The Turtlebot platforms are used during the lab sessions of the mobile robotics course. Students receive three such lab sessions. In order to make the most of the available time, students are asked to prepare each session beforehand. And receive a few questions to turn in at the beginning of each session. The majority of their grade is based on these preparations. Labs are organized as follows:

1. **Introduction and dead reckoning:** In the very first session, students get acquainted with the robots and software. Basic tasks such as driving in a straight line or an ellipse are to be performed by the robots. Additionally, the concept of dead-reckoning is treated in a number of simulation and practical exercises. The disadvantages of drift are discussed. Finally, students gain experience with so called 'error ellipses', which are visual representations of uncertainty on a position estimate.

2. **Kalman filter:** Drift is one of the main disadvantages of dead-reckoning based localization. The Kalman filter and its derivatives can be used to include more sensor data into the position estimate, and thus compensate for this drift. In the second session, students design a linear Kalman filter, derive state

transition and observation models for an Extended Kalman Filter (EKF) and experiment with the effects of different types of sensor noise on pose estimation.

3. **Path planning:** In this session, students test different path planning algorithms that were discussed in the lectures. They are then asked to compare their advantages and disadvantages for certain situations. Finally, students design a program that makes the Turtlebot follow a wall. Braitenberg vehicle type behavior is used to achieve this task [21].

5   Example Use Case: Embedded Control Systems Experience

In this second use case, we discuss how we made use of the robots in a guided project that was part of an Embedded Control Systems (ECS) course. We refer to this project as an 'experience'. Students learn the theoretical concepts in a lecture format. Several control strategies are covered in these lectures. The more classical approach of PID controllers and design with the Ziegler-Nichols method is discussed. Additionally, more modern control schemes in the Z-domain or in state space are also covered [22,23]. In exercise sessions, students practice designing the controllers discussed in the lectures on paper or in a simulation environment. MATLAB is used for these simulations. Finally, the experiences require students to put concepts from the lectures and exercise sessions into practice. The aim of these experiences is to independently design a controller for a physical system. In groups of 2 or 3 students, different parts of the control loop are built according to a milestone system. This means that the overall assignment is broken up into parts, each of which can grant a percentage of the total grade. After completing one of these parts, teaching staff checks progress, and gives a pass/fail on the current milestone. A pass allows students to progress to the next milestone, a fail means that current work is not sufficient, and more time on the current milestone is needed. This system provides students with regular feedback, and allows major flaws to be detected early. At the end of the project, students present their work with a demonstration and turn in a written report.

The campus building is constructed with a spiral walkway in the center (see Fig. 3), the overall assignment of the experience sessions is to follow this spiral with the Turtlebot. The robot should keep a constant distance relative to the wall whilst driving. The overall control loop is shown in Fig. 4, the following sections will explain the different parts in further detail.

M1: Motor model identification: The servo motors of the Turtlebot have built-in PID controllers, which regulate the speed of each individual wheel. In the very first sub-assignment, a model for the DC motors with PID controllers is identified. From the step response, students can derive the transfer function.

M2: Speed control: Using the transfer function (**M1**), students are asked to design a controller, which takes wheel speed as an input, and outputs an

Fig. 3. Spiral walkway inside the Group T campus

actuator command such that the speed follows a given reference. Next, individual wheel speeds are translated to a forward (v) and rotational (w) speed of the robot. The overall controller therefore takes v and w as an input, and internally translates these to velocity commands for the left and right wheel. By writing the controller inputs this way, we found it easier to design the distance controller (**M3**). Students are allowed to use any control scheme they know of. However, all groups designed some form of PID controller.

M3: Distance control: The final part of the overall control loop is a controller that allows the robot to drive at a constant distance relative to the wall of the spiral. It will take LIDAR measurements as an input (which may have to be processed first) and return a forward and a rotational speed as an output. These speeds are then used as an input for the speed controller (**M2**). In this stage, students are asked to use a state feedback controller, such that they gain hands-on experience in designing this type of controller as well.

M4: State estimator: It is possible to control the speed of the motors based on just encoder measurements. However, sensors are not perfect. In order to reduce inaccuracies, students are asked to include a state estimator into the control loop. An estimator can filter noisy sensor values and include multiple sources of information, thus improving accuracy and robustness.

6 Evaluation of Learning Outcomes

Table 2 summarizes performance indicators from the academic years before (2017) and after (2018) the introduction of the Turtlebots. In the academic year 2017 and before, robot platforms were already used. Therefore, the change between 2017 and 2018 is not the introduction of physical platforms, rather,

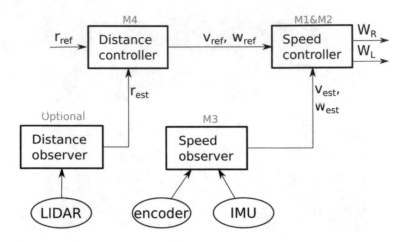

Fig. 4. Overview of the control loop and milestones of the project

the platforms used are now much better suited for teaching. For the Embedded Control systems course, other factors changed as well between the academic years (different teacher, course content and project). Therefore, only data for the mobile robotics course is included. Table 2 shows the median final scores per academic year. These scores are based on the exercises of the exam, as the theory questions should not be influenced by the use of robot platforms. Scores for Mobile robotics increased by 2%. However, the correlation between the lab scores and the exam scores is very low. While the correlation increases in 2018, the low value may indicate that better lab performance may not necessarily influence exam scores. Therefore, we can not make any definitive conclusions about the impact of these new robot platforms on students' grades.

Table 2 also shows the passing rate of students. That is, the percentage of students who pass the course on their first examination attempt. For both courses, the passing rate increased by approximately 4%, reaching almost 90% for the Mobile Robotics course. However, this increased passing rate might not be completely be attributed to the introduction of new robots. Small fluctuations are always possible. Therefore, a longer time period should be observed before drawing conclusions.

6.1   Mobile Robotics

Through an anonymous feedback form, students were asked to rate the course based on a number of questions. 10 students responded to this request. With the maximum score being 6, the question "The things I learned in this course are relevant for my education." scored 4.9. The question "I was satisfied with the quality of teaching in this course." received an average score of 4.8. Figure 5 shows the distribution of the received responses. When asked for feedback in person, many students requested more sessions to expand on the discussed topics.

Table 2. Learning outcomes before and after Turtlebots

Academic year	2017	2018
Number of students	60	65
Median score	60%	62%
Exam participation	100.00%	96.92%
Correlation lab/exam	0.13	0.20
Success rate	85.00%	89.23%

Contrary to previous years (when Turtlebots were not used), students had no negative comments on the platforms themselves.

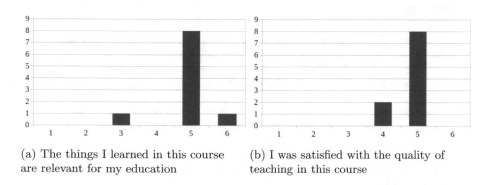

(a) The things I learned in this course are relevant for my education

(b) I was satisfied with the quality of teaching in this course

Fig. 5. Student responses to anonymous survey

6.2   Embedded Control Systems

No anonymous feedback forms are available for this course. The university organizes course evaluation, and not every course is evaluated every year. Nonetheless, students were still asked for their feedback in person. This feedback was also positive. Students appreciated the hands-on experience, as they found implementing a control loop on a physical system to be quite different from designing one based on a theoretical model. The project was, however, deemed more challenging than initially estimated. Particularly the design of the state feedback controller was the most difficult milestone.

7   Conclusion

In this paper, we discussed our experiences in using the Turtlebot 3 as a robotics platform in higher education. We propose a number improvements to the software, and discuss how we used the robots in two different courses. We present performance indicators before and after the introduction of the new platforms. It is challenging to draw a definitive conclusion about the effect of these platforms on student performance. Feedback from students and teaching staff has, however, been overwhelmingly positive. Their experience in using the robots has significantly improved compared to previous years. In the future, we would like to improve the platforms further by implementing a system to automatically update the software. Such a system would make the platforms easier to maintain. Additionally, for the control systems course, we would like to offer students different kinds of projects to choose from, instead of one common track.

Acknowledgments. Robin Amsters is an SB fellow of the Research Foundation Flanders (FWO) under grant agreement 1S57718N.

References

1. Stohr-hunt, P.M.: An analysis of frequency of hands-on experience and science achievement. J. Res. Sci. Teach. **33**(1), 101–109 (1996)
2. Ma, J., Nickerson, J.V.: Hands-on, simulated, and remote laboratories. ACM Comput. Surv. **38**(3) (2006). 7–es. ISSN 03600300. https://doi.org/10.1145/1132960.1132961
3. Sauter, M., et al.: Getting real: the authenticity of remote labs and simulations for science learning. Distance Educ. **34**(1), 37–47 (2013). ISSN 01587919. https://doi.org/10.1080/01587919.2013.770431
4. CYBERBOTICS Ltd. Webots documentation: Adept's Pioneer 3-DX (2019). https://www.cyberbotics.com/doc/guide/pioneer-3dx?version=master. Accessed 01 Sept 2019
5. Mondada, F., Franzi, E., Guignard, A.: The development of Khepera. In: Proceedings of the First International Khepera Workshop, Paderborn, pp. 1–21 (2015)
6. Sanco Middle East LLC. AmigoBot - Sanco Middle East LLC. http://sanco-me.net/amigobot/. Accessed 01 Mar 2019
7. Parallax. Scribbler 3 (S3) Robot—28333—Parallax Inc. (2018). https://www.parallax.com/product/28333. Accessed 01 Mar 2019
8. Yujin robotics. TurtleBot2. https://www.turtlebot.com/turtlebot2/. Accessed 01 Mar 2019
9. Microbric. Edison Programmable Robot - Ideal for school classroom education. https://meetedison.com/. Accessed 01 Mar 2019
10. Anki. *Cozmo—Meet Cozmo* (2018). https://www.anki.com/enus/cozmo. Accessed 01 Mar 2019
11. Robotis. TurtleBot3. http://emanual.robotis.com/docs/en/platform/turtlebot3/overview/. Accessed 01 Mar 2019
12. Husarion. Husarion - Robot development made simple (2019). https://husarion.com/. Accessed 01 Mar 2019
13. ROBOTIS. TurtleBot 3. https://robots.ros.org/turtlebot3/. Accessed 01 Aug 2019
14. Ubuntu Mate. About Ubuntu Mate (2015). https://ubuntu-mate.org/about/. Accessed 24 Dec 2018
15. Edo Scalafiotti. Turn a RaspBerryPi 3 into a WiFi router-hotspot—Edo Scalafiotti–Medium (2016). https://medium.com/@edoardo849/turna-raspberrypi-3-into-a-wifi-router-hotspot-41b03500080e. Accessed 10 Apr 2018

16. Martin, P.: Using your new Raspberry Pi 3 as a WiFi access point with hostapd. https://frillip.com/using-your-raspberry-pi-3-as-a-wiFi-access-point-with-hostapd/. Accessed 10 Apr 2018
17. O'Kane, J.M.: A gentle introduction to ROS (2014). http://www.cse.sc.edu/~jokane/agitr/agitr-small.pdf
18. Mathworks. Features - Robotics System Toolbox - MATLAB & Simulink. https://nl.mathworks.com/products/robotics/features.htmlfn#grobot-operating-system-%28ros%29. Accessed 01 Sept 2019
19. Corke, P.: Robotics, Vision and Control: Fundamental Algorithms in MATLAB, vol. 73. Springer, Heidelberg (2011). ISBN 978-3-642-20143-1. https://doi.org/10.1007/978-3-642-20144-8. arXiv:0509398v1 [arXiv:cond-mat]
20. Corke, P.I.: A robotics toolbox for MATLAB. IEEE Robot. Autom. Mag. **3**(1), 24–32 (1996). ISSN 1070-9932. https://doi.org/10.1109/100.486658
21. Braitenberg, V.: Vehicles: Experiments in Synthetic Psychology. MIT Press, Cambridge (1986)
22. Aström, K.J., Murray, R.M.: Feedback Systems: An Introduction for Scientists and Engineers. Princeton University Press, Princeton (2010)
23. Leondes, C.T.: Digital control systems implementation and computational techniques (1996)

SLIM - A Scalable and Lightweight Indoor-Navigation MAV as Research and Education Platform

Werner Alexander Isop$^{(\boxtimes)}$ and Friedrich Fraundorfer

Institute of Computer Graphics and Vision (ICG), Graz, Austria
isop.alexander@gmail.com, {isop,fraundorfer}@icg.tugraz.at,
https://sites.google.com/view/w-a-isop/home/education/slim

Abstract. Indoor navigation with micro aerial vehicles (MAVs) is of growing importance nowadays. State of the art flight management controllers provide extensive interfaces for control and navigation, but most commonly aim for performing in outdoor navigation scenarios. Indoor navigation with MAVs is challenging, because of spatial constraints and lack of drift-free positioning systems like GPS. Instead, vision and/or inertial-based methods are used to localize the MAV against the environment. For educational purposes and moreover to test and develop such algorithms, since 2015 the so called *droneSpace* was established at the Institute of Computer Graphics and Vision at Graz University of Technology. It consists of a flight arena which is equipped with a highly accurate motion tracking system and further holds an extensive robotics framework for semi-autonomous MAV navigation. A core component of the *droneSpace* is a Scalable and Lightweight Indoor-navigation MAV design, which we call the *SLIM* (A detailed description of the *SLIM* and related projects can be found at our website: https://sites.google.com/view/w-a-isop/home/education/slim). It allows flexible vision-sensor setups and moreover provides interfaces to inject accurate pose measurements form external tracking sources to achieve stable indoor hover-flights. With this work we present capabilities of the framework and its flexibility, especially with regards to research and education at university level. We present use cases from research projects but also courses at the Graz University of Technology, whereas we discuss results and potential future work on the platform.

Keywords: SLIM · Dronespace · Micro aerial vehicle ·
Mobile robotics · Indoor navigation · Mapping ·
Research and education

1 Introduction

During the past decade, research and development on aerial robotic platforms, especially MAVs, became increasingly popular. With different types of phys-

© Springer Nature Switzerland AG 2020
M. Merdan et al. (Eds.): RiE 2019, AISC 1023, pp. 182–195, 2020.
https://doi.org/10.1007/978-3-030-26945-6_17

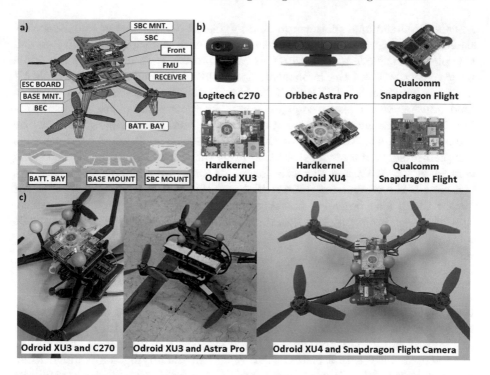

Fig. 1. Overview of the *SLIM*. (a) Principle layout of main components. Custom mounts are shown in blue and electronic components are represented in green colors. (b) Utilized sensors ([1,2]) and single-board-computers ([3,4]). (c) Overview of the different sensors (monocular camera, RGBD-sensor and high-resolution monocular camera integrated into the flight controller) and single-board-computer setups onboard the *SLIM*.

ical MAV setups and underlying control framework, state of the art applications widely aim for autonomous, or at least semi-autonomous navigation. Though, commercial off-the-shelf systems are most commonly designed for acting autonomously in outdoor environments, whereas typical use cases involve agriculture [5], parcel-delivery [6] or even search and rescue missions [7]. Clearly, such use cases put different demands on the MAV, compared to indoor flights. In outdoor environments spatial constraints play a subsidiary role and very often GPS-based localization and navigation is possible. Only few MAV platform designs exist which are focusing on indoor navigation mainly and more importantly even provide access to the appropriate interfaces to enable drift-free localization and navigation in indoor environments. This is especially important for educational purposes, if development and testing of vision based tracking or reconstruction algorithms is required. To this purpose, an indoor flying-arena for MAVs, including physical MAV design and software-framework, was established at the Institute of Computer Graphics and Vision (ICG) at Graz University of Technology. Besides of serving for research projects, the goal was to achieve a scalable educational platform for indoor flights, where Master and Bachelor-students are able to gain first hands-on experience in terms of MAVs. We call this platform the *SLIM* and, with this work, present main design aspects, while we give details about the implementation of hardware and software. The MAV design is based on the PIXHAWK [8] open source flight controller and able to hold multiple vision sensor configurations. For educational purposes, also drift-free pose measurements from a highly accurate motion tracking system can be used to localize the MAV against the environment. The physical MAV setup is complemented with an extensive ROS-framework [9] which serves for high-level control of the MAV and visualization of important data. This work gives an overview of the current capabilities of the platform. It was successfully used for indoor flight, also providing tracking-based navigation and online reconstruction of the environment. In a last step we discuss potential future work on the platform as an outlook.

2   Related Work

Design, implementation and evaluation of MAVs for indoor use was widely investigated on during the last two decades. Bouabdallah et al. [10] present design and control of an MAV for indoor flights, whereas they do not provide results in free-flight but create a testbench to test their underlying control algorithm. Further, How et al. [11] introduce a real-time indoor autonomous vehicle test environment which they call RAVEN. They discuss design of their framework and evaluate performance of their vehicles during flight operation. More recent work addresses MAV design, control and localization based on vision algorithms. For example, Vempati et al. [12], in general discuss design and control of an MAV, specifically discussing non-linearities, to achieve a highly accurate model. In addition they suggest localization based on SLAM algorithms like PTAM. However, they use a considerably larger and heavier MAV frame with an all-up weight of about 1 kg without any additional sensory setup. Also, they do not focus on a full educational framework with flexible sensor setup. Moreover, their design does not allow for injection of ground-truth measurements from external sources, like motion trackers. A more lightweight design is presented by Loianno et al. [13]. Their concept aims for an overall weight of below 250 g, also including a monocular vision camera for localization and navigation. However, their full setup does not include an RGBD-sensor which can be beneficial for mapping tasks in spatially constraint indoor environments, if rich and more detailed maps are required [14]. Kushleyev et al. [15] introduce a very small and lightweight design as part of a swarm of agile MAVs. The ready-to-fly (RTF) weight of their design is below 100 g and they also localize their MAVs with an external motion tracker. However, their setup does not provide any computational unit onboard for processing more expensive tasks like online mapping or any additional vision-sensor. In comparison, a lightweight solution below 100 g is also available from DJI called Ryze Tello [16], whereas a monocular camera is attached onboard that enables live video streams. On the other hand, this solution provides only very limited pay-loads and is not capable of carrying more advanced vision sensors like RGBD. Besides, various off-the-shelf products aim for small-sized flight control solutions and MAVs ([17,18]), with fully integrated vision sensors [4], also emphasizing an educational context. On the other hand they do not provide a full educational framework, including software solutions for higher-level robotic tasks. Finally, none of these solutions discuss a flexible and scalable design for multiple configurations of computational units and sensors.

3   The SLIM

In this section we present motivation and main aspects, which were considered when designing the *SLIM* as a research- and education-platform. Besides of achieving the challenging task of enabling indoor flight in a flexible setup, also easiness of use by unexperienced students played an important role. Additionally, the goal was to make experimental flights as safe as possible.

3.1   Design-Aspects and Constraints of the Physical MAV-Setup

Considering the general constraints of MAVs with regards to limited payload and flight times, finding a well balanced design is challenging. On one hand, larger MAVs with higher-all-up weight are capable of carrying more payload and typically have increased flight times. On the other hand, their produced thrust, size of propellers and outer dimensions increase accordingly. Very often this is not practical, specifically if flights in more narrow indoor environments is concerned. Moreover, if flying close to humans is required or unavoidable for educational purposes, too large platforms could be too dangerous for practical use.

In general, well-selecting appropriate components for the given design requirements was challenging, since flexibility and maximum accessability of the individual interfaces excluded any off-the-shelf solution.

Platforms like the well known M100 from DJI or smaller sized MAVs like the DJI Mavic 2 [16] are capable in terms of flight time, easiness of steering and quality of the attached monocular vision camera. In addition, the M100 is provided with an extensive ROS framework, whereas the Mavic 2 is provided with a mobile SDK to enable development of custom applications. Still, with a diagonal wheelbase of 650 mm and 2.4 kg of all-up weight the M100 is considerably larger and was not assessed as practical for the indoor flights in narrow space. In comparison, the Mavic 2, with a diagonal wheelbase of 354 mm and 907 g of all-up-weight, is of smaller size but unfortunately does not provide a ROS API directly, which would make it considerably more handy for educational purposes. In addition, the Mavic 2 is not easily hardware-customizable regarding more capable vision sensors like RGBD.

Additionally, there exist a variety of very small sized off-the-shelf platforms like the DJI Spark [16] or the crazy-fly from BitCraze [19], whereas their design aims for frame-wheelbases about the size of a humans hand. However, these platforms do not provide more powerful vision-based onboard sensors like RGBD and there is no easy access to hardware and flight controller functions. Also taking into account the very limited payload in general, this ultimately makes it more difficult to achieve position-stabilized hover-flight or environmental reconstruction in indoor environments.

To address the afore discussed problems, the main design goals for the *SLIM* platform were to consider a **scalable hardware configuration** with components that are **easily accessible** in the physical setup. Further, to include **open- and accessible** hardware interfaces, compatible with **open source** software frameworks. Moreover, a **lightweight** setup was aimed for, allowing for **adequate flight times** and **safe operation**.

Scalability

To provide maximum scalability in terms of supporting multiple sensor types and according computational hardware, a middle-sized frame design from off-the-shelf hardware was selected. The frame setup is based on the Parrot Bebop 2 [20] and has a diagonal wheelbase of 335 mm, whereas it is extended by custom mounts (Fig. 1a) to hold various sensor types and single-board-computers (SBCs), shown in Fig. 1b. The sensor systems, which were attached to the *SLIM* and also successfully used during lectures were a Logitech C270 Pro Monocular Camera, an ORBBEC Astra Pro RGBD-sensor and Snapdragon Flight Monocular Camera. Supplementary, the Hardkernel Odroid XU3, the Odroid XU4 and the Qualcomm Snapdragon Flight onboard SBC were used as small-sized computational units (Fig. 1b).

Easy-To-Access Hardware Components

The basic layout of the individual components to achieve stable hover flight is centered around the Bebop 2 frame. Use of the Bebop 2 frame is motivated in the following subsections, whereas placement of the components in principal is shown in Fig. 1a. The overall goal was to make as many components accessible for replacement or repair. As a result the *SLIM* was designed based on multiple component layers, which are arranged in a stack-like structure. For example, if access to the receiver or battery emitting circuit (BEC) inside the frame is required, only the base mount has to be removed. The base mount is simply "clipped" into the stock base-frame and can be quickly removed. If connector cables of the engines are unplugged, it is further possible to remove all layers above the base-frame including ESC, flight controller, RC-receiver, SBC and potentially connected vision sensors. The individual layers are separated by the custom mounts and connected via hexagonal plastic spacers and screws. Subsequently, removal of each layer is simple and time effective. Details about the assembly of the *SLIM* can be found via our website.

Open Hardware Interfaces

Access to hardware interfaces was of great important for our design, especially with regards to research and education. In general it can be stated, that closed off-the-shelf components might provide basic and robust functionality until some degree. Though, there are clear drawbacks in terms of letting students gather knowledge and experience, as customizable interfaces enrich educational and research potential. To this purpose, the main hardware components like flight controller, SBC and sensors were selected so that they provide maximum access to interfaces on hardware level.

Connected to this, another important role plays open source compatibility, because of its great flexibility and richness available solutions and support from the community. Accordingly, also the software frameworks were selected, whereas details are outlined in Sect. 3.3.

Lightweight Setup

For safety reasons, especially when flying close to unexperienced students, designing the MAV as lightweight as possible, was a crucial aspect. On the other hand, like discussed in Sect. 3.1, a too small-sized MAV design significantly reduces its capabilities. In addition, given a constrained flight height, the maximum flight velocity plays an important role. For better comparison, Falanga et al. [21] and Loianno et al. [13] presented aggressive MAV flight maneuvers with a max. velocity of up to 3.0 m$\cdot$s^{-1} and 4.5 m$\cdot$s^{-1} respectively. Whereas the design of Falanga et al. includes an all-up-weight of 0.83 kg, Loianno et al. introduce a lightweight design below 250 g and aim for similar capabilities like the *SLIM*, although both works include a monocular vision camera only. From design perspective, achieving online mapping during flight based on the richer RGBD-sensor data is still challenging in the weight category below 250 g. As a result trade-offs have to be considered. To first of all define a baseline for weight- and velocity constraints, the maximum energy of MAVs for outdoor use was selected under consideration of the aviation law in Austria. According to the Austrian Aviation Act [22], the maximum overall energy of a MAV during operation must not exceed

$$E = E_{kin} + E_{pot} \leq 79 \text{ J} \tag{1}$$

before it is subject to authorization. Since the height of the tracked flying space is below 3 m and considering the discussed works, the maximum MAV velocity during experimentation was defined to not exceed $v_{max} = 5.0$ m$\cdot$s^{-1}. In addition a weight of 0.7 kg was defined as a competitive maximum all-up-weight, compared to the afore mentioned works. Based on this weight constraint a maximum tolerable flight velocity $v_{tol.}$ was calculated for better comparison and is given in Eq. 3.

$$E_{kin} + E_{pot} \leq 79 \text{ J}, E_{kin} \leq 79 \text{ J} - 0.7 \text{ kg} \cdot 9.81 \text{ m} \cdot \text{s}^{-2} \cdot 3.0 \text{ m}, E_{kin} \leq 58.399 \text{ J} \tag{2}$$

whereas the maximum flight velocity is then given with

$$E_{kin} = 0.5 \cdot m \cdot v^2 \Rightarrow v_{tol.} \leq \sqrt{\frac{2 \cdot E_{kin}}{m_{AUW}}} \leq \sqrt{\frac{2 \cdot 58.399 J}{0.7 \text{ kg}}} \leq 12.917 \text{ m} \cdot \text{s}^{-1} \tag{3}$$

Regarding safety considerations, the *SLIM*'s maximum tolerable flight velocity $v_{tol.}$ (Eq. 3), under the given weight constraint was estimated to be far above the defined maximum of $v_{max} = 5.0$ m $\cdot$ s^{-1}.

As a result of the maximum all-up-weight and to start from a basic working setup, a semi-customized design was established. In stock configuration, the off-the-shelf Parrot Bebop 2 MAV [20] can provide flight times of up to 25 min and has an all-up weight of 0.5 kg. Thus, it is far below the required maximum of 0.7 kg regarding the need of mounting additional computational or sensor-units. Further, its frame and propellers provide enough stiffness and comparably low weight (43 g and 3 g each), compared to other carbon fiber frames of even smaller size (e.g. 190 mm-DroneArt MAV X-Frame with 26 g [23]). In addition, the frame provides enough hollow space in the center to potentially hold smaller hardware components. Moreover, its cheap price and availability makes it an attractive base for a semi-customized setup.

Based on the maximum all-up weight of $m_{AUW} = 0.7$ kg, in the following also the maximum flight times should be discussed. As basic requirement for educational- but also for research purposes, a minimum of $t_{hover} \geq 10$ min hover-flight time was defined. To maximize flight times, an efficient battery type was selected considering the energy to weight ratio. Evaluation and comparison of a wide range of battery types resulted in selection of the "DroneTec HD Power Battery For Parrot AR Drone 2.0". As it provides 3 cells (same as the stock Bebop 2), 2300 mA · h of capacity and due to its small size ($71 \times 37 \times 34$mm) and low weight (146 g), it was selected as the best fit.

Considering these boundary conditions, in a next step the hover-flight time was estimated. In general, the hover-flight times of a Lithium-Polymer (LiPo) battery powered quadcopter, can be calculated based on the following parameters, whereas details can be found on our website: C defines the battery capacity, V_n defines the nominal battery voltage, P_m defines the electrical power for the remaining electrical components, P_e defines the electrical power and η defines an efficiency-factor, taking into account energy losses from the electric components.

A crucial aspect for estimating hover-flight times is P_m, since it is the required power of each engine to let the MAV hover. An estimate $P_{m,est.}$ for one single engine given in Watts can be derived based on a mathematical thrust-model for the static case [24], and is given in the following equation.

$$P_{m,est.} = K_{ad} \cdot \frac{\sqrt{F^3}}{r} = 0.3636 \cdot \frac{\sqrt{6.867^3}}{0.0762} = 10.7321 \text{ W} \tag{4}$$

whereas $K_{ad} = 0.3636$ is the air-density coefficient given with the nominal value under the assumption that the air pressure is 1atm and the room temperature is at 20 °C. $F = m \cdot g = \frac{1}{4} \cdot 0.7$ kg $\cdot 9.81$ m·s^{-2} = 6.867 N is the thrust required for hover expressed in Newtons and $r = \frac{1}{2} \cdot 6 \cdot 0.0254 = 0.0762$ m is the radius of the propeller considering the 6 in. propellers of the Bebop 2 model.

Noticeable is a deviation of true measured power (P_m empirically measured with 12 W) from the estimated power, required for hover-flight. Typically they are a result of unmodeled power losses, which can occur due to non-perfect stiffness of rotor blades, efficiency of engines, power electronics and the fact that in our model the propelled air follows a perfect cylindrical shape, which is not true in the real world case. Although it is mentionable that deviations were still in an acceptable range still, also considering the resulting hover-flight times.

Based on the afore mentioned parameters and values, the estimated hover-flight time of the *SLIM* is then given with

$$t_{hover} = \eta \cdot \frac{60}{1000} \cdot \frac{C \cdot V_n}{4 \cdot P_m + P_e} = 0.8 \cdot \frac{60}{1000} \cdot \frac{2300 \cdot 11.1}{4 \cdot 12 + 30} = 15.711 \text{ min} \quad (5)$$

As a result, the estimated flight time was considered to be above the required minimum of $t_{hover} = 15.711$ min ≥ 10 min.

Safety

Concerning safety, the *SLIM* also provides basic features enabling operation and close-flight to unexperienced persons. One major aspect are the propeller guards, which were also customized and reduced in weight for the *SLIM* design. In addition, the PIXHAWK flight controller offers a safety-switch which enables the MAV's engines only after they were manually armed by an operator.

3.2 Modelling and Control of the MAV

The *SLIM* is setup in a common 4 propeller X-configuration including 4 engines that are individually positioned at the end of the stock Bebop 2 frame. The direction of motors is set in such a way that a pair of motors rotate clockwise while the other pair rotates counter clockwise. This arrangement of motors is set to generate vertical lifting force to raise the MAV up in the air. Choi et al. [25] describes two frames of reference which are the fixed world-frame F_W and the relative body frame F_B of the *SLIM*. The transformation of the *SLIM*'s body frame F_B relative to the world frame F_W is defined with $^W T_B$. All frames are right handed with the x-axis pointing in forward-flight direction and the z-axis pointing upwards. They are defined to derive the equations of motion for a 6 degree of freedom (DOF) configuration of the *SLIM*.

Degrees of Freedom

Like described by Choi et al., the *SLIM* in the current configuration can be controlled via angular movements along 3 different axis, which are roll-angle θ (rotation around the x-axis), pitch-angle ϕ (rotation around the y-axis) and yaw-angle ψ (rotation around the z-axis). Thus, the following in-air maneuvers can be performed by the MAV: hovering, pitching forward or backwards, rolling left or right and turning around the z-axis. Hovering can be achieved if all engines turn at the same speed and produce the same vertical thrust. Roll- and pitch movements can be achieved if the speed of one pair of motors is changed, while the other pairs turning speed remains constant. Turning around the z-axis is achieved by altering the speed of the two pairs of engines. The orientation angles θ, ϕ and ψ are expressed as Euler angles.

Model of the MAV

The model of the MAV can be defined with the equations of motion. They are expressed as Newton-Euler equations, whereas they reflect the combined translational and rotational dynamics of a rigid body. Assuming a simplified point mass model of the MAV the equations are given in Eqs. 6 and 7 and reflect the MAV's 6DOF,

$$\ddot{\phi} = \dot{\theta}\dot{\psi}\left(\frac{J_y - J_z}{J_x}\right) + \frac{U_2}{J_x}, \ddot{\theta} = \dot{\phi}\dot{\psi}\left(\frac{J_z - J_x}{J_y}\right) + \frac{U_3}{J_y}, \ddot{\psi} = \dot{\phi}\dot{\theta}\left(\frac{J_x - J_y}{J_z}\right) + \frac{U_4}{J_z} \quad (6)$$

$$\ddot{z} = \frac{U_1}{m}\cos\phi\cos\theta - g, \ddot{x} = \frac{U_1}{m}(\cos\phi\sin\theta\cos\psi + \sin\phi\sin\psi)$$
$$\ddot{y} = \frac{U_1}{m}(\cos\phi\sin\theta\sin\psi - \sin\phi\cos\psi) \quad (7)$$

whereas J_x, J_y and J_z reflect the inertia terms on the main diagonal of the inertia matrix J, m parametrizes the point mass and g is the earths gravity constant. The equations are simplified, based on Beard et al. [26], whereas coriolis terms and time-derivates of the rotational component of the transformation matrix $^W T_B$ are neglected. Accordingly, the 4 control inputs U_i of the model can be defined as given in Eq. 8.

$$U_1 = b\left(\omega_1^2 + \omega_2^2 + \omega_3^2 + \omega_4^2\right), U_2 = b\left(\omega_4^2 - \omega_2^2\right)$$
$$U_3 = b\left(\omega_3^2 - \omega_1^2\right), U_4 = d\left(\omega_4^2 + \omega_2^2 - \omega_3^2 - \omega_1^2\right) \quad (8)$$

whereas ω_i is the angular velocity of each rotor, b is the thrust coefficient and d is the drag coefficient. U_1 can be interpreted as the overall thrust force applied to the MAV along the z-axis in the center of its body-frame F_B. U_2 and U_3 lead to pitch- and roll torques respectively, while U_4 is leading to the torque around the z-axis.

Control of the MAV
Since the *SLIM* is designed to fly indoors mainly, turbulences were expected to disturb flight performance. For sakes of robustness and due to the already existing architecture of the PIXHAWK the integrated linear controllers were utilized to achieve stabilized flight. They consist of nested PID controllers, which can be expressed in general by

$$C_{PID}(s) = K_p + K_i \cdot \frac{1}{s} + K_d \cdot s \quad (9)$$

whereas K_p, K_i and K_d are the proportional-, integral- and differential gain parameters of the controller and typically tuned based on the Ziegler-Nichols method [27]. The linear position control approach for the MAV can be separated into altitude control (z-axis) and control of the horizontal movement (x/y-axis). Control of the angular position around the z-axis (yaw-angle ψ) is directly achieved by the PIXHAWK's inner attitude controller, whereas control of x,y and z-position is achieved by the integrated position control loop. Details are discussed in the following.

First of all, the MAV's rotor dynamics play a crucial role, whereas they can be approximated with a first order system, including a linear coefficient K_M and a time constant τ_M resulting from the inertia of the rotors and engines. In the Laplace domain the dynamics are then given with

$$G(s) = \frac{K_M}{\tau_M s + 1} \tag{10}$$

Laplace transformation of the translational equations of motion ($\ddot{x}$, $\ddot{y}$ and $\ddot{z}$ in Eq. 7) and combination with the linearized rotor dynamics (Eq. 10) results in the transfer functions for displacement in x-,y- and z-direction. Different modelling approaches exist here, whereas a common approach is to approximate the transfer functions for displacements by a double integrator combined with first- and second-order systems (Seidel [28], Joyo et al. [29]). Thus, assuming hovering condition ($\frac{U_1}{m} - g = 0 = const.$ with $\phi = \theta = 0$) for vertical displacement and considering a small-angle approximation for the horizontal displacement, we can express the according transfer functions given in Eq. 11. Noticeable are the dynamics for attitude stabilization G_{xy} for pitch- and roll-angles, which are approximated by a second order system [30]. The according gain parameters can then be defined with $K_{xy} = g$ and $K_z = \frac{1}{m}$.

$$G_{xy}(s) = \frac{K_{xy}}{s^2}\left(\frac{\omega_0^2}{s^2 + 2D\omega_0 s + \omega_0^2}\right), G_z(s) = \frac{K_z}{s^2}\left(\frac{K_M}{\tau_M s + 1}\right) \tag{11}$$

Combining these transfer functions with the PID-control approach expressed in Eq. 9, the transfer functions of the closed loop system can then be expressed with

$$T_{xyz}(s) = \frac{G_{xyz}(s) \cdot C_{PID}(s)}{1 + G_{xyz}(s) \cdot C_{PID}(s)} \tag{12}$$

Experiments for Position Stabilization

For sakes of completeness, performance of the position stabilization of the *SLIM* was evaluated in relation to the model and control approach discussed in the previous section. The setup used for experimenting included the C270 camera and the Odroid XU3, whereas details about the different *SLIM* configurations are shown in Fig. 1. In a first step, datasets necessary for evaluation and further controller tuning were taken during flight and it is remarkable that the proportional gain of the position control loop of the PIXHAWK in the first step was set to $MPC\_XY\_P = MPC\_Z\_P = 1$. No additional integral or derivative gain was used ($K_p = 1, K_i = K_d = 0 \rightarrow C_{PID}(s) = 1$) and all other gain parameters (e.g. for the feed-forward path and state estimator) were left at default configuration. The procedure for taking the datasets is described in the following. First the MAV was commanded to achieve stable hover flight at 1 m with zero heading. After the MAV stabilized around the given setpoint, step inputs with a relative change of 1 m were applied in x, y and z-direction accordingly. Data for the applied position command and resulting actual position of the MAV was recorded at 60 Hz. The step responses with tuned controller gains applied to the system described by Eq. 12 are shown in Fig. 2 and in good approximation show PT2 behaviour. They reflect adequate closed loop control performance of the fully modelled MAV, including nested control loops of the internal system structure.

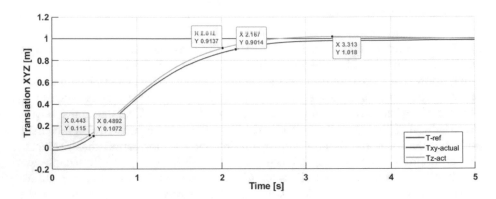

Fig. 2. Shown are step responses with tuned controller gains for translation in x/y- and z-direction. Remarkable are adequate rise times ($t_{r,x} \simeq t_{r,y} \simeq 1.57$ s and $t_{r,z} \simeq 1.68$), small overshoot ($PO < 5\%$) and sufficiently small steady-state error ($e_\infty < 5\%$).

3.3   Design of the Software Framework

The main motivation of the software design was compatibility to the selected hardware components in Sect. 3.1, which also provide open and accessible interfaces for the *SLIM* platform. This involves typical components used for research with MAVs, like the open source PIXHAWK flight controller, but also computers with more computational power like the Odroid SBCs from Hardkernel (XU4, XU3). They well support UNIX-based open source operating systems (Ubuntu), whereas support for commercial operating systems like Microsoft Windows is not as extensive. Consequently, with the aspects of scalability and open source compatibility in mind, the Robot Operating System (ROS) [9] was selected as state of the art framework. Not only because it is open source and well supported under UNIX based operating systems, but also because of its rich ecosystem.

3.4   Architectural Overview and Utilized Methods

The main aspect for design of the software architecture was to provide maximum flexibility with regards to research and educational projects. Consequently, a minimum set of core-functionalities were integrated as individual components in a ROS framework, such as low-level flight control, localization, high-level flight control (path-planning, exploration), environmental mapping and object detection. More details about the framework's components and the distributed ROS-messages can be found via our website.

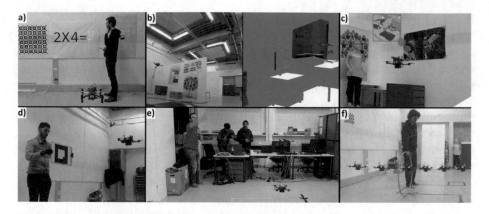

Fig. 3. Example use cases of the *SLIM*. (a) An earlier version of the *SLIM* acting as teaching assistant [31]. (b) The *SLIM* during experimentation for the DAHV [32]. (c) Victim detection during the camera drones' rescue challenge. (d) Marker-based visual servoing. (e) Avoidance of a thrown reflective marker. (f) Hula-Hoop visual tracking and passing through.

4   Use Cases in Research and Education

Since 2015, the *SLIM* was used in various projects at the ICG. These were either research projects but also lecture courses and students projects. Research contributions which were utilizing the *droneSpace* and the *SLIM* platform were namely the Micro Aerial Projector (MAP) [31] and the Drone Augmented Human Vision (DAHV) [32].

Additionally, a lecture course where the *SLIM* platform was extensively used is called *Camera Drones* and was established in Winter Term 2016/2017 at Graz University of Technology. During this term students had to work on individual projects based on a reference implementation. Each student received a *SLIM*, equipped with different vision sensors and SBCs. Amongst others, projects included visual servoing based on fiducial markers (Fig. 3d), collision avoidance when throwing a reflective marker towards the *SLIM* (Fig. 3e) and a hula-hoop flight (Fig. 3f).

In Winter Term 2017/2018 for the first time students had to compete in an indoor drone rescue challenge. A maze was set up in the *droneSpace* as part of an artificial disaster scenario, whereas the task for each team was to explore the environment and report back the 3D-positions of found victims. Based on a reference implementation, the teams had to solve individual sub-tasks which included localization, path-planning and navigation, 3D-mapping and detection of fiducial markers using the onboard RGBD-sensor of the *SLIM* (Fig. 3c).

For more details about research projects, lectures and the *droneSpace* in general please also refer to the droneSpace website[1].

---

[1] A detailed description of the *droneSpace*, the *SLIM* and related projects can be found at our website: https://www.tugraz.at/institutes/icg/education/the-dronespace/.

5   Conclusion and Future Work

In conclusion, the *SLIM* served as versatile platform during the recent years. It was successfully used for research projects and moreover for education, providing a flexible MAV setup and an extensive ROS-based framework. On one hand, students gained valuable experience in **control, sensing, navigation** and also **environmental mapping** with MAVs. On the other hand the *SLIM* could also serve as versatile educational platform for robotics lectures in general. Further, it is remarkable that work is in progress and the platform is constantly improved and extended, whereas a brief overview on future work is given in the following.

One improvement which can be seen as a future work is to achieve trajectory tracking by feeding higher-derivative reference signals into the PIXHAWK flight controller. Based on a position trajectory, discretized velocity- and acceleration inputs could be derived and directly fed into the inner control architecture. Ultimately, the tracking accuracy could be improved. However, it was found that during experimentation for research purposes, but also during lectures, the *SLIM* provided adequate flight dynamics and tracking errors during experimentation.

References

1. ORBBEC: Orbbec Astra Pro (2017). https://orbbec3d.com/
2. Logitech: C270 HD Webcam (2010). https://www.logitech.com/
3. Hardkernel: Odroid XU3/XU4 (2014). https://www.hardkernel.com/
4. Qualcomm: Snapdragon Flight (2014). https://shop.intrinsyc.com/products/snapdragon-flight-dev-kit
5. Tripicchio, P., Satler, M., Dabisias, G., Ruffaldi, E., Avizzano, C.A.: Towards smart farming and sustainable agriculture with drones. In: 2015 International Conference on Intelligent Environments (IE), pp. 140–143. IEEE, July 2015
6. Puerta, J.P., Maurer, M., Muschick, D., Adlakha, D., Bischof, H., Fraundorfer, F.: Package Delivery Experiments with a Camera Drone (2017)
7. Silvagni, M., Tonoli, A., Zenerino, E., Chiaberge, M.: Multipurpose UAV for search and rescue operations in mountain avalanche events. Geomat. Nat. Hazards Risk **8**(1), 18–33 (2017)
8. Meier, L., Tanskanen, P., Heng, L., Lee, G.H., Fraundorfer, F., Pollefeys, M.: PIXHAWK: a micro aerial vehicle design for autonomous flight using onboard computer vision. Auton. Robot. **33**(1–2), 21–39 (2012)
9. Quigley, M., Conley, K., Gerkey, B., Faust, J., Foote, T., Leibs, J., Wheeler, R., Ng, A.Y.: ROS: an open-source robot operating system. In: ICRA Workshop on Open Source Software, vol. 3, no. 3.2, p. 5, May 2009
10. Bouabdallah, S., Murrieri, P., Siegwart, R.: Design and control of an indoor micro quadrotor. In: 2004 IEEE International Conference on Robotics and Automation, Proceedings, ICRA 2004, vol. 5, pp. 4393–4398. IEEE, April 2004
11. How, J.P., Behihke, B., Frank, A., Dale, D., Vian, J.: Real-time indoor autonomous vehicle test environment. IEEE Control Syst. **28**(2), 51–64 (2008)
12. Vempati, A.S., Choudhary, V., Behera, L.: Quadrotor: design, control and vision based localization. IFAC Proc. Vol. **47**(1), 1104–1110 (2014)

13. Loianno, G., Brunner, C., McGrath, G., Kumar, V.: Estimation, control, and planning for aggressive flight with a small quadrotor with a single camera and IMU. IEEE Robot. Autom. Lett. **2**(2), 404–411 (2017)
14. Henry, P., Krainin, M., Herbst, E., Ren, X., Fox, D.: RGB-D mapping: using depth cameras for dense 3D modeling of indoor environments. In: The 12th International Symposium on Experimental Robotics (ISER) (2010)
15. Kushleyev, A., Mellinger, D., Powers, C., Kumar, V.: Towards a swarm of agile micro quadrotors. Auton. Robots **35**(4), 287–300 (2013)
16. DJI: Ryze Tello, Mavic 2 and Spark. (2017). https://www.dji.com/at
17. Infineon: Educopter (2018). https://www.infineon.com/cms/en/applications/consumer/multicopters-and-drones/
18. Intel: Intel Aero (2016). https://software.intel.com/en-us/aero
19. Bitcraze: Bitcraze Crazyflie 2.0 (2014). https://www.bitcraze.io/
20. Parrot, S.A.: Parrot Bebop 2 (2016). http://www.parrot.com/
21. Falanga, D., Mueggler, E., Faessler, M., Scaramuzza, D.: Aggressive quadrotor flight through narrow gaps with onboard sensing and computing using active vision. In: 2017 IEEE International Conference on Robotics and Automation (ICRA), pp. 5774–5781. IEEE, May 2017
22. Austro Control: Regulations for Unmanned Aerial Vehicles (2014). https://www.austrocontrol.at/drohnen
23. DroneArt: DroneArt Aeon X-Frame (2018). https://redbee.de/DRONEART-Aeon-HighEnd-Series-X-Frame-UL12
24. Ramamurthy, S.: Thrust models (2018). http://www.dept.aoe.vt.edu/~lutze/AOE3104/thrustmodels.pdf
25. Choi, Y.C., Ahn, H.S.: Nonlinear control of quadrotor for point tracking: actual implementation and experimental tests. IEEE/ASME Transactions Mechatron. **20**(3), 1179–1192 (2015)
26. Beard, R.: Quadrotor dynamics and control rev 0.1 (2008)
27. Ziegler, J.G., Nichols, N.B.: Optimum settings for automatic controllers. Trans. ASME, **64**(11) (1942)
28. Seidel, M.S.C.: Entwurf und Stabilitätsanalyse der Höhenregelung und Wandvermeidung des FINken II Quadrokopters
29. Joyo, M.K., Hazry, D., Ahmed, S.F., Tanveer, M.H., Warsi, F.A., Hussain, A.T.: Altitude and horizontal motion control of quadrotor UAV in the presence of air turbulence. In: 2013 IEEE Conference on Systems, Process and Control (ICSPC), pp. 16–20. IEEE, December 2013
30. Andreas, R.: Dynamics identification & validation, and position control for a quadrotor. Swiss Federal Institute of Technology Zurich, Spring Term (2010)
31. Isop, W.A., Pestana, J., Ermacora, G., Fraundorfer, F., Schmalstieg, D.: Micro aerial projector-stabilizing projected images of an airborne robotics projection platform. In: 2016 IEEE/RSJ International Conference on Intelligent Robots and Systems (IROS), pp. 5618–5625. IEEE, October 2016
32. Erat, O., Isop, W.A., Kalkofen, D., Schmalstieg, D.: Drone-augmented human vision: exocentric control for drones exploring hidden areas. IEEE Trans. Vis. Comput. Graph. **24**(4), 1437–1446 (2018)

An Open Solution for a Low-Cost Educational Toy

Pavel Petrovič[1](✉) and Jozef Vaško[2]

[1] Comenius University, Bratislava, Slovakia
ppetrovic@acm.com
[2] Fablab, Bratislava, Slovakia
jozef.vasko@fablab.sk

Abstract. In the summer of 2018 we organized two 5-day summer camps each for 20 children aged 11-15 in Fablab Bratislava. In addition to learning about 2D and 3D modelling and Arduino programming, every child has built and experimented with a humanoid toy robot. For this purpose, we have developed a flexible and extensible software solution that easily transitions and scales up to any other toy or even advanced robot controlled by Arduino. In this article, we describe our experiences from the camp as well as some other makers events and efforts and give details on the respective framework and discuss the role of makers movement in the educational process.

Keywords: Dtdt · Otto · Fablab · Makers · Arduino

1 Introduction and Related Work

Young people growing up in this decade benefit from an easy access to and availability of information and technologies of all kinds thanks to the development of Internet. Youngsters who have a deep interest in a particular hobby, subject, a challenge or a specific project, and who have the time, enthusiasm, energy, and a goal mindedness are in a much better starting position to enter and advance along the learning trajectory in a fast pace than ever before. Yet, paradoxically, the amount of information available works counter-productive when the children are facing a difficult question of where to start, how to digest information that is too complex to start with and how to find in the sea of the options – tasks, challenges, projects, and ideas – those that are suitable for their actual knowledge and skill levels. Parents and other adults who face their own challenges in a demanding world are seldom able to follow up on the developments and fail to provide the youngsters the shared time and the opportunities to let them grow from experiences while working with an appropriate learning material.

One possibility – often used by the parents and sometimes schools – is to rely on good quality solutions, example of which are LEGO robotics educational sets and programs. Their price, however, makes it inaccessible for many. Furthermore, the marketing strategies of such producers are often somewhat limiting in terms of

M. Merdan et al. (Eds.): RiE 2019, AISC 1023, pp. 196–208, 2020.
https://doi.org/10.1007/978-3-030-26945-6_18

variability, good modularity, interoperability, portability, open-hardware and software, and the software stability consequently. We feel that the development cycle of 6-8 years[1] is too slow to reflect on the fast technology developments and educators' needs. Thus meanwhile they are seconded by copyright breaching, but technologically superior sets such as KAZI EV5 sold for a quarter price (although possibly not in the same production quality). There is never enough of the credit to give to the robotics set producers for the valuable work they do on these useful educational tools, it is hard to believe that in the highly flexible production factory processes of today it is so difficult to provide a broader selection of different options that would better fit various needs given by the educational goals of the customers. Many alternate solutions are fragmenting the user base meanwhile, an interesting example of such being [9].

Educational programs such as FIRST LEGO League or RoboCup Junior provide an excellent escape from the lack of opportunities. However, peeking at the results from the regional tournaments, we see too many teams achieving too low score. For example, in 2018, from 98 teams participating in 7 regional tournaments of FLL in Slovakia only 30 achieved more than 100 points and 29 achieved less than 50 points suggesting that many teams do not allocate enough time or efforts needed for the learning transfer to be effective. Team performance often depends on the presence of a necessary catalyst, in this case a skilled team-leader providing the technical, motivational, pedagogical guidance. In addition, relatively short meetings duration combined with low meeting frequency further prevents efficient progress. The same material, skills and ideas have to be re-acquired multiple times after having been lost. Sometimes it takes tens of minutes to be able to follow-upon on the previous meeting. Especially when the meeting club must tidy up the room after every single meeting and stow the equipment and the work in progress somewhere in a storage cabinet.

Somewhat more effective regime with faster, yet longer-lasting learning progress can be achieved in summer schools and summer camps. The positive impact of summer schools on the future skill level and performance of participants is well known [8].

For more than 30-years a group of enthusiastic students, professionals, and training staff members organizes a two-week electronics summer camp for talented children (LSTME – Letné sústredenie talentovanej mládeže v elektronike, lstme.sk), where both authors have participated several times as leaders or instructors. From our personal experience, working both as a leader in an afternoon robot-club and as an instructor in this summer camp, we claim with confidence that skills of a typical strongly motivated child advance further in the camp than during a full year of active and periodic participation in an afterschool robot club. Intensity of the learning in the camp is extremely accelerated by the presence of about 15 leaders and instructors - experienced technicians who have answers to all the questions of the curious young mind. Adding up to that they also bring various hardware and workshop equipment to the camp making it possible to demonstrate the ideas in practical realizations: well-defined or even open-ended projects. Disposition of the camp typically includes a full electronics workshop for producing PCBs, LEGO robotics workshop, 3D printing workshop,

---

[1] MINDSTORMS releases: RCX: 1998, NXT: 2006, EV3: 2013.

computer room and audio-video studio. It takes place in the heart of nature and thus interleaving laboratory work with a relaxing time and sports in the beautiful environment.

One of the authors is a leader and manager of Fablab Bratislava – a fabulous laboratory where everybody can come to realize his or her dreams using workshop equipment such as 3D printers, laser cutters, vinyl cutter, automatic sewing machine, miller machine, and more. Inspired by our experience from LSTME, we decided to organize a day-camp in the space of Fablab Bratislava. We named it Denný tábor digitálnych technológií (DT)^2 – a day camp of digital technologies. We had several goals in mind when preparing the camp: (1) to let the children have a hands-on learning experience with digital technologies so that they will understand their principles, purpose and use and that they will be capable of fabricating various designs, potentially coming back later and extending the Fablab users family, (2) show them the whole process of completing a full project resulting in a real product that they take home and can continue using and tinkering with it later on, (3) give them sufficient background on 2D and 3D modelling and Arduino programming so that they understand the complete process of a design and development of a novel prototype.

Further sections of this article describe some related work, the individual activities of children in the camp, the robot that the children built and the framework that we have developed for the robot and finally summarizes our thoughts on the role of such activities in the educational process.

2 Organization

Children were divided into two groups of 10 based on their age and skills. These two groups alternated between two workshops: (1) 2D and 3D modelling and (2) Electronics and programming. The day was started with a social warm-up game while we waited for everybody to arrive, followed by the morning session lasting about 4 h. After the lunch, we spent some time outside, playing games and easy sport such as discgolf. In particular, we have used the activities from the Systems thinking playbook [1] that helps people develop a systems thinking perspective when observing the world or solving problems in a funny and gentle way. In the afternoon, the groups have exchanged the rooms, and continued in another workshop session of about 4 h. In the middle of the camp, we visited the laboratories of the National Centre for Robotics with live demos of manipulators, 3D scanners, large mobile robots and industrial applications. At the end of the camp, all children presented their results and products to the whole group, see Fig. 1. We feel the sharing, and enjoying the sharing is among the most important principles in educational activities.

Fig. 1. A happy participant in a final presentation and children playing with Ottos.

3 2D and 3D Modelling

The first group of the participants learned about 2D modelling by creating various designs in Inkscape open-source software. Every participant drew a picture that was printed on vinyl cutter and ironed on a T-shirt, and they also produced fabric bags and tiny items such as jewelry and souvenirs. They were introduced into the world of 3D design using the TinkerCAD software. Every participant has designed a fairy-tale 3D scene (Fig. 2).

Fig. 2. Example designs from the 2D workshop.

4 Introduction to Arduino Platform

One of our primary goals was to show the children a technology they would be able to use at home or their clubs after they will have returned from the camp. Arduino platform is available at very reasonable prices, has a huge user-base and a community with solutions to almost any challenge that an electronic hobbyist typically could encounter. Every participant received the following equipment for experimentation:

Arduino Nano board, Arduino Nano Expansion Board, USB mini cable, and a set of sensors and servo-motors. We have prepared a set of challenges – little projects on which we demonstrated the various elementary features of the Arduino platform – such as digital and analog inputs and outputs, PWM control as well as the language features – expressions, variables, arrays, statements, conditions and loops.

A typical project consisted of a challenge that we solved together, while explaining the principles. Next, the participants were asked to make modifications, improvements, and solve similar challenges. These included – flashing LEDs on LED panel, alarms, reacting to sound intensity, remembering and replaying a clapping rhythm. In some activities, we combined LEGO parts with Arduino and servo. For example, designing a ticking clock, or making the FLL mission models to move on their own.

In addition, we presented a set of additional optional projects to the participants that demonstrated further features and that were meant for those who are confident with the acquired or previous experience and were seeking more information. We have used some of these ideas in our framework as well. This included: using interrupts to respond to ultrasonic sensor, harmless software serial communication (see below), playing more complex melodies described in easy sequence, playing tones and melodies in the background (Fig. 3).

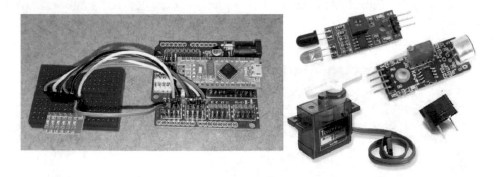

Fig. 3. Parts used in the Arduino challenges.

5   Building Otto

We used a design that was inspired by [2], but we wished to make it more human-like. We added the arms of the same shape as the legs, except of making the arms a little bit wider in order to increase their distance from the body. We have produced both a version for the 3D printer and a version for laser cutter. The main difference from our point of view was the time required for producing the parts – about 12 h for 3D printing compared to about 30 min of plywood laser cutting. The difference increases if the production process fails for any unspecified reason. Since we really wanted to prepare parts for all the 40 participants, we chose the laser-cut version to save the time. Another difference between the two designs is the way the joints are attached to the servo-motors. In the 3D printed version, the motor is attached by a single screw, whereas in the plywood version, servo horn is attached to the motor, and the horn fits inside a wooden part to smoothly control the turning. The original author of Otto has later improved the 3D version too. The plywood version contains more parts and requires more advanced manual skills. We have therefore used the 3D printed version in the group with younger and less experienced students (Fig. 4).

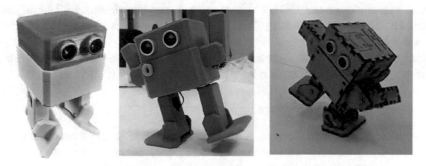

Fig. 4. The original [2] and our two versions of robot Otto.

Once the participants acquired the elementary programming skills and began to be fluent in using Arduino platform and programming in its C++ language, they received a bag containing laser-cut plywood or 3D printed parts, screws, battery pack, wires, and electronic parts and they began to build their own robot Otto. The building process took the whole 4-h session, but every participant managed to build it. The process is documented at the website of the day camp [3]. For the wooden version, participants used hammer to align the parts properly, and a glue just for a very few connections with otherwise loose contact (Fig. 5).

Fig. 5. Building robot Otto from parts.

Otto has been prepared for use of the Arduino Nano with its expansion board, which fit nicely into its head. Apart from that, we chose to design the electronics in our own custom way and write all the software completely on our own.

The following parts have been used (in total including the material worth about 20 Eur): 6 pcs. SG-90 micro-servo, Arduino Nano with ATmega328 (and mini/micro USB cable), Arduino Nano Shield I/O Extension Board, 4 AA batteries holder, passive buzzer, ultrasonic sensor HC-SR04P, at least 10pcs female-female 10 cm jumper cables, HC-05 Bluetooth module, SB550A Schottky diode – or similar with about 0.5–0.7 voltage drop, minimum voltage 10 V, and current 3A, 1000 uF capacitor – or similar to filter out high current demands, KCD11 power switch, 4pcs screws M3/5 mm, optionally: DFPlayer mini mp3 player, 8 Ohm 1 W speaker.

5.1 Powering and Wiring the System

A very nice feature of the Arduino Nano Expansion Board is that it provides multiple pins with GND and 5 V connections, allowing easy connection of many sensors, servos, and other devices. Unfortunately, all these 5 V pins rely on a single power regulator, which is capable of delivering 1.5 A current. The power consumption of 6 servo-motors, Bluetooth module, ultrasonic sensor, and other devices summed up indicates higher demands. Therefore, we ought to skip the DC power input of the expansion board, its power regulator and built-in capacitor and connect the batteries directly to the 5 V pins, ensuring the highest possible current supply. Unfortunately, that would breach another limitation – the absolute maximum ratings of the Arduino board (and in fact most TTL electronics) voltage, which is stated as 6 V in the data-sheet. A pack of 4 AA alkaline batteries when fresh (each producing 1.65 V) gives 6.6 V in total, which is unacceptable. We have therefore inserted a Schottky diode, with about 0.5–0.7 voltage drop to clamp the maximum within the limits. Since the original capacitor was also circumvented, we added another high-capacity 1000 uF electrolytic capacitor. This part of the process as well as installation of the ultrasonic sensor in a more space-efficient way required a little bit of soldering. Most of the participants had no soldering experience, and this was a nice opportunity for them to see it and try it for the first time, while we observed the process and made sure the resulting connections are fine.

The connections to the buzzer, Bluetooth module, ultrasonic sensor and optional devices were made with usual jumper wires.

6 Open Software Framework

To satisfy one of the goals stated in the introduction, we intended to create a more sophisticated control framework for the Otto robot so that the participants do not have to invent their own complete code for the Otto robot. The latter option was not possible due to a short time remaining in the stage of the camp after they have built their robots. Yet, with the background they have acquired in the first part of the week, they should be able to make modifications in the code and tune it to their needs and desires. We wanted to give the children the ability to create their own choreographies in a simple manner so that they could also be easily shared with others.

6.1 Calibration

When building the robot, servos are screwed in at some particular angular position while the leg or arm is turned to a particular configuration. The builder has to observe that the full range of movement can be achieved for each degree of freedom. Yet, it is very difficult to mount the motors at the same exact position on every robot and the servo motors themselves can exhibit somewhat different behavior as such. To compensate for that and to produce the very same results when dancing according to some shared choreography, each robot should be calibrated. The semi-automatic calibration procedure involves manually tuning the position of each leg to a standardized position and then storing the calibrated values into EEPROM memory so that the robot will remember the calibration even when powered off and on again. In addition, some children have made a mistake when mounting the servo motor on the leg, turning it around – getting a fully functional robot, but with one (or more) servos reversed. To deal with the situation, the calibration also stores the orientation of each servo. When dancing according to some choreography, the actual value is recomputed to match both the servo shift and the servo reversal. The calibration also stores the allowed range for each degree of freedom, which is enforced when controlling the robot movement manually.

6.2 Bluetooth Communication

Arduino boards are typically programmed in a very convenient way through built-in USB to serial converter and using a built-in bootloader program, and we like to keep up with this standard. However, unfortunately, the ATmega328 microcontroller only has a single USART hardware port. Connecting the Bluetooth module to its Tx, Rx pins interferes with the programming and the module must be disconnected each time before the programming takes places. That is a no-option, since the robot would have to be opened each time for the program download. In addition, we would like to keep the Bluetooth communication in a separate channel so that the wireless communication link does not have to be re-established every time we want to update the program with a new version. A possible solution is provided by the SoftwareSerial library for Arduino. Unfortunately, this library is implemented in a poor way: during the communication on the serial line, the global interrupt flag is disabled, meaning the timer-generated PWM signals by the Servo library for the movements of the legs and arms get extremely distorted, resulting in chaotic and unpredictable movements. We have therefore implemented our own "software serial" communication that relies only on the pin-change interrupt, which is available on most Arduino pins. The challenge is that every byte transmitted may have a different number of pin change interrupts and often even after the last pin-change interrupt occurs, we do not yet know, which byte is being transmitted. Consider, for instance, two different situations: receiving byte $191 = (10111111)_2$ vs. receiving byte $175 = (10101111)_2$, see Fig. 6.

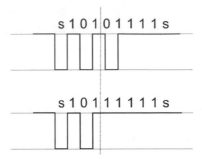

Fig. 6. Ambiguity within the pin-change interrupt routine.

After reading bits 0-2 (101), we will see no more pin change interrupts in the case of 175. We must leave the buffer in a "quantum" undecided state and infer that the received byte is 175 only after we will have received a request to read the next byte from the port. If that request, however, comes before the expected time of the stop bit, we may not be sure, and we must report that no new bytes are available yet. Otherwise, we can conclude that the received byte must be 175. The full implementation of this efficient and non-disturbing software serial algorithm is demarked in the source code [4].

6.3   Playing Melodies in the Background

A simple passive buzzer is a very low-cost solution to add sound to an Arduino project, and even though the sound is a bit loud, it can play nice melodies. Arduino has the built-in *tone()* function for producing sounds at specified frequencies of a specified duration in the background, using timer 2. Unfortunately, this plays only a single tone, and we need to play the full melodies. And since all Arduino timers are already occupied (0 – system time, 1 – servos, 2 – tone), there is no remedy. We chose to phase-out the standard *tone()* function and write our own that allows playing full melodies. A melody is a sequence of bytes, where each typically byte represents a single tone – its duration (full, half, quarter, eighth, sixteenth), and octave (1–5). It allows also rests, dotted notes, playing sounds of arbitrary melody, and in the latest version also a repetition sign. The tricky part for this implementation was that the timer pre-scaler has to change depending of the note frequency since humans have quite a large audible frequency range, see again the source code for details.

6.4   Choreographies

The most important feature of the framework is the ability to design choreographies with no programming skills. A choreography is a plain text file consisting of a sequence of movements, one per line. Each movement is a triple: the time in milliseconds (from the start of the movement, or the latest time reset command), the degree of freedom to move, and the target position to reach. Each degree of freedom can move using a different speed, which can be changed anywhere within the choreography. Everything on a line behind a # character is treated as a comment. Instead of a movement command, the line can contain a control command. These include: starting a specified melody, playing a sound effect, "goto" – continuing the choreography from a specified line, resetting the time clock, which is important because the time is specified only using a 16-bit integer, which would prevent specifying longer choreographies, setting a total time of the choreography – even when looped, the dance will be stopped after that time, and in the most recent version, also the possibility to define procedures (i.e. sequences of movements) that can be inserted at any other location by a single command. This involves recalculating the times for the movement commands, meaning the time specification inside of a procedure is relative to the time of the procedure start. In the framework that we have presented to the participants, there was a limitation of one movement at a time (with the exception of full-speed movements – they could be triggered in parallel). In the most recent version, we have implemented the required data structures and procedures that allow simultaneous movements. This, however, changed the semantics – times in the original version meant the time to wait since the last movement command, whereas times in the latest version are times relative to the choreography start, or the most recent clock reset command. Every choreography is terminated by a triple "0 0 0".

Children in the camp used the last day to get their robot working, and they created a couple of little dances as well as the sequence that produced a bipedal walking. To upload the choreography to the robot volatile SRAM memory, it can be simply copy-pasted to the terminal window, preceded by the '@' character. The framework allows to store up to three choreographies to the non-volatile EEPROM memory and recalling them after the robot is powered up again. It is also possible to set the autostart flag, which means the robot will start a selected choreography automatically when turned on.

6.5   Starting Programs Using Ultrasonic Sensor

We wanted to implement a convenient interaction with the robot using the ultrasonic sensor and timing. Placing a hand, or some more hard-surface object (such as mobile phone) triggers interaction. After that, the robot observes carefully the movements in front of its sensor. By repeating the obstacle/no-obstacle sequence of gestures, the robot counts the "program" to be started. When the sequence stops, it will know how much it has counted, and starts the respective program. To make the interaction easier, the whole scenario is accompanied by sounds.

The robot is equipped with a single sensor – a forward pointing low-cost ultrasonic sensor, which occasionally suffers from freezing when it receives no echo. It can be reset by grounding its output ECHO signal for a short time, but we have to take caution to detect this behavior with proper timing. Also, the measured distance in this case would be replaced by a magic constant and it has to be distinguished from the real distance readings. The robot may be operated in a large room, or in a small space, and thus to program the above-described procedure took some noticeable efforts, see again the sources for details.

6.6 Controlling the Robot with Android Devices

The interaction with the robot can take place both at the USB serial communication line and on the Bluetooth serial line. All information is printed to both and inputs from both are being responded to. Most commands are triggered immediately by pressing a single key. This communication interface allows for a direct control of any degree of freedom, changing the speed of the movement, testing all the sound effects, melodies, saving and loading the choreographies and calibration parameters, entering the calibration procedure, but also triggering special choreographies. We have tuned walking behavior in four directions: forward and backward walking, and turning at a spot left/right, and these could be triggered by sending a single character through the communication line. Thus it is possible to easily control the robot from Android after being connected through Bluetooth, using a free BT control application, such as Arduino Bluetooth Controller [5].

7 Example Choreographies

We have implemented two example choreographies, which attracted some attention to this robot and our summer activity at various national fora: a cancan dance based on the melody of Offenbach, and walking accompanied by Jarre's Popcorn melody. Both can be found at the camp's website [3]. Here is one version of infinite walking choreography with music in the background

```
@1 11 1                 #change hands          100 3 62
#lean to the left       1 6 90                 1 4 69
100 1 48                1 5 0                   # -
1 2 69                  100 1 111              100 1 48
#move right hand        1 2 146                1 2 69
1 6 180                 1 6 0                  #end of steps
100 4 104               1 5 90                 1 9 2
1 3 104                 # second step          0 0 0
```

:

8 Framework Scalability

This framework for the robot Otto that we have developed is a generic framework that can in fact be used with any other toy controlled by Arduino. We have done just that after we have built a larger humanoid robot Lilli with 25 degrees of freedom [7]. Servos are controlled by two external boards connected through I2C bus, but this required a very little work. Furthermore, we have added a very low-cost mp3 player that interacts with Arduino and added commands that start or stop playing a specified sound sample on a connected speaker.

8.1   Multiple Arduino Units

Sometimes a single Arduino Nano is not sufficient for the project needs. One possibility is to leave the platform and move to more advanced solutions such as the family of the STM32 boards, however, nothing can really beat the high availability of solutions and libraries that are available to Arduino. One option is to connect several Arduino computers together. We have done just that in one of our outdoor robots, where we have developed a proprietary fast communication protocol using standard GPIO lines (since the serial lines, I2C and SPI are often used by other connected devices) [6]. We are currently working on extending Otto's framework with the ability to distribute the control over multiple Arduinos by forwarding the communication traffic to all the Arduinos in the chain.

9   Makers Movement in Educational Process

Our efforts are very deeply founded in the makers movement. A prototype of Lilli, the more advanced of the robots controlled by our framework has first been presented at Maker Faire in Vienna in May 2018 by Per Salkowitsch. We have encountered Otto robot for the first in June 2018 at Robotic Day in Prague. Makers movement provides unlimited sources of inspiration. It is in conformance with our very strong belief in the joy of sharing. Fablab's also are very central to the Makers movement – for instance Maker Faire in Vienna is organized by Viennese Fablab. However, what is the role of all this technology – 3D modelling, printing laser, vinyl cutter, automated sewing machine, water ray production, 3D scanning, and other in the educational process?

With the Industry 4.0, there is a clear shift into very high versatility in production, very large degree of automation and customization. Prototyping is a skill that becomes essential for all parties involved in the new organization of the development and production cycle. Some schools (with the help of EU funding) have invested large resources into establishing technical workshops, where all pupils spend several hours per week learning about this revolutionary technology. We believe that significant efforts are needed to make it easier for the educators to select in the sea of available information suitable activities, projects, and technologies to make this transition succeed. Our little contribution has been realized with this idea in mind. We also believe in the change of the organization of school education. Learning 1 lesson of history per week might contribute to a reasonable and regular work habit, but it is not suitable for efficient learning. Instead, spending more hours a week until a particular educational unit is completed gives stronger and longer lasting experience and allows for some courses to take place outside of the schools, in regional centers similar to Fablabs.

10   Future Directions and Conclusions

We have received only a positive feedback from both the participants and the parents. Several participants have attended "Otto service days" in the weeks after the camp. Some of them are now regularly attending a club for children in Fablab

Bratislava, others joined robotics clubs around the city. We are planning a second $(DT)^2$ in the forthcoming summer, considering our own reconfigurable robotics as the target platform to work on. Since the last summer, "our" version of Otto has been built by several groups around the country and we often use it at various public presentations. The framework has been used multiple times in seminars with the students of Applied Information at Comenius University. Links to the Github repositories with the software and all the details about the camp and our version of Otto robot can be found at the $(DT)^2$ website [3].

References

1. Sweeney, L.B., Meadows, D. (eds.): The Systems Thinking Playbook: Exercises to Stretch and Build Learning and Systems Thinking Capabilities. Chelsea Green Publishing, White River Junction (2010)
2. Otto DIY (2019). www.ottodiy.com
3. $(DT)^2$ website (2018). dtdt.fablab.sk
4. $(DT)^2$ Otto Github repo. (2019). github.com/Robotics-DAI-FMFI-UK/dtdt-otto/
5. Ioannis Tzanellis: Arduino Bluetooth Controller. play.google.com/store/apps/details?id=eu.jahnestacado.arduinorc
6. Smelý Zajko Github rep. github.com/Robotics-DAI-FMFI-UK/smely-zajko-ros
7. Cu-lilli robot home. kempelen.dai.fmph.uniba.sk/lilli/
8. Markowitz, D.G.: J. Sci. Educ. Technol. **13**, 395 (2004)
9. Klein, M., et al.: Hedgehog: a versatile controller for educational robotics. In: Constructionism 2018 Conference Proceedings (2018)

Environment Virtualization for Visual Localization and Mapping

David Valiente[1]($\boxtimes$), Yerai Berenguer[1], Luis Payá[1], Nuno M. Fonseca Ferreira[2],
and Oscar Reinoso[1]

[1] Department of Systems Engineering and Automation,
Miguel Hernández University, Av. Universidad sn, 03202 Elche (Alicante), Spain
{dvaliente,lpaya,o.reinoso}@umh.es, yerai.berenguer@graduado.umh.es
[2] Department of Electrical Engineering (DEE),
Engineering Institute of Coimbra (ISEC), Polytechnic Institute of Coimbra (IPC),
Coimbra, Portugal
nunomig@isec.pt

Abstract. Mobile robotics has become an essential content in many
subjects within most Bachelor's and Master's degrees in engineering.
Visual sensors have emerged as a powerful tool to perform reliable localization and mapping tasks for a mobile robot. Moreover, the use of
images permits achieving other high level tasks such as object and people detection, recognition, or tracking. Nonetheless, our teaching experience confirms that students encounter many difficulties before dealing
with visual localization and mapping algorithms. Initial stages such as
data acquisition (images and trajectory), preprocessing or visual feature
extraction, usually imply a considerable effort for many students. Consequently, the teaching process is prolonged, whereas the active learning
and the students' achievement are certainly affected. Considering these
facts, we have implemented a Matlab software tool to generate an open
variety of virtual environments. This allows students to easily obtain
synthetic raw data, according to a predefined robot trajectory inside
the designed environment. The virtualization software also produces a
set of images along the trajectory for performing visual localization and
mapping experiments. As a result, the overall testing procedure is alleviated and students report to take better advantage of the lectures and
the practical sessions, thus demonstrating higher achievement in terms
of comprehension of fundamental mobile robotics concepts. Comparison
results regarding the achievement of students, engagement, satisfaction
and attitude to the use of the tool, are presented.

Keywords: Virtual environments · Visual localization ·
Omnidirectional image · Simulation

1 Introduction

During the last decade mobile robots applications have substantially increased in
many industrial and quotidian scopes, being responsible for a wide range of tasks.

M. Merdan et al. (Eds.): RiE 2019, AISC 1023, pp. 209–221, 2020.
https://doi.org/10.1007/978-3-030-26945-6_19

To that end, two requisites are fundamental for mobile robots when operating autonomously in an unknown environment: (a) obtaining a map representation of the environment; and (b) estimating its current pose and orientation within that map. This whole procedure entails a non-trivial problem, namely SLAM (Simultaneous Localization and Mapping), which has been widely studied so far [1], and became an essential content in many engineering degrees at university.

The manner in which the robot interacts with the environment can be modeled by the use of a wide range of sensory data. These would be utilized to generate the map representation but also to estimate the pose of the robot with respect to such map [2]. In this sense, visual sensors provide advantages in terms of low cost, lightness, and good amount of information of the environment, in contrast to traditionally acknowledged sensors like laser [3] or sonar [4]. Thus they also turned to be essential in the learning programs [5,6] in engineering. Regarding the image data processing, students get introduced to basic projection systems such as the pinhole camera model and the stereo pair model. However, it is also worth presenting the most up-to-date projection systems for full-view scene acquisition, such as the omnidirectional, the fish-eye, and the panoramic projection.

Regarding the academic scope, it has been largely reported that science, and engineering students find several difficulties on the learning of scientific and technology contents [7]. Usually this comes associated to old methodologies which led the students to construct wrong misconceptions. Nevertheless, these days the tendency has evolved towards more active learning approaches by the introduction of innovative learning programs [8,9]. Some significant interventions to promote active learning in robotics are represented by simulation [10]. And more specifically, by virtualization of laboratories, which have been extensively accepted as a beneficial tool to empower students towards autonomous learning [11,12]. To our understanding, this kind of solutions should be exploited under a careful guidance of the instructors, who need to find the proper balance between theory lessons, practical and hands-on sessions supported by software and virtualization tools.

In this sense, we conduct several courses in a Master's degree on Robotics, in the Miguel Hernandez University (Spain). Our goal is to integrate software tools for virtualization within the academic plans, for an active learning. In particular, the students learn how to extract and process visual information from the environment, by means of a set of images captured by the robot. They are also taught how to produce robust estimates for the robot localization and the map building. To that purpose, they are provided with a consistent theoretical background, despite the fact that these subjects are eminently practical. The practical and hands-on sessions take place in the robotics laboratory. Firstly, during the initial design stages, the students start testing with simple and static environments in order to assess the performance of different algorithms. Secondly, they move on to work with further extended environments, with real data under realistic conditions. Our experience confirms that the acquisition of real environments is time-consuming and it is highly likely to cause issues to the students. Besides this, we have to consider noisy elements associated to real conditions. The control and motion data, but more particularly, the images acquired by the robot, are commonly affected by occlusions, changing lightning conditions, presence of dynamic objects, etc. This results in corrupted data which directly compromise the final mapping and localization estimation. Therefore the learning process is highly compromised at the first stages.

There are some well-acknowledged tools for this aim, such as *Gazebo*, *Morse* or *Webots*. However, due to the time limitation and the diverse background of the students, we are not able to dedicate the proper time to introduce such tools. That is the main reason why we opted for the custom development of a simply tool, as a first solution. According to this, we aim at reducing the appearance of these difficulties that students may encounter, at the same time that we contribute to their active and autonomous learning. To that end, we have implemented a virtualization software tool in Matlab to ease the data acquisition and image processing stages. This tool permits generating virtual environments, in which a robot equipped with a specific camera sensor can traverse a predefined trajectory. The data extracted from the different poses, are also available by user configuration. Thus this tool simplifies the initial procedure, by easily generating sets of images along the trajectory, from which specific information can be also extracted in order to test robotic algorithms under idealistic conditions. Hence we help students concentrate on higher objectives regarding the learning and comprehension of advanced robotic concepts, such as mapping and localization. Moreover, students have at their disposal, a powerful tool for their own purposes, in case they need to extend their algorithm developments in a synthetic environment.

The remainder of this paper is: Sect. 2 presents an overview to the contents of the course; Sect. 3 describes the vision system embedded in the virtualization software; Sect. 4 provides details about the software design and the generated environments; Sect. 5 states a final discussion with the derived conclusions.

2 Learning Contents

This section summarizes the main contents regarding visual mobile robotics, which are taught during the Master's degree on Robotics, in the Miguel Hernández University. An initial subject comprises the main theory essentials of visual mapping and localization. Students are introduced to general approaches of visual mobile robotics, according to the classification into three aspects: (i) the map model; (ii) the mapping estimation algorithm; (iii) the visual techniques to process the information extracted from the environment.

Traditional map representation models produced with laser range finder and sonar sensors, are briefly presented. The main exponent to these is the occupancy map [13]. However, the course concentrates on the evolution towards discrete mapping, such as landmark maps with feature points [14] and hybrid topological maps [15]. This has been possible thanks to the emergence of visual sensors, and to the improvement of the processing techniques. More precisely, students are introduced to the main distinctions between specific visual techniques: appearance-based methods and feature-based methods. Both coincide in pursuing a discrete visual representation of a scene, namely visual descriptor. On the one side, the introduction to appearance-based methods concentrates on the processing of the pixel intensity, treating the entire set of pixels of an image as a unique representation, by means of specific computation and metrics. A set of reliable examples are explained: the Fourier Signature [16], HOG [17] and Gist [18]. On the other side, feature-based methods are explained in order to be used in more challenging situations, so that distinctive and robust physical points of the environment can be acquired under harmful scene circumstances. Some well acknowledged descriptors are described during the course, such as Harris [19], SIFT [20], SURF [21] and ORB [22]. All these methods allows the students to understand the final composition of the estimated map model. Contrarily to the previous approaches, based on occupancy areas, visual landmark maps can be only conformed by a reduced set of appearance-based or feature-based descriptors for a single image. Consequently, the students perceive the considerable improvement on the efficiency in terms of data acquisition, processing, and estimating the map and the localization of the robot.

2.1   Visual Mapping and Localization

The structure of the course is divided into four main blocks of content, which are subsequently addressed. Each one, is in turn, subdivided into theory lectures, practical and hands-on sessions at the robotics laboratory.

1. Perception system: in this block the students learn how to acquire data from a mobile robot equipped with cameras, a laser range finder, and wheels' encoders. They practice with the virtualization software under simulated conditions that emulate the presence of noise. The robot traverses the predefined trajectory while the software provides a log data for each sensor.
2. Image processing: once each set of images is captured along the different poses of the robot, then certain visual descriptors can be computed on the images. As a result, the students can choose between several appearance-based and feature-based descriptors.
3. Observation measurement. After the visual descriptors are generated, next the observation measurement stage makes use of these descriptors to input the mapping and localization algorithms and eventually to produce a valid estimation. The process is achieved by comparing visual descriptors between images acquired by the robot, and images stored as part of the current map. In the case of appearance-based methods, the difference between descriptors is tabulated from their actual dimensionality to a metric reference system. As for the feature-based methods, assuming a well known calibration, several metric relations can be straightforward extracted between the image frame and the robot reference system.
4. Re-estimation. Once the robot localizes itself within the map, then it is necessary to update the last estimation of the map, but also to back-propagate the estimation update to the trajectory followed by the robot.

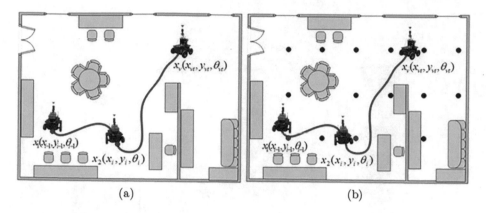

(a)   (b)

Fig. 1. (a) Map composed by images (red dots) acquired along the trajectory (blue). (b) Map composed by images (red dots) pre-acquired on a given grid basis (black dots).

5. Mapping update. Several approaches of incremental mapping update can be tested. The robot may initiate new parts in the map, as long as its decision module advises that the uncertainty is high and thus a new image in the map has to be stored. Figure 1 shows two examples. Figure 1(a), presents a synthetic trajectory in which the robot has been initializing some images in the map, as long as it has been traversing the path. By contrast, Fig. 1(b) shows an example with the map as a given distribution of a set of grid images, which where pre-acquired. With this approach, the robot decides whether an image of the grid has to be included in the current representation of the map. Red circles indicate poses where images were stored as part of the map.

In the end, the set of images, their visual descriptors and the topological and geometrical relationships between them, eventually conform the final estimate of the map of the environment. Overall, the students receive an in-depth theoretical and practical instruction to the formulation of the mapping and localization problem, which is normally expressed by an augmented state vector, $\bar{x}(t)$. This state vector comprises the current pose of the robot, x_v, at each time step, and the pose of the set of images stored in the map, x_n. This set ultimately determines the final map estimate. Notice that the specific visual descriptors are also stored, and linked to each associated image. In consequence, the state vector that contains the different variables of the map is expressed as:

$$\bar{x}(t) = \begin{bmatrix} x_v\ x_1 \cdots x_n \end{bmatrix}^T \tag{1}$$

with each view $n \in [1, \ldots, N]$. Then the state vector encodes a map constituted by a total number of N views.

The current pose of the robot and the pose of the images, are expressed in the a global reference system, with the addition of the orientation. In a 2D planar reference system, the elements of the state vector, with orientation, can be assumed as:

$$x_v = (x_{vt}, y_{vt}, \theta_{vt}) \tag{2}$$

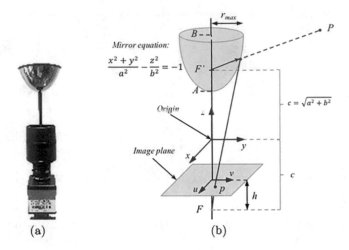

Fig. 2. (a) Catadioptric (omnidirectional) vision system constituted by a hyperbolic mirror and a CCD digital camera. (b) Projection model of the vision system.

$$x_n = (x_i, y_i, \theta_i) \tag{3}$$

The visual descriptors related to each image in the map, d_n, can be calculated in a multidimensional subspace, according to the particular visual descriptor technique and its dimensionality (NxM), that is to say, $d_n \in \mathbb{R}^{NxM}$. Additionally, the uncertainty on the map and the localization estimations has to be registered. That is, a record in the form of the square matrix, $P(t)_{AxB}$, with $A = B = 3n + 3$.

3   Vision System

This section concentrates on the description of the virtual vision system implemented within the virtualization software tool. The model corresponds with a catadioptric vision system [23], consisting of a hyperbolic mirror jointly coupled to a digital CCD firewire camera, as it may be observed in Fig. 2(a). Figure 2(b) represents the projection model associated to the vision system by which a ray coming from a certain 3D point, P, projects onto the surface of the mirror and then directs towards the focus of the hyperbolic, F. The path followed by the ray finally intersects in the image frame, generating an image pixel, $p(u, v)$. Note that the centre of projection coincides with the focus of the hyperboloid so that to focalize the 3D scene properly. The dimension of the mirror are determined by the lengths a and b. The result of the image generation can be observed in Fig. 3 with real examples. Figure 3(a) and (b), show the difference between the omnidirectional and the panoramic projections, respectively.

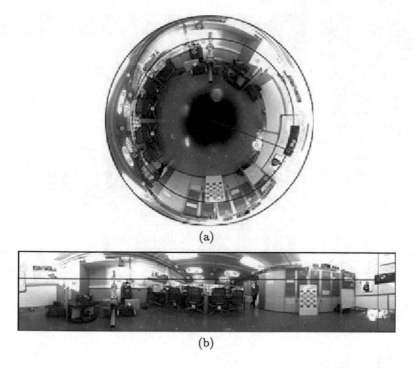

Fig. 3. (a) Original omnidirectional image. (b) Panoramic conversion.

4 Visual Environment Generation

Once the vision system has been described, here it is presented the main aspects regarding the generation of visual environments by using the Matlab software.

4.1 Virtual Objects

First of all it is necessary to define a virtual layout in either a 2D or a 3D reference system. A set of objects has been developed through a custom library. The basis for their construction consists of a cluster, which results from the intersections of different 3D planes. This strategy permits constructing a wide set of objects. The students can modify their dimensions and to decide their position along the virtual environment.

4.2 Virtual Layout

The main menu window of the software (Fig. 4) permits configuring the entire virtual environment. Students can set: (a) the dimensions of the 3D virtual environment; (b) the parameters of the vision system; (c) the sort of image projection; (d) the poses of the trajectory (in blue); (e) the arrangement and characteristics of the virtual objects. Note that any specific orientation (rotation) for the 3D coordinates of each pose, can be also configured.

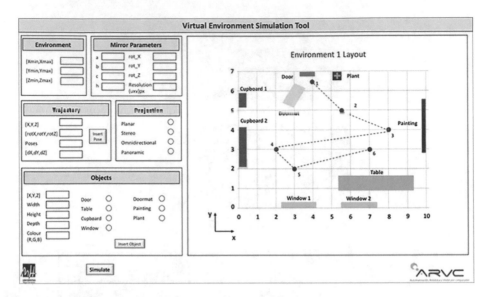

Fig. 4. Main menu window. The left side options permits setting the virtual environment. The right side options show the virtual environment layout.

4.3   Virtual Images

The images associated to the trajectory poses are generated once the virtual layout with specific objects has been configured. Then, a back-projection algorithm is run by following these steps:

- Each pixel in the image is the origin of a back-projected ray, which translates in a Matlab vector that represents the path of the ray from the focus to the mirror surface.
- The set of vectors (one for each pixel) is directed from the mirror towards the 3D world.
- The rays emerging from the mirror surface (vectors) intersect with the virtual objects situated in the environment. This means that the rays can intersect with the planes that conform the arranged objects in the 3D world. In case an intersection occurs with a plane, the resulting intersection point (part of an object), must appear in the image. The algorithm returns the RGB value for that point and it is registered as the color of the image pixel in the image. Once all the pixels receive their color value as the result of the intersection with the 3D virtual world, the image is completely generated.

4.4   Virtual Output

After configuring and defining all the desired parameters, then the students can click the *Simulate* button at the bottom of the menu. The software generates all the trajectory data, associated images in each pose along the trajectory, and

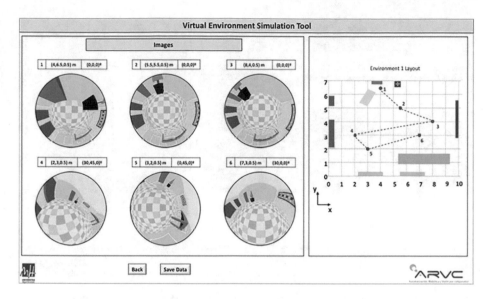

Fig. 5. Results window. The left side presents the omnidirectional virtual images with their position and orientation indicated. The right side presents the virtual environment layout, with the trajectory and poses where the images are generated.

a complete set of several visual descriptor for each image. Figure 5 presents the results window with a sample of the synthetic omnidirectional images virtualized

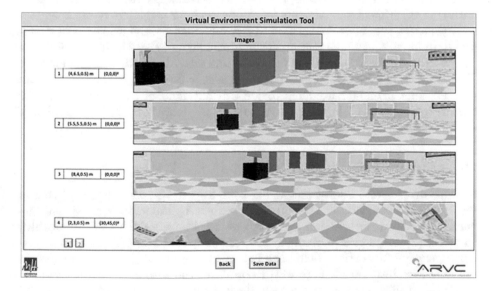

Fig. 6. Panoramic virtual images converted from the omnidirectional virtual images, presented in Fig. 5.

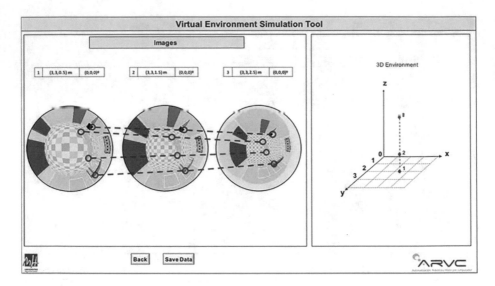

Fig. 7. Omnidirectional virtual images for estimating height. Feature matching points are indicated in blue.

in each predefined pose along the trajectory. Moreover, a secondary windows is also prompted, in order to present additional data such as those relating to other selected image projections, as shown in Fig. 6. Another example may be observed in Fig. 7, where a 3D trajectory along the height coordinate is generated. In addition, some feature matching points are indicated between images, in blue. With this experiment the students learn how to infer height from the images.

Ultimately, the students can save in a compressed file the whole set of generated data for this virtual environment, maintaining a list with the logs for each sort of data and specific folders for each content. All in all, the students are allowed to easily generate a virtual environment in few minutes. Later, they can load the same virtual environment to edit any elements, or they can just use the data to input specific mapping and localization algorithms that are taught in several subjects of this Master's degree on Robotics.

4.5   Assessment

The introduction of the software tool has been assessed from two points of view in Fig. 8: (*i*) achievement results as grades obtained by students; and (*ii*) engagement, motivation, satisfaction and attitude to the use of the tool. Figure 8(a) compares the mean grades obtained in three subjects during the last three academic years. It is worth noticing the better grades in 2018, when the software tool was first introduced. Figure 8(b) compares the responses of the students according to their experience before and after they used the software tool. The scale for the responses varies from *1-totally disagree* to *5-totally agree*.

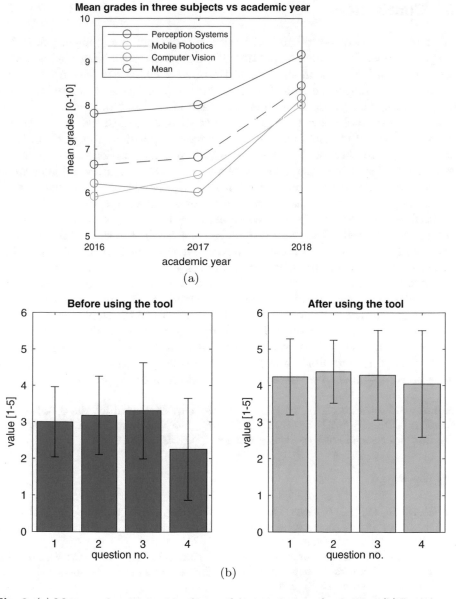

Fig. 8. (a) Mean grades obtained in three subjects versus academic year. (b) Responses of the students to questions regarding engagement (question no. 1), motivation (question no. 2), satisfaction (question no. 3) and attitude to the use of the software tools (question no.4), before (left) and after (right) the software tool was introduced.

5 Conclusions

We have presented a software tool to generate virtual environments, aimed at supporting teaching of mobile robotics. The design has been integrated in a Master degree on Robotics, in the Miguel Hernández University (Spain). Students traditionally confirmed certain difficulties on the procedures associated to initial stages during the practical and hands-on sessions at the laboratory of robotics. Therefore this tool alleviates the procedure of data acquisition and data processing, in terms of visual data and trajectory acquisition. But it also aids in the testing of further robotics algorithms for visual localization and mapping for mobile robotics. The virtualization tool also permits configuring different vision systems (camera, mirror and projection model), and the sort of visual descriptor extracted from the images (appearance-based or feature-based). Moreover, realistic experiments can be also configured, by adding noise data, accordingly represented by well acknowledged models. As a result, the time of the practical and hands-on sessions is be efficiently dedicated to deal with tasks regarding localization and mapping, since the students can easily obtain a set of synthetic data, ready to be used in such tasks. Our experience demonstrate that this contribution enhances the active and autonomous learning of the students, who report positive satisfaction and motivation toward the combined use of this software tool with the lectures. In addition, their achievement is proven to be better, in terms of their conceptual understanding, but also in their practical skills. Comparison results in terms of achievement, engagement and satisfaction, have been presented to validate the contributions of the approach.

Acknowledgements. This work has been partially supported by: the Spanish Government (DPI2016-78361-R, AEI/FEDER, UE); the Valencian Research Council and the European Social Fund (post-doctoral grant APOSTD/2017/028).

References

1. Payá, L., Gil, A., Reinoso, O.: A state-of-the-art review on mapping and localization of mobile robots using omnidirectional vision sensors. J. Sens. **2017**, 1–20 (2017)
2. Valiente, D., Payá, L., Jiménez, L.M., Sebastián, J.M., Reinoso, O.: Visual information fusion through bayesian inference for adaptive probability-oriented feature matching. Sensors **18**(7), 2041 (2018)
3. Cole, D., Newman, P.: Using laser range data for 3D SLAM in outdoor environments. In: IEEE ICRA, U.S.A., pp. 1556–1563 (2006)
4. Lee, S.J., Song, J.B.: A new sonar salient feature structure for EKF-based SLAM. In: IEEE IROS, pp. 5966–5971 (2010)
5. Ferreira, N.M.F., Freitas, E.D.C.: Robotics as multi-disciplinary learning: a summer course perspective. In: 2018 IEEE 16th International Conference on Industrial Informatics (INDIN), pp. 536–543, July 2018
6. Oliver, J., Toledo, R., Valderrama, E.: A learning approach based on robotics in computer science and computer engineering. In: IEEE EDUCON 2010 Conference, pp. 1343–1347, April 2010

7. López-Gay, R., Martínez Sáez, J., Martínez Torregrosa, J.: Obstacles to mathematization in physics: the case of the differential. Sci. Educ. **24**, 591–613 (2015)
8. Ferrerira, N.M.F., Freitas, E.D.C.: Computer applications for education on industrial robotic systems. Comput. Appl. Eng. Educ. **26**(5), 1186–1194 (2018)
9. Xuemei, L., Jiashu, C., Jinhu, L., Gang, X.: Innovative education activities with vision based robot navigation system. In: 2010 International Conference on Optics, Photonics and Energy Engineering (OPEE), vol. 2, pp. 505–507, May 2010
10. Gil, A., Juliá, M., Reinoso, O.: MRXT: the multi-robot exploration tool. Int. J. Adv. Robot. Syst. **12**(29), 1–10 (2015)
11. Tosello, E., Michieletto, S., Pagello, E.: Training master students to program both virtual and real autonomous robots in a teaching laboratory. In: 2016 IEEE Global Engineering Education Conference (EDUCON), pp. 621–630, April 2016
12. Saraiva, A.A., Barros, M.P., Nogueira, A.T., Fonseca Ferreira, N.M., Valente, A.: Virtual interactive environment for low-cost treatment of mechanical strabismus and amblyopia. Information **9**(7), 175 (2018)
13. Grisetti, G., Stachniss, C., Burgard, W.: Improved techniques for grid mapping with rao-blackwellized particle filters. IEEE Trans. Rob. **23**(1), 34–46 (2007)
14. Mur, R., Tards, J.D.: ORB-SLAM2: an open-source SLAM system for monocular, stereo, and RGB-d cameras. IEEE Trans. Rob. **33**(5), 1255–1262 (2017)
15. Liu, M., Siegwart, R.: Topological mapping and scene recognition with lightweight color descriptors for an omnidirectional camera. IEEE Trans. Rob. **30**(2), 310–324 (2014)
16. Menegatti, E., Maeda, T., Ishiguro, H.: Image-based memory for robot navigation using properties of omnidirectional images. Robot. Auton. Syst. **47**(4), 251–267 (2004)
17. Dalal, N., Triggs, B.: Histograms of oriented gradients for human detection. In: 2005 IEEE Computer Society Conference on Computer Vision and Pattern Recognition (CVPR 2005), vol. 1, pp. 886–893, June 2005
18. Friedman, A.: Framing pictures: the role of knowledge in automatized encoding and memory for gist. J. Exp. Psychol. Gen. **108**, 316–55 (1979)
19. Harris, C.G., Stephens, M.: A combined corner and edge detector. In: Proceedings of Alvey Vision Conference, Manchester, UK (1988)
20. Lowe, D.: Distinctive image features from scale-invariant keypoints. Int. J. Comput. Vis. **60**, 91–110 (2004)
21. Bay, H., Ess, A., Tuytelaars, T., Van Gool, L.: SURF: speeded up robust features. In: Proceedings of the European Conference on Computer Vision, Graz, Austria (2006)
22. Rublee, E., Rabaud, V., Konolige, K., Bradski, G.: ORB: an efficient alternative to SIFT or SURF. In: Proceedings of the 2011 International Conference on Computer Vision, Washington, DC, USA, pp. 2564–2571 (2011)
23. Valiente, D., Gil, A., Reinoso, O., Juliá, M., Holloway, M.: Improved omnidirectional odometry for a view-based mapping approach. Sensors **17**(2), 325 (2017)

Programming a Humanoid Robot with the Scratch Language

Sílvia Moros[1]([✉]), Luke Wood[1], Ben Robins[1], Kerstin Dautenhahn[1],
and Álvaro Castro-González[2]

[1] School of Computer Science, University of Hertfordshire, Hatfield AL9 10AB, UK
`s.moros@herts.ac.uk`
[2] Ingeniería de Sistemas y Automática, Universidad Carlos III, 29811 Leganés,
Madrid, Spain
`https://www.herts.ac.uk/`

Abstract. In this paper we present a novel approach to programme Kaspar, a 22 DOF humanoid robot used for robot-assisted therapy with children with Autism Spectrum Disorder (ASD). The original software used to programme Kaspar was developed to primarily be used in research. However, Kaspar is now used increasingly in other environments, operated by non-roboticists. While Kaspar has a user-friendly interface to be operated by non-programmers, new games or behaviours were mainly created by the research team; thus, we needed to develop an interface that would allow non-roboticists to programme Kaspar.

As a solution, we used the Scratch programming language. We tested the Scratch interface with over 170 school children aged 7 to 10, who had the chance to programme Kaspar and give their feedback. In general terms, Scratch was thought to be a fun, useful and easy way to programme Kaspar, and the majority of the children were willing to use it again.

Keywords: BabyRobot project · Scratch ·
Human-robot interaction · UI · User Interface ·
Block programming

1 Introduction

From the very beginning, the Kaspar robot was designed to be able to interact and operate in the real world [1]. It is specifically designed to work with children with Autism Spectrum Disorder (ASD); the imperative need here is that it must be able to work and be useful outside the laboratory.

In the early stages of the project, a researcher of the Kaspar project was always present but, as the Kaspar project expands, more Kaspars are being built and more facilities and private houses are interested in having a Kaspar for a certain period. This trend highlights that our team, as developers, need to provide all users (teachers, therapists, parents and children) with an easy way

© Springer Nature Switzerland AG 2020
M. Merdan et al. (Eds.): RiE 2019, AISC 1023, pp. 222–233, 2020.
https://doi.org/10.1007/978-3-030-26945-6_20

to programme Kaspar. There has been a substantial effort to provide the end user with user-friendly interfaces, such as a keypad, RFID cards, etc, with which they can easily use to switch between games and play with Kaspar. However, those systems do not allow users to easily change the content of the games or to programme new ones; the content has already been preprogrammed by the research team. To edit or create new games, the system provides a GUI, but it is slightly complicated and past experience showed that only very few non-technical users have used it themselves to change and adapt existing games or create new game behaviours. The majority of users relied on the research team to carry out these tasks; therefore, the main objective is to develop a system that is easier to programme so that users can easily adapt the games, create new ones or tailor them according to the children's individual preferences.

With all of this in mind, our team considered the point of view of a person who wants to programme Kaspar with minimal technical knowledge. For users, the most interesting feature underlying Kaspar's behaviour is what our team calls "sequences". Sequences are defined as a list of poses that Kaspar will execute in order; a pose is any specific configuration of the motors that is considered interesting, e.g. a meaningful gesture or expression, and thus is saved into a file. From that point on, the Kaspar programme will remember this particular configuration, and it can be used for building new sequences.

In this paper, then, we outline an approach whereby we provide an easy way to programme these sequences using the Scratch programming language. Section 2 describes the components of the system; Kaspar the robot and the Scratch programming language. Section 3 explains the new architecture developed to create the unified system. Section 4 presents the results of a test in a school, while Sect. 6 outlines the conclusions. To conclude this paper, Sect. 7 highlights the future work to do in this project.

2   Current Architecture

2.1   Kaspar

Kaspar is a fully programmable, 22 DOF humanoid robot designed and built in the Adaptive Systems Group at the University of Hertfordshire [2]. This robot has the size of a small child and it is specifically designed to help children with autism to interact and communicate with other people. The first version of Kaspar was built in 2005, but since then many improvements have been made and features have been refined. The robot has been part of international projects, and also has been used in schools as well as in private houses and other research facilities ([3–5]) (Fig. 1).

Kaspar is controlled via a Java system, which launches a Graphic User Interface (GUI) with different tabs, as presented in Fig. 2. Some of the tabs allow the person operating the robot to fine-tune Kaspar's motors and save the current pose Kaspar is in (tab "Pose"), to play sequences using an array of different poses and add one sound file per sequence (tab "Sequences"), or to have an overview of the Force Sensitive Resistor (FSR) sensors (tab "FSR Sensors"), etc.

Fig. 1. The Kaspar robot.

When it comes to the interaction with the robot in a real environment, the most frequently used tab, and for that reason a focus of this article, is the "Control" tab, as shown in Fig. 2. In this tab the user can select any of the pre-constructed sequences and play it, making Kaspar execute the sequence and therefore move.

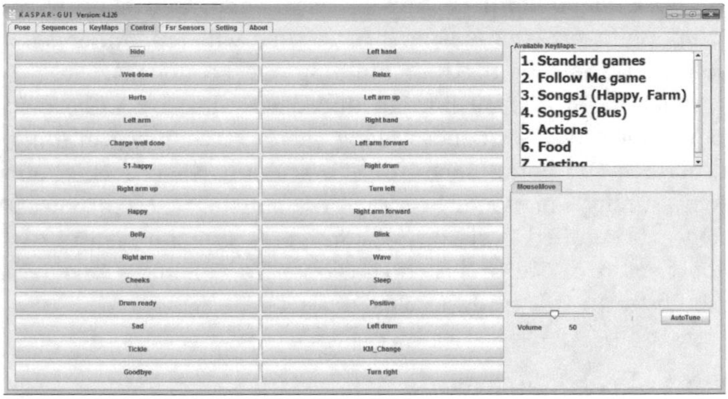

Fig. 2. Current GUI from which to control Kaspar.

Every sequence presented in the "Control" tab in Fig. 2 is created on the "Sequences" tab, shown in Fig. 3. In this tab, all the poses available can be selected and ordered in what we call a "sequence". The resulting sequence can then be saved and played again at will, without having to configure it again from scratch. From this it follows that, for us, a sequence is defined as "an orderly set of poses that are reproduced one after the other and have interest and entity enough as a whole to be saved for future uses".

However, this system is designed to be primarily used by researchers. To reach a wider audience and in order to empower families, teachers and carers to use Kaspar as a tool without requiring the support of the research team, some easy to use system is needed that acts as a front-end.

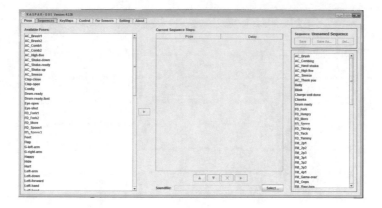

Fig. 3. "Sequences" tab from the current GUI.

How Kaspar Works: A wi-fi emitter is located inside Kaspar; this means that each of the Kaspars, once powered up, emits its own wi-fi, making it possible to work with many of them at the same time without interferences, each of them from its own laptop. To operate Kaspar, after powering it up a laptop must connect to Kaspar's wi-fi and then run the program that controls Kaspar.

This program, coded in Java, has direct access to the motors, the speaker, FSR sensors, the microphone and other components inside Kaspar. The program has been widely used for many years in numerous experiments, demonstrations, summer schools, etc., works reliably, and is thus the foundation of every possible modification. This program will always retain the control of the motors and will always run in the background.

2.2   Scratch Programming Language

Scratch is a block programming language designed by the Massachusetts Institute of Technology (MIT), originally for children from economically disadvantaged backgrounds to help them engage with programming activities in the context of Computer Clubhouses ([6,7]). The Scratch website states that it is especially targeting children between 8 and 16 years old, and data shows a peak of new 'Scratchers' at the age of 12, as seen in Fig. 4.

The Scratch programming language created has become very popular in primary and high schools, and studies highlight its potential to help the children learn maths [9], programming [10,11] or creative arts [12], due to the appealing interface and easy way to programme.

In this work we will study how the Scratch system can be used for non-programmers to programme a humanoid robot, a complex mechatronic system used in research. To accomplish this, we will use the experimental version of the Scratch programming language, called "ScratchX", which allows all users to create their own blocks. This version of Scratch allows the user to be offline

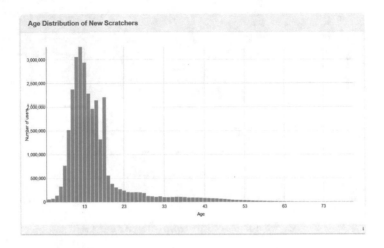

Fig. 4. Age distribution of new Scratchers [8].

as well. In this paper, then, we will generally say "Scratch" to refer to the environment, be it Scratch or ScratchX version.

3 The Scratch Front-End Architecture

The Scratch layer is meant to be the front-end of this system; this means that Scratch is the interface and it does not communicate with the motors itself. Control of the motors is done via the Java system, which has been developed over the years to be robust and reliable, and it has been widely and extensively used to operate Kaspar in many studies. It is therefore sensible to use it as the backbone for new developments.

The ScratchX interface communicates with the Application Programming Interface (API) via a JavaScript program, called "extension", that has to be loaded in the ScratchX website. The API then calls the Java programme, which executes the poses and gives instructions to the motors. The communication is unidirectional and in open loop, to make it simpler and easier for the first version of the system, as seen in Fig. 5.

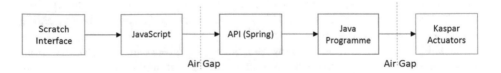

Fig. 5. Diagram of the ScratchX architecture.

3.1   Scratch Interface

The Scratch interface is a website that usually is launched via a web browser. It can be launched via any browser, but we used Google Chrome since it has better support for Adobe Flash player, which Scratch uses. This means that the computer needs to have access to the Internet to be able to launch Scratch; however, in this project it is an important requirement that the system can be activated offline, since Internet access cannot be guaranteed in all the different real-world environments that Kaspar is being used in.

The way our team chose to work around this was to download the code from the ScratchX website, called "Scratch Offline Editor". It is available to be freely downloaded on the Scratch website, and can be served directly from the controlling computer via a "localhost:8080" service. This means that the ScratchX screen that we see does not have all the functionalities that can be found on the usual ScratchX website. However, this loss of functionality is not important in this project: since we are not creating sprites, the missing "Motion" and "Looks" tabs are not needed.

Once the ScratchX screen is launched, to add our customized blocks it offers the options to open an existing project (with a .sbx file) or to paste an URL, which is done via a JavaScript script, explained in the next subsection. Next, the screen allows all users to use the blocks that move Kaspar, as seen in Fig. 6. This has to be done every time that ScratchX is launched.

Fig. 6. ScratchX screen with customized blocks for Kaspar.

3.2   JavaScript

JavaScript is the way to create new blocks in ScratchX [13]. Inside the JavaScript file, a block descriptor allows the developer to create new blocks and assign some features to them. Each block is assigned to its own function, that will make a call to a specific resource of the API. HTTP calls facilitate the communication between JavaScript and the API. This is because JavaScript is executed via the browser; it means it can only access browser resources, but not the Operating System (OS) resources. The resource against which the JavaScript is making the calls has to be agreed with the API.

The Scratch system does not receive any feedback and JavaScript is an asynchronous language; this means that some of the processes are launched all together, and may not arrive in the original order to the API. In spite of that, Kaspar has to move according to the sequence it is programmed to do; that is, the poses have to follow a specific sequence. To avoid executing poses in an incorrect order a pause was hardcoded into the script.

3.3   The API (Spring)

The API receives the information that the JavaScript sends in the form of HTTP calls, and connects each call with the specific function in the original Java program that will handle it.

The API itself acts as a translator, but the part which retains the full control of the movement is the original Java program. The execution of all the poses and sequences is activated via the Java program.

To open the communication channel there is a library called Spring [14], which is used in this project because is easy and quick to integrate with Java.

4   User Evaluation Study

4.1   Ethics Statement

This research was approved by the University of Hertfordshire's ethics committee for studies involving human participants, protocol number: COM/SF/UH/03320. Informed consent was obtained in writing from all parents of the children participating in the study.

4.2   Study

The Scratch system was tested with 175 primary school children from a mainstream school in Hatfield (Hertfordshire, United Kingdom). The children were aged 7 to 10 (MD $= 8.87$; SD $= 0.85$), 53% males and 47% females, and all had previous experience using Scratch, as they had regular lessons on it every Friday.

All of the children had the same experience with Kaspar, namely an hour and twenty minutes of programming it, in pairs. First of all, the children were given 10 min to familiarise themselves with the system. Next, they had to programme 5 "emotions": happy, sad, surprised, angry and silly. They were given 10 min to express each emotion by making Kaspar move. Once the movement was created they could also add sounds to it. Next, they were asked to prepare a small story using some or all the emotions created and explain it to the other children, while Kaspar was acting along. At the end of the session, a questionnaire was run, which contained the following items:

- Fun while programming Kaspar.
- Easiness of programming Kaspar.
- Willingness of programming Kaspar again.
- Perceived usefulness of Scratch when programming Kaspar.

Each question was answered with a Likert scale from 1 to 5, depicted with happy/sad faces to make the choice easier for the children, as in Fig. 7.

Fig. 7. Answering scale from 1 to 5 with happy/sad faces.

4.3 Results

Fun: As seen in Fig. 8, approximately 95% of the students thought programming Kaspar was Fun or Very Fun. Less than 2% thought it was Boring or Very Boring, distributed in two males from the older grade and one female from the younger grade, while 3.41% thought it was OK. The black line represents the aggregated percentage.

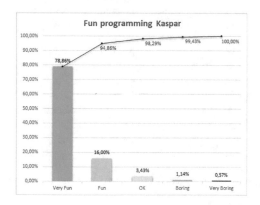

Fig. 8. Fun expressed by the children on programming Kaspar.

Easiness: Over 75% of the children thought programming Kaspar was Easy or Very Easy. Two children left this question blank, and slightly over 6% thought it was Hard or Very Hard. All the children that found it "Very Hard" were from the younger grade and female. For the remaining 17% of the children, they rated the easiness as "OK", as shown in Fig. 9.

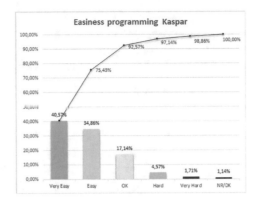

Fig. 9. Easiness expressed by the children on programming Kaspar.

Willingness to Programme Kaspar Again: About 87% of the children were willing to programme Kaspar again, while only two children said "No" and none of them was definitively against it. The children who said "No", however, expressed that they thought Kaspar was "Fun", and were from the middle grade. Roughly 12% of them answered "Maybe", as shown in Fig. 10.

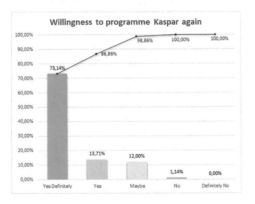

Fig. 10. Willingness expressed by the children to be able to programme Kaspar again.

Perceived Usefulness: As seen in Fig. 11, 84% of the children thought that Scratch was "Very Useful" or "Useful" in order to programme Kaspar. Only around 3.5% of the pupils thought it was not useful for that purpose, and again all of them thought Kaspar was "Fun" or "Very Fun", and 12.5% thought it was "OK".

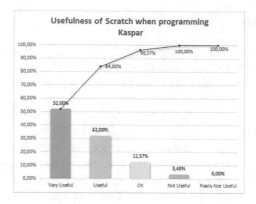

Fig. 11. Perceived usefulness of the Scratch system when programming Kaspar.

5   Discussion

This study had two separate aims. The first one was to investigate how a Scratch-based system would work with our robot Kaspar, and what complications could arise after extensively using this system with children programmers. Considering all the sessions, just once we needed to conduct a repair on one Kaspar, which was a basic repair concerning motors' overheating. The Scratch system performed well during all the sessions, only having problems when the Scratch was giving two different commands simultaneously to Kaspar's program, because Kaspar works sequentially.

The second aim was to investigate if the children could profit of a different take in learning to program. The usual way to teach programming in schools is via a computer, in a screen-based approach, but the children seldom have the opportunity to program outside the computer something in the real world that moves and changes behaviour according to the programming on the fly. We chose to use the robot named Kaspar and created a Scratch interface for it, so the children were more or less comfortable with the GUI and knew how it worked beforehand. When this system was placed into a real world environment, i.e. a class full of children where they are taking control and programming the robot, the Scratch system has proven to be fun (according to $\approx$95% of the students), easy to use ($\approx$75%) and made the children willing to repeat the experience and programme Kaspar again ($\approx$87%). The fact that the great majority of them ($\approx$84%) thought that Scratch is an adequate and useful tool to programme a humanoid robot like Kaspar points out that they are interested in it, and therefore we can see the potential and feasibility of this system. These results show that this approach leads to children eager to continue programming with a real robot.

6   Conclusions

We developed a novel way for non-programmers to programme Kaspar, using Scratch and integrating it with the robot's existing control software. The system, which we call the "Scratch system", has been thoroughly tested during sessions at a primary school. In each session the Scratch system was up and running for more than 1.5 hours, so we conclude that it works well and allows non-programmers to programme Kaspar's behaviours in real time. The system has some issues with simultaneity though; Scratch allows processes to be run simultaneously, but Kaspar only allows commands to be executed sequentially. When Kaspar tries to do some sequences simultaneously, its motors will overwork, causing them to overheat and eventually stop. If this happens, a total reboot of Kaspar is needed in order to get the motors working again. The Scratch system we developed is also able to save the sequences and moves created, in both the Scratch "projects" (with .sbx extension) and in the Java system as sequences (with a .seq extension), and it is able to reproduce them later.

During the session, some children started developing intricate stories using Kaspar and its abilities; in some cases, they created a background for Kaspar using sounds and others; for example, they stated that Kaspar was "so angry it broke the piano", using one of the sounds provided which was a piano; or that it was surprised because aliens were coming, using an UFO sound. They were able to make Kaspar act consistently with the emotion created plus adding context, which indicates that as well as being able to program the emotion, they were able to explore their creativity as well. We consider that by learning to programme in this way, children will be more interested in programming in general; as we could see in the sessions, even children who were not very skilled in programming and did not consider the session as an easy one had generally a good and fun time, and wanted to repeat the experience.

7   Future Work

Our next steps will be making the system more reliable by adding a feedback loop; the system as it is works, but the Scratch system does not receive any confirmation from the motors or the Java programme indicating wether a movement was successfully executed or failed in its execution.

Another future step is to bring the Scratch system into the hands of therapists, parents and teachers to have their insight on the system as well.

Acknowledgements. The authors thank Dr. Gabriella Lakatos, Dr. Patrick Holthaus and Alessandra Rossi for their assistance in the schools' robotics sessions.

This work has been funded by the BabyRobot project, supported by the EU Horizon 2020 Programme, under grant 687831.

References

1. Dautenhahn, K., Nehaniv, C.L., Walters, M.L., Robins, B., Kose-Bagci, H., Mirza, N.A., Blow, M.: KASPAR-a minimally expressive humanoid robot for human-robot interaction research. Appl. Bionics Biomech. **6**(3–4), 369–397 (2009). https://doi.org/10.1080/11762320903123567
2. Wood, L.J., Zaraki, A., Walters, M.L., Novanda, O., Robins, B., Dautenhahn, K.: The iterative development of the humanoid robot Kaspar: an assistive robot for children with autism. In: International Conference on Social Robotics, pp. 53–63. Springer, Cham (2017)
3. Costa, S., Lehmann, H., Dautenhahn, K., Robins, B., Soares, F.: Using a humanoid robot to elicit body awareness and appropriate physical interaction in children with autism. Int. J. Soc. Robot. **7**(2), 265–278 (2015). https://doi.org/10.1007/s12369-014-0250-2
4. Huijnen, C.A., Lexis, M.A., de Witte, L.P.: Matching robot KASPAR to autism spectrum disorder (ASD) therapy and educational goals. Int. J. Soc. Robot. **8**(4), 445–455 (2016). https://doi.org/10.1007/s12369-016-0369-4
5. Wainer, J., Robins, B., Amirabdollahian, F., Dautenhahn, K.: Using the humanoid robot KASPAR to autonomously play triadic games and facilitate collaborative play among children with autism. IEEE Trans. Auton. Ment. Dev. **6**(3), 183–199 (2014)
6. Maloney, J., Burd, L., Kafai, Y., Rusk, N., Silverman, B., Resnick, M.: Scratch: a sneak preview [education]. In: Second International Conference on Creating, Connecting and Collaborating Through Computing, Proceedings, pp. 104–109. IEEE, January 2004
7. Maloney, J.H., Peppler, K., Kafai, Y., Resnick, M., Rusk, N.: Programming by choice: urban youth learning programming with scratch, vol. 40, no. 1, pp. 367–371. ACM (2008)
8. Scratch Statistics (n.d.). https://scratch.mit.edu/statistics/. Accessed 27 Nov 2018
9. Calao, L.A., Moreno-León, J., Correa, H.E., Robles, G.: Developing mathematical thinking with scratch. In: Design for Teaching and Learning in a Networked World, pp. 17–27. Springer, Cham (2015)
10. Wilson, A., Hainey, T., Connolly, T.: Evaluation of computer games developed by primary school children to gauge understanding of programming concepts. In: 6th European Conference on Games-Based Learning (ECGBL), pp. 4–5, October 2012
11. Moreno-León, J., Robles, G.: Code to learn with Scratch? A systematic literature review. In: 2016 IEEE Global Engineering Education Conference (EDUCON), pp. 150–156. IEEE, April 2016
12. Resnick, M.: Sowing the seeds for a more creative society. Learn. Lead. Technol. **35**(4), 18–22 (2008)
13. Home: ScratchX Wiki, January 2018. https://github.com/llk/scratchx/wiki. Accessed 23 Aug 2018
14. Spring Boot Reference Manual. (n.d.). https://docs.spring.io/spring-boot/docs/current-SNAPSHOT/reference/htmlsingle/. Accessed 23 Aug 2018

Integrating Robotics with School Subjects

Bringing an Educational Robot into a Basic Education Math Lesson

Janika Leoste[(⊠)] and Mati Heidmets

Tallinn University, 10120 Tallinn, Estonia
leoste@tlu.ee

Abstract. The teaching practices of STEAM subjects, especially mathematics, need to be modernized in order to successfully prepare today's students for the future jobs. Using technology, in particular educational robotics, is considered as one of the ways of meeting this demand. While there are several studies that explore the effects of robot-supported teaching on students' math learning outcome, these studies do not give us a clear picture of what would be the prerequisites and the outcome if regular math teachers would use robot-supported teaching as a supplementary part of the regular curriculum during a longer period of time. The authors of this paper are currently conducting a multi-stage research on this topic in Estonia. The paper describes the design of the research's first stage, including the overview of the practical problems of designing lesson plans for robot-supported teaching in regular classrooms and of our experience in solving these problems. Finally, the outlines of the main problem areas and further recommendations are discussed.

Keywords: Educational robotics · Mathematics · Math · Basic education · Technology

1 Introduction

Modern technologies have enormously changed the world that is available for human perception. There have been huge changes in "what" and "how" we consume and in "what" and "how" we produce. Artificial intelligence combined with advanced robotics will automate a lot of jobs that used to require heavy human intervention. Many new jobs will be created for satisfying our need for products that were unheard of only a few generations ago. These jobs are often similar in their nature in that the features inherent to specific fields will disappear and workers are able to move freely between different domains [1, 2]. Also, in the modern world both schools and interactive content available through personal smart devices are competing for the students' attention. The digital social interaction, videos and compulsive gaming activate the same dopaminergic reward circuits that are responsible for associative learning and outcome evaluation [3]. In the USA, 95% of teens have access to a smartphone and 45% of them are almost constantly online, consuming for example YouTube videos (32% of teens use it "most often") [4].

These developments are creating immense pressure on the educational systems of different countries for changing the learning and teaching practices [2]. As one of the

© Springer Nature Switzerland AG 2020
M. Merdan et al. (Eds.): RiE 2019, AISC 1023, pp. 237–247, 2020.
https://doi.org/10.1007/978-3-030-26945-6_21

key priorities of the educational systems is the production of graduates with well-developed 21st century skills, including STEAM skills [5–8] then it is necessary to find meaningful and engaging ways of teaching the subjects that on the one hand are traditional and require good concentration abilities, and on the other hand are still relevant [2, 9, 10] and inevitably necessary for most of the 21st century jobs [7, 11, 12].

The change of teaching and learning is especially required in the subjects that are still seeing high levels (almost 30%) of learner dislike, for example in STEAM subjects, particularly in math, at the higher grades [8]. Mathematics is often considered as the thirdly important basic academic survival skill besides reading and writing. This skill accumulates knowledge with each passing year and acquiring new knowledge requires good understanding of fundamental math knowledge. As the fundamental knowledge in mathematics is mostly based on abstract concepts and formulas then children may perceive learning these as boring and incomprehensible, causing gradual loss of their interest towards the subject. In fact, using traditional methods for improving math learning outcomes seems to increase the number of students who dislike math. The correlation coefficient for the percentage of TIMSS (Trends in International Mathematics and Science Study)[1] mathematics achievement in 2015 and students in the same countries who dislike math was statistically significant $r = 0.50$ ($p < 0.05$) [8].

One of the ways of making math more interesting and engaging to students is to bring into curriculum constructivist teaching methods that help students to build their understanding of the subject by allowing collaborative learning and providing more student independence. Accumulated evidence points that compared to the lecture-based approaches more active forms of collaborative group work, inquiry and problem solving will achieve better educational outcomes, and most of these active forms include using some form of technology in the classroom [13]. This trend is actually already seen in the mathematics curricula of TIMSS countries: in the past two decades incorporation of digital devices into educational settings has become widespread practice (about 90% of TIMSS 2015 countries reported initiatives in this field) while most large scale efforts to integrate technology and education are relatively recent [8]. This is not always painless process as technology cannot be just brought into classroom – instead there should be an adaption of mathematics teaching to the needs of information technology in modern society [14], meaning that there should be fusion of mathematics teaching and technology that is suitable for classroom use.

One of the suitable technologies for helping students to construct their math knowledge is educational robotics. These robots are especially designed for younger students and have been used in classroom for the purposes of rising students' interest towards learning process [15–18]. Bringing educational robots into classroom leans on the idea of constructivism that describes how student builds meaning through behavior of physical objects [14, 15] and it is assumed that the integration of robots supports also the change of the teacher's teaching practices [19, 20]. While these robots can be used at the lower grades as tools that enable students to explore and visualize abstract concepts, they are also suggested as means of increasing older students' engagement, employability skills such as creativity, teamwork, problem-solving and communication [14].

---

[1] The list of the TIMSS countries is available on https://nces.ed.gov/timss/countries.asp.

Although the field of educational robotics is relatively new there are already many works studying its use in classrooms. Yet, there are only a few works that evaluate the influence of robots on students' math skills [21]. Of these, the works of Iturrizaga [22], Lindh et Holgersson [16], and Hussain, Lindh et Shukur [23] come forward as having relatively large sample sizes (more than 600 students) and relatively long duration of experiments (at least one school year). The overall results of conducted studies refer that using educational robots will almost always improve the students' motivation and sometimes also their learning outcome.

An educational robot can be brought into classroom as a teacher's assistant (robot-assisted learning and teaching) or as a learning tool (robot-supported learning). Although there is a body of research for using robot as a teacher's helper [24, 25], the meaningful use of robotics as a full substitution of a teacher requires advanced artificial intelligence that is not available for mass use as of today [26]. Another approach, robot-supported learning, is using a robot as a learning tool for applying the student's skills about the learned subject and as an object for introducing accompanying teaching and learning methods (for example collaborative learning) into traditional subjects [27].

2 Study of Robot-Supported Math Lessons

The previous research has indicated that using educational robots in math classroom may have several positive outcomes for students. However, the existing works have various shortcomings. For example, most of the works did not evaluate the long term intervention impact on students' learning motivation or learning outcomes, and had too small sample size for making generalizations. Also, the studies with large samples and durations [16, 22, 23] did not use strict integration of robots into ordinary mathematics teaching, allowing either integration with technology teaching or no integration at all.

Our main goal was to evaluate the effects of robot-supported math learning when the lessons were conducted by regular math teachers as a supplement of a regular math curriculum. For this we planned to conduct a multi-stage research. The first stage was carried out on teacher's teaching approach, on students' motivation and on students' engagement of mathematics learning during the second half of the school year 2017/2018. The additional aim of this study was also to prepare the ground for the more comprehensive next stage of the research, and therefore this study also searched for answers to the practical questions of integrating the robot-supported teaching into regular classroom of basic education stage 1 [28]. The purpose of the current paper is to present the findings related to these practical questions:

1. What are the suitable educational robotics platforms to be used in robot-supported math lessons in basic education stage 1?
2. How to address teachers' lack of robotics and coding skills when starting to use educational robots in regular lessons in basic education stage 1?
3. How to connect mathematics with robotics in a meaningful way in robot-supported math lessons in basic education stage 1?

4. How to design a robot-supported basic education math lesson in basic education stage 1?
5. Can robot-supported math lessons in basic education stage 1 be used with SEN students?

2.1 The Design of the Study

In Estonia the compulsory school start age is 7. The ISCED basic education stage 1 corresponds to the stages 1 and 2 of Estonian basic school as follows: the grades 1–3; and the grades 4–6. The ISCED basic education stage 2 corresponds to the local basic school stage 3 (classes 7–9) [29]. The national standardized tests for measuring mathematics knowledge are conducted at the grades 3 and 6 that are the last grades of the local basic school stages 1 and 2, respectively, while both belong to the ISCED basic education stage 1. We chose these grades to focus to in our study for the following reasons:

1. Possibility of using the results of the national standardized tests for comparing the students' development.
2. At the last grade of any Estonian basic school stage students should have had acquired the concepts that are important for that stage.
3. We wanted to compare the influence of educational robotics on math learning on at least 2 different stages of Estonian basic school.

The sample included 208 students (106 in the 3rd grade and 102 in the 6th grade). The students were taught by 4 class teachers, 6 math teachers and they were supported by 5 educational technologists.

The study took place from December, 2017 to May, 2018 while the robot-supported math lessons were conducted from February, 2018 to May, 2018. Each participating class was requested to take at least 10 to 20 robot-supported math lessons during that period.

The teaching material for the study was worked out in collaboration with participating teachers, while paying special attention to collaborative learning approach and student independence. The mathematical content was based on the regular curriculum of the respective grade. The selection of robotics platforms took into consideration the educational robots that were available in the participating schools. There was no special scripting for the lessons, teachers were encouraged to try different approaches in order to find out what approach would yield the best results. During the experiment, researchers were available to support teachers by using e-mail and phone calls. On-site support was provided by educational technologists.

At the end of the study comprehensive interviews were conducted with teachers and educational technologists to collect data about the strengths and weaknesses of the method. Additional information was also gathered through teachers' lesson diaries and through online questionnaires that were filled by the students and teachers at the end of each robot-supported lesson.

2.2   Results

We were interested in creating an understanding about the practical difficulties of bringing an educational robot into a basic education math classroom. Before the start of the study we had defined 5 questions that were answered as a result of the study.

What Are the Suitable Educational Robotics Platforms to Be Used in Robot-Supported Math Lessons in Basic Education Stage 1? Most of the studies that explore the influence of educational robots on students' math learning outcomes have been based on (as of today) obsolete LEGO Dacta/Educational platform robots [21] that could not be used for the purposes of our study. Based on the initial feedback from participating teachers we defined following criteria for choosing educational robotics platforms for our study:

1. The battery life of a robot had to allow conducting at least 2 consecutive lessons.
2. The size of a robot had to afford easy logistics of classroom sets inside the school-house.
3. A robot had to be able to move a predetermined distance (with the help of color sensor if necessary).
4. There had to be intuitive block-based iPad or Android tablet apps available for robots to allow students with little or no English language skills to work with robots in a regular math classroom.
5. The robotic platform had to be currently supported by the manufacturer.
6. The robotic platform had to be available in the participating school.
7. The robotic platform had to be common in Estonian schools [30, 31].

Based on these criteria we chose following educational robotics platforms for the study (the description of the robots is available on a separate link[2]):

- The LEGO Mindstorms EV3 robot for the grades 3 and 6.
- The LEGO WeDo 2.0 robot for the grade 3.
- The Edison robot for the grade 3.

The feedback from the study indicated that there was no need to add additional platforms or to remove existing ones. However, teachers considered it important that robot should allow accurate sensor readings and movement combined with good reliability, making the LEGO MindStorms EV3 robot as the most suitable choice. The sensors of other used robotics platforms lack accuracy, the LEGO WeDo 2.0 robot is not able to turn and the Edison robot is usually not able to drive a perfectly straight line.

How to Address Teachers' Lack of Robotics and Coding Skills When Starting to Use Educational Robots in Regular Lessons in Basic Education Stage 1? The results of the study, based on the data gathered from the teachers' interviews and questionnaires, indicated that both math teachers and class teachers do not usually have any previous experience with educational robots or with their programming apps. It was common for teachers to feel themselves insecure when conducting a robot-supported math lesson, especially during the first lessons. However, this insecurity had

---

[2] The description of the robots is available on the following link: http://bit.ly/2AvwZpB.

additional basis, including technical reasons (possible robot malfunctions or connection problems) and having to use the new teaching/learning methods that were introduced by the robot-supported approach (students working in teams, students solving exercises independently, some students actively coaching the members of their own teams but also of other teams, also students moving freely around the classroom while solving robotics exercises). During the study these problems were solved by providing teachers with technical assistance from their schools' educational technologists and by allowing them to directly contact the researchers via any suitable channel (e-mail, phone, messenger apps, etc.) at any time.

How to Connect Mathematics with Robotics in a Meaningful Way in Robot-Supported Math Lessons in Basic Education Stage 1? We considered it of utmost importance to ensure that the connection between mathematics and robotics was real and perceivable. To this end, before the study, we observed two different math teachers conducting experimental math lessons using educational robots for visualizing certain math constructs. We also examined the findings of the existing studies that described using robots as aiding tools for visualizing math concepts [14], constructing math knowledge [15], or as embodied agents for getting deeper perception of math [20]. Based on this information and on existing Estonian math curriculum for the selected grades we, in collaboration with the participating teachers of the study, created a number of math word problems. Of these we selected 20 that were reasonably easy to program by a 3rd grade student.

The study confirmed that the chosen word problems were mostly suitable. However, the feedback from the teachers indicated that the wording had to be revised in some cases in order to become more easily understandable for the students with poor functional reading skills. The examples of the worksheets are available on a separate link[3].

How to Design a Robot-Supported Basic Education Math Lesson in Basic Education Stage 1? For the study we scripted relatively loose lesson plans that consisted of creating teams of 2 (or in justified cases of up to 4) students, solving a math word problem and then solving up to 3 robotic exercises that were based on the math word problem. The duration of the lesson was supposed to be one school hour (45 min) but teachers were allowed to use 2 school hours or to move the math word problem into previous regular math lesson or to have it solved as a homework. If solved in a robot-supported lesson then the math word problem could be solved individually or in teams. Teams were formed by students themselves but more often by teachers. Teachers were provided with printable worksheets that contained both math word problem and robotics exercises. These worksheets were sometimes modified by some teachers with the goal of aiding students to understand texts more easily. The teachers who did not modify worksheets pointed out that functional reading skills of students had improved during the experiment although during the first lessons there were difficulties in understanding the exercises, especially in the grade 3. It was also pointed out that worksheets should be in an electronic format in order to improve students' digital competence.

---

[3] The example worksheets are available on the link http://bit.ly/2Q1n9AS.

Can Robot-Supported Math Lessons Be Used with SEN Students in Basic Education Stage 1? The study included one class of the SEN students (in the 6th grade). There were also individual SEN students in some of the other participating classes. The interviews with teachers and their lesson diaries revealed that SEN students enjoyed robot-supported math lessons significantly more than regular lessons. Robots were able to engage even those students who in the regular math lessons "were just looking out of the window". However, these students were initially strongly reluctant to working in teams and, in classes of mostly regular students, SEN students were sometimes working at slower pace than other students although they still wanted to finish all of the 3 robotics experiments.

The collected data showed that SEN students were able to take part in the robot-supported lessons, that in these lessons they were able to grasp some of the math concepts more easily than in regular lessons but these students also needed more guidance in respect of collaboration and extra time to finish all of the robotics exercises (either during the break or after hours).

3 Conclusions and Discussion

Digital technologies have remarkably changed many of the ways we use for processing information, creating pressure on the national educational systems to assimilate these changes in order to form students' knowledge and thinking according to the contemporary requirements of society. One of the ways of making education more relevant is to integrate the elements of modern technology, for example, educational robotics, into appropriate teaching subjects. However, although introduction of technology into, for example, math lessons, could seem to be one of the most logical and, perhaps, easiest things to do, in practice it can involve a lot of complications.

We are currently conducting a multi-stage research that is examining the different aspects of robot-supported math teaching and learning. This paper explored the experience about the practical aspects of conducting a robot-supported math lesson, based on the data that was collected during the first stage of our research. This collected data revealed that there are several areas that require careful approach when designing a robot-supported math curriculum. The answers that we got for the research questions of this study allow following conclusions.

3.1 Conclusions

The choice of the robotics platform must take into consideration the battery life, availability of intuitive programming interface, and the size of individual robots, allowing easy logistics inside the school house. Also, the robot has to be generally easy to use and to be budget-friendly. However, the robot has to work reliably, meaning that the performance should be error free, its sensors should be as accurate as possible and the movement should allow exact control (i.e. drive a straight line and turn exactly 90°). Based on these criteria, the LEGO Mindstorms EV3 robot or its analogues (i.e. VEX IQ robot) should be preferred. The Edison robot could be used as a low budget alternative. However, if the school already has a reasonable quantity of other robots, then it should be evaluated whether the existing robots can be used for robot-supported math teaching.

The majority of teachers need scaffolding as they lack technical knowledge that is required for programming and managing educational robots and tablets, and knowledge about associated modern teaching practices. In reality, at least in the beginning, educational technologists should help in conducting the robot-supported math lessons, providing technical support and acting as an assistant teacher. This need is also pointed out by the previous research work [32]. Besides that, an additional scaffolding structure should be created with the goal of helping teachers to become self-sufficient, while considering the following notions:

The experience of the researchers that have studied Technology Enhanced Learning suggests that for successful integration of a new technology into actual lessons, teachers would need to perceive the positive influence of the technology on the learning outcome while being certain that this technology would be comfortable enough to use. One of the suitable ways of ensuring this is to use the Technology Acceptance Model [33–35]. Research also points out that scaffolding teachers in adopting new teaching practices requires understanding about how innovative information becomes appropriated at the individual, organizational and domain level. We think that using a tool similar to the Knowledge Appropriation Model (KAM), proposed by Ley et al. [36, 37] would be useful in order to ensure the sustainability of the innovation.

Robot-supported lesson's design requires connecting meaningful math content with robotics exercises [14, 15, 20], while providing teachers with lesson scripting that directs them to use modern teaching practices. Robot-supported math lessons require and encourage more student independence, collaborative working, and peer tutoring, compared to regular math lessons. Similar conclusions are found also in previous studies [10, 11, 20, 21]. Studies about similar innovations in education point out that lesson designs for such innovations should be collaboratively created by School-University Partnerships (SUP), ensuring thus that theoretical knowledge of researchers would be combined with practical know-how of teachers [38, 39].

For SEN students the robot-supported math lessons, compared to regular math lessons, could prove to be more developing and engaging, provided that teachers are able to allocate more time for these students.

In conclusion we estimate that robot-supported teaching requires more effort and time from teachers and is also more demanding for students, at least in the beginning. However, after becoming accustomed to this teaching method it will start supporting both teacher's professional development and student's appropriation of math knowledge [22, 36].

3.2   Limitations

Our study has several shortcomings. For example, the sample size is relatively small and the duration of the study is short. We also designed all of the lesson plans for the students with no previous robotics and programming skills. However, if the intervention were longer, then the accumulation robotics skills should be taken into consideration and the lesson plans should progressively include more math content. In addition, our study was providing math teachers with ready-to-use learning material.

We feel that there is a need to examine if it would be viable to teach and encourage math teachers to design their own robotics material to complement their standard curricula. Also, we focused on the grades 3 and 6, leaving the higher grades uncovered, while it could be extremely important to find out if using robots to support math learning in the grade 9 could stop or even reverse the fade of students' interest towards math. Another area that has to be examined more closely is finding out the mathematics topics where robot-supported learning would be most meaningful.

3.3   Further Developments

Based on the results of the study, discussed in this paper, we have designed the second stage of the research (Study 2). The purpose of this stage is to conduct a larger, more reliable study on the effects of using robot to support math teaching.

The sample of the Study 2 involves 67 participating schools with 137 classes (98 in the grade 3 and 39 in the grade 6) with more than 2000 students (1450+ in the grade 3 and 550+ in the grade 6). The Study 2 is taking place from autumn 2018 to spring 2019.

The Study 2 stage is focused on measuring students' math motivation and engagement. For this we are using pre- and post-testing of students and comparing the results of students' national standardized tests to national averages. We are also measuring teachers, namely the change of their teaching practices. For this we are using pre-test and post-test surveys. There will be an additional semi-structured questionnaire for measuring teachers' adaption of innovation.

For scaffolding purposes we use a KAM based SUP that is focused on collaborative creation of learning designs, establishing a support network of peers, and formalizing the knowledge collected and created during the study.

The results of this study can be used as recommendations for a researcher who is going to conduct a similar study or for a teacher who is going to design and conduct robot-supported lessons. Also, the results can be useful in preparing teacher training or in planning scaffolding for robot-supported lessons.

The conclusions can be generalized and used for conducting lessons of other STEM subjects.

Acknowledgments. This project has received funding from the European Union's Horizon 2020 research and innovation programme under grant agreement No. 669074.

We are grateful to all of the students and teachers participating in the research.

European
Commission

References

1. Manyika, J., Lund, S., Chui, M., Bughin, J., Woetzel, J., Batra, P., Ko, R., Sanhvi, S.: Jobs lost, jobs gained: workforce transitions in a time of automation. McKinsey Global Institute (2017)
2. OECD: Enabling the next production revolution: the future of manufacturing and services – Interim report. OECD Publishing (2016)
3. Veissière, S.P.L., Stendel, M.: Hypernatural monitoring: a social rehearsal account of smartphone addiction. Front. Psychol. **9**, 141 (2018)
4. Anderson, M., Jiang, J.: Teens, social media technology 2018. Pew Research Center (2018)
5. Matson, E., DeLoach, S., Pauly, R.: Building interest in math and science for rural and underserved elementary school children using robots. J. STEM Educ.: Innov. Res. **5**(3/4), 35–46 (2004)
6. Dede, C.: Comparing Frameworks for "21st Century Skills". Harvard Graduate School of Education (2009)
7. OECD: Strengthening Education for Innovation (Science, Technology and Industry e-Outlook). OECD Publishing (2012)
8. Mullis, I.V.S., Martin, M.O., Loveless, T.: 20 Years of TIMSS: International Trends in Mathematics and Science Achievement, Curriculum, and Instruction. Boston College, Chestnut Hill (2016)
9. Prensky, M.: Digital natives, digital immigrants part 1. Horizon **9**(5), 1–6 (2001)
10. Gerretson, H., Howes, E., Campbell, S., Thompson, D.: Interdisciplinary mathematics and science education through robotics technology: its potential for education for sustainable development (a case study from the USA). J. Teach. Educ. Sustain. **10**(1), 32–41 (2008)
11. Ribeiro, C., Coutinho, C., Costa, M.F.: Educational robotics as a pedagogical tool for approaching problem solving skills in mathematics within elementary education. In: 6th Iberian Conference on Information Systems and Technologies (CISTI 2011), pp. 1–6 (2011)
12. Savard, A., Freiman, V.: Investigating Complexity to Assess Student Learning from a Robotics-Based Task. Digit. Exp. Math. Educ. **2**, 93–114 (2016)
13. Acosta, A., Slotta, J.: CKBiology: an active learning curriculum design for secondary biology. Front. Educ. **3**, 52 (2018)
14. Samuels, P., Haapasalo, L.: Real and virtual robotics in mathematics education at the school–university transition. Int. J. Math. Educ. **43**, 285–301 (2012)
15. Papert, S.: Mindstorms: Children, Computers, and Powerful Ideas. Basic Books, New York (1980)
16. Lindh, J., Holgersson, T.: Does lego training stimulate pupils' ability to solve logical problems? Comput. Educ. **49**, 1097–1111 (2007)
17. Highfield, K., Mulligan, J., Hedberg, J.: Early mathematics learning through exploration with programmable toys. In: Figueras, O., Cortina, J.L., Alatorre, S., Rojano, T., Sepulveda, A. (eds.) Proceedings of the Joint Meeting of Pme 32 and Pme-Na Xxx, PME Conference Proceedings, vol. 3, pp. 169–176. Cinvestav-UMSNH, Mexico (2008)
18. Barker, B., Ansorge, J.: Robotics as means to increase achievement scores in an informal learning environment. J. Res. Technol. Educ. **39**, 229–243 (2007)
19. Kopcha, T.J., McGregor, J., Shin, S., Qian, Y., Choi, J., Hill, R., Mativo, J., Choi, I.: Developing an integrative STEM curriculum for robotics education through educational design research. J. Form. Des. Learn. **1**, 31–44 (2017)
20. Werfel, J.: Embodied teachable agents: learning by teaching robots. In: Conference Proceedings (2014). http://people.seas.harvard.edu/~jkwerfel/nrfias14.pdf. Accessed 08 Nov 2018

21. Leoste, J., Heidmets, M.: The impact of educational robots as learning tools on mathematics learning outcomes in basic education. In: Digital Turn in Schools - Research, Policy, Practice, Conference Proceedings. Manuscript submitted for publication (2018)
22. Iturrizaga, I.M.: Study of educational impact of the LEGO Dacta materials – InfoEscuela - MED. Final Report, Infoescuela (2000)
23. Hussain, S., Lindh, J., Shukur, G.: The effect of LEGO training on pupils' school performance in mathematics, problem solving ability and attitude: Swedish data. Educ. Technol. Soc. **9**(3), 182–194 (2006)
24. Shamsuddin, S., Yussof, H., Hanapiah, F.A., Mohamed, S., Jamil, N.F.F., Yunus, F.W.: Robot-assisted learning for communication-care in autism intervention. In: 2015 IEEE International Conference on Rehabilitation Robotics (ICORR), Singapore, pp. 822–827 (2015)
25. Hemminki, J., Erkinheimo-Kyllonen, A.: A humanoid robot as a language tutor - a case study from Helsinki skills center. In: Proceedings of R4L HRI2017, Wien, Austria (2017)
26. Smith, C.: Artificial intelligence that can teach? It's already happening. ABC Science (2018)
27. Kennedy, J., Baxter, P., Belpaeme, T.: Comparing robot embodiments in a guided discovery learning interaction with children. Int. J. of Soc. Robot. **7**, 293–308 (2015)
28. UNESCO: International Standard Classification of Education ISCED 2011. UNESCO Institute for Statistics (2012)
29. Statistics Estonia: Mõisted ja metoodika (2018). http://pub.stat.ee/px-web.2001/Database/ RAHVASTIK/01RAHVASTIKUNAITAJAD_JA_KOOSSEIS/04RAHVAARV_JA_RAHV ASTIKU_KOOSSEIS/RV_0231.htm. Accessed 15 Mar 2019
30. Leppik, C., Haaristo, H.S., Mägi, E.: IKT-haridus: digioskuste õpetamine, hoiakud ja võimalused üldhariduskoolis ja lasteaias. Poliitikauuringute Keskus Praxis (2017)
31. HITSA: ProgeTiiger programmis toetuse saanud haridusasutused 2014–2018 (2018). https:// www.hitsa.ee/ikt-haridus/progetiiger. Accessed 08 Nov 2018
32. Leoste, J., Heidmets, M.: Õpperobot matemaatikatunnis. Estonian Research Council (2019). http://www.miks.ee/opetajale/uudised/opperobot-matemaatikatunnis. Accessed 15 Mar 2019
33. Aypay, A., Çelik, H.C., Sever, M.: Technology acceptance in education: a study of pre-service teachers in Turkey. Turk. Online J. Educ. Technol. **11**, 264–272 (2012)
34. Miller, M.D., Rainer, R.K., Corley, J.K.: Predictors of engagement and participation in an on-line course. Online J. Distance Learn. Adm. **6**, 1–13 (2003)
35. Davis, F.D., Bagozzi, R.P., Warshaw, P.R.: User acceptance of computer technology: a comparison of two theoretical models. Manag. Sci. **35**(8), 982–1003 (1989)
36. Ley, T., Leoste, J., Poom-Valickis, K., Rodríguez-Triana, M.J., Gillet, D., Väljataga, T.: CEUR Workshop Proceedings (2018). http://ceur-ws.org/Vol-2190/CC-TEL_2018_paper_1. pdf. Accessed 08 Nov 2018
37. Ley, T., Maier, R., Waizenegger, L., Manhart, M., Pata, K., Treasure-Jones, T., Sargianni, C., Thalmann, S.: Knowledge appropriation in informal workplace learning (2017). http:// results.learning-layers.eu/scenarios/knowledge-appropriation/. Accessed 08 Nov 2018
38. Korthagen, F.: The gap between research and practice revisited. Educ. Res. Eval. **13**(3), 303–310 (2007)
39. Coburn, C.E., Penuel, W.R.: Research-practice partnerships in education: outcomes, dynamics, and open questions. Educ. Res. **45**(1), 48–54 (2016)

Inviting Teachers to Use Educational Robotics to Foster Mathematical Problem-Solving

Vladimir Estivill-Castro[✉]

Griffith University, Nathan 4111, Australia
v.estivill-castro@griffith.edu.au

Abstract. We have developed three lessons supported by the principles of inquiry-based learning (IBL) and problem-based learning (PBL) in educational robotics with the aim of steering and emphasising the mathematics aspects of the curriculum and the role of mathematics in STEM, while also touching on the social context and impact of STEM. Our goal is to inspire and prompt the curiosity in the participants to seek further understanding in mathematics, to develop mathematical thinking and problem-solving skills, and to see applicability in the emerging world where artificial intelligence and automation are transforming the skills learners will use as professionals. Moreover, we have delivered our ideas to educators in high-school who indicated they would incorporate our challenges and tools to cross-pollinate different areas of STEM.

1 Introduction

As part of our university's outreach activities, and as others previously [18,31, 38], we aim at establishing motivating links with the STEM curriculum using hands-on experiences with robots. More importantly, we hope to inspire the so much needed *curiosity* [15] and assist talented individuals to seek explorations beyond our activities and encourage their teachers to develop lessons around our proposed challenges. At a minimum, we provide some exposure to our participants for the usefulness of mathematical thinking in problem-solving.

The context also derives from the need to ensure high-school pupils meet prerequisites from our programs that offer professional outcomes in STEM. In Queensland, for example, the senior Mathematics (Year-11 and Year-12) curriculum is structured into Maths A, Maths B, and Maths C. Maths A covers more practical topics than Maths B and Math C, and although it is OP eligible (part of the weighted average for ranking for a government-funded place in university), it is insufficient for admission into Software Engineering (or other engineering disciplines) or the Bachelor of Computer Science. At a minimum, candidates require Maths B as it demands more advanced algebra skills than Maths A and more sophisticated mathematical thinking. Maths C goes beyond Maths B, and covers additional pure-maths topics (including complex numbers, matrices, vectors, further calculus and number theory). We note that topics like Control

M. Merdan et al. (Eds.): RiE 2019, AISC 1023, pp. 248–261, 2020.
https://doi.org/10.1007/978-3-030-26945-6_22

Theory [1] would require to deal with feedback-loop control and core notions of automation need complex numbers, matrices, vectors, calculus and differential equations. Thus, ideally, we aim at inspiring students to pursue Maths C. A study covering 10 years (2007–2016) by the Australian Mathematical Science Institute reports that across Australia participation in "Advanced Mathematics (Maths C for Queensland) has steadily dropped from 10.2% in 2007 to 9.5% in 2017. A very similar decreasing trend has been recorded for "Intermediate Maths" (Maths B in Queensland): from 21.2% in 2007 to 19.4% in 2016 [3].

We have developed 3 lessons supported by the principles of inquiry-based learning (IBL), problem-based learning (PBL) with the aim of steering and emphasising the mathematics aspects of the curriculum and the role of mathematics in STEM, while also touching on the social and context impact of STEM. We hope that topics such as vectors and matrices emerge from the IBL and PBL exploration as useful mathematics modelling tools, and alleviate the stigma that Math C topics are pure-mathematics with little application. Hence, our activities are aimed at the later years of the high-school curriculum and offering opportunities for teachers of those senior years for interaction in the development of high-school lessons or projects involving STEM and using robots. We hope to bridge the teachers from the mathematics department with those of the science departments, those of new STEM initiatives, those on technology and IT.

Our goal is to prompt the curiosity in the participants to seek further understanding in mathematics, to develop mathematical thinking and problem-solving skills, and to see applicability in the emerging world of where artificial intelligence and automation are radically transforming the skills learners will use as professionals. We note that problem-solving (highly regarded as a 21st-century skill [33] and core to STEM skills [12]) cannot be achieved without conceptual understanding, especially for mathematics [40].

In Sect. 2 we provide some more details regarding the context in which our educational experiences take place. We will provide some information on our chosen infrastructure, and then we will briefly describe our first challenge in Sect. 3. While our first challenge is perhaps more IBL and less PBL (although IBL is the core framework for PBL [24]), our second challenge is probably more PBL than IBL. We will provide a few more details for the second challenge in Sect. 4. Because of space we only touch briefly on our third challenge. We have conducted sessions describing the activities to three groups of high-school teachers as part of two professional-development conferences to teachers. We aimed to evaluate whether they would adopt these ideas into more elaborate series of lessons and project. Section 6 will present the results of the survey.

2   Context

We regularly receive at the university campus students from high school as part of our STEM educational activities. The students have very different backgrounds because they can be as young as Year 6 (± 12 y.o.), or already in Year 12. The may have (but maybe not because they are not even in high-school) an initiation to technology, and in particular (coding) programming. But they typically see little value in the mathematics they are studying, or the implications in society by developments in STEM.

Although in some cases using robots was not effective [14], there is now strong evidence that involving learners in robotics, and in particular, robotics competitions and informal camps, stimulates interest for STEM [7,8,11,12,21–23,20]. For example, interest in STEM professions is increased among middle-school students when the activities are perceived as *fun*, and involve hands-on experiences [19]. Some have identified that learning with robots provides a positive social experience, access to an engaged community and feelings of success [16]. Because of this impact, several projects to enhance educational robotics as a catalyst for curiosity in STEM have appeared. In Europe, one example of an educational platform is the Orbital educational framework [6], while an example of a large project is ER4STEM [17].

However, our context is different from competitions and camps. We are constrained to a two-hour interval. Thus, we adopt a path that includes inquiry-based learning (IBL) and problem-based learning (PBL) [24]. We exclude project-based learning because of the time limitations. But, as explained, we are regularly engaging teachers in high-schools to motivate them to develop project-proposals from extending our activities.

Each activity is, therefore, based on a challenge, accompanied with a lesson plan (enabling different facilitators to conduct the activity) and a booklet of problems. Solving these problems would guide to the solution of the challenges, and we do not expect pupils to solve them during the sessions. The second half of the booklet has some sample solution (not all problems have a unique solution). The booklet is a souvenir for the learners to prompt their teachers or their parents about the solutions, perhaps requesting a further explanation or attempting the problem without knowing an answer and use it for later reflection.

The challenges are, to a certain extent, ill-defined. The first characteristic aspects of our approach that align with IBL is that we expect the participants to play and explore with LEGO®'s EV3s and their programming environment (or with SoftBank's Pepper and Nao and the programming environment of Choregraphe) before being introduced to the formal argumentation or the scientific concepts and principles in mathematics and robotics we aim to introduce (the booklet contains a glossary at the end of terminology and nomenclature). Second, our learners are guided to explore, compare, investigate, and repeat their actions in an attempt to discover. Third, we emphasise that *"inquiry is the art of questioning and the art of raising questions"* [24].

Our presentation of problem-solving follows one of the fundamental strategies for mathematical thinking stated as Polya's First Principle [26]: *Understand the problem.* Scrutinising the problem by inquiry (looking at simpler cases, exploring the context, and even changing the question) is part of the learning behaviours and skills we expect to foster.

All of our lessons work for LEGO®'s EV3s built under the standard shape for the differential-drive model named *TRAC3R* (see Fig. 1). All programming can be completed using the EV3 kit software (LABVIEW). LEGO®'s EV3s

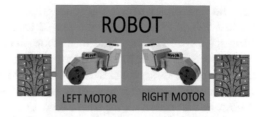

(a) *TRAC3R* as built   (b) Main components of a differential drive

Fig. 1. The LEGO®'s EV3 *TRAC3R* standard construction.

are robust, they hardly break, and they represent an affordable choice for many educational institutions, including some parents (even with the separately sold rechargeable pack). LEGO-Mindstorms have been used to stimulate STEM topics for more than 15 years [39], and its different versions have been widely used [7,14,19,20,29,39]. However, we shift emphasis from the coding skills to the mathematics skills, because we maintain than algorithm understanding (the proof of correctness of an algorithm) is fundamental to relational understanding [34] in mathematics, and central to the quality of software development.

LEGO® produced a series of "EV3 and STEM Curriculum Grids" for different countries, including Australia, where learning objectives of several aspects of the curricula are cross-referenced against LEGO®'s activity packs. For example, there is a guide for mathematics curriculum [35]. LEGO® has also sponsored the First LEGO League® and develop the *Australian Curriculum Links* that cross-reference objective such as "General Capabilities - Literacy", "General Capabilities – Personal and Social Capability", and "General Capabilities – Information and Communication Technology (ICT) Capability" to the opportunities to accommodate such objectives with the involvement with the "First LEGO League"®. Our activities must inspire learners to seek the reasoning and mathematical problem solving that is developed in high-school and tertiary education. However, our links to mathematical concepts and skills are significantly more specific regarding the Mathematics curriculum by Australian Curriculum, Assessment and Reporting Authority (ACARA). We are inspired by the hands-on, exploratory and interactive approaches of others [38] for Lower Secondary School, but we offer significantly more complex challenges.

Our approach follows the design of activities inspired by the activities suggested by Mayerová and Veselovská [18]. However, our approach is more into Problem-Based Learning [33] and fostering creativity to apply lateral thinking and cognitive skills that sit high in the Bloom taxonomy. Some of these correspond to so-called *21st-century skills* [33]: Critical Thinking Skills, Creativity Skills, Communication Skills, Collaboration Skills. In robotics education, Inquiry-based Learning [29] and Problem-Based Learning [33] have been contrasted with Project-Based Learning [24]. We follow the proposal of Samuels and Poppa [31] for "*Themed challenges*: Students were briefed on a number of specific challenges around an engaging theme for which they were required to build a robot". Our lessons developed from contact with high-school clubs using robotics for STEM is a similar fashion that theirs [31].

We aim for closer links between mathematical thinking, computational thinking and problem-solving. It is important to analyse the concept or robot, or the constructs of programming. But, we seek the fundamental understanding of the notion of algorithm as a generic solution for all the valid inputs. We emphasise the notion of algorithm correctness in the true spirit of relational understanding by Skemp [34]. We share the hope to recognise and guide STEM talent [25]. Most of the topics of our activities seem to be innovative with respect to the multi-grade curriculum of the project *READY* [28]. Also, there seems to be significant use of the links between educational robotics and topics such as control theory (feedback-loop control and related matters of sensor and actuators) and programming at the university level [36]. Similarly, some service units for educational robotics focused on programming skills and efficacy skills, and cooperative learning![20]. Interest is increased for STEM [20], but the topics remain distant to mathematical concepts. In some cases, the aim for trans-disciplinary curriculum has reached computational thinking, coding, and engineering [11], it still falls short of crossing mathematical thinking and societal impact. We attempt to inspire discussion of technology's impact, but we admit we do not aim to reach the current level of debate [37]. Similarly, we are aware that mathematics provides perhaps the simple models for the physical world, directly propelling the progress of automation and robotics. We use some examples to inspire debate on the successes of artificial intelligence in emulating human cognition [4]; for example, we use Apple's Siri (a voice-controlled personal assistant) to find out who was Winston Churchill, or to convert radians to degrees. These demonstrations of semantic understanding illustrate the impact on recommender systems is no longer merely syntactic matching by search engines and convincingly argues that memorisation and *instrumental understanding* [34] is becoming less valuable. However, we barely touch on the treatment of values, for example, the potential implementation of systems with emotions [9].

3 First Challenge

Our first challenge is to program the robotic champion of "paper, scissors, rock". It enables the practice of understanding the problem. This task requires to build a program, so the robot moves its arm in a specific position. This sub-problem is rather simple for the configuration of the robot, as the required program is small: It only requires one block. By using the vast amount of resources about LEGO-Mindstorms or the guidance of the instructor, we have seen every Year 6 child or older complete this task. Thus, every child succeeds with the first problem. Here is where we bring the other aspects of PBL. Although considered a variant of IBL [24], PBL starts from real and meaningful problems. We link the "paper, Scissors, rock" challenge to the current academic, economic and social debate of ethical decisions by autonomous vehicles. We bring to the attention the reality of moral dilemmas and the suitability of machines selecting the fate of some humans over others. We point to the recent media reports and the controversial and contradictory views different people have on this regard.

Another aspect for adopting a PBL philosophy is that we want to create the need to know about some mathematical tools. The virtue of our problems is that they are easy to formulate, and in principle, they are understood across the diversity of the students. However, they are also at the frontier of current research, (recall that the first challenge is an abstraction of the issues around ethical machines). Satisfactory solutions (usually raise more question) can be attempted quite satisfactorily by experimentation. Thus, we immediately suggest creating (coding) behaviour, so the robot can engage, or compete against other similar robots in a tournament of the game of "scissors, paper, rock". Facilitators participate also illustrating the inquiry and problem-solving investigation asking the student question such as "Why?", "How do you know for sure?" "Is that the best possible? " Depending on the progression of the Socratic debate, we have been able to introduce concepts such as Nash equilibrium and mixed strategies. Students have reinforced their notions of probability to calculate the expected wins of the program that chooses randomly among the three options.

While guiding learners for this challenge, we have found opportunities to introduce similar triangles, and as a result, units for bearings, such as degrees and radians. Just debating where is left, right, ahead or behind enables the discussion frames of reference. Converting between polar-coordinates and Cartesian coordinates results in motivation for trigonometric functions (although polar-coordinates has been removed from Math A and Math B in the Queensland curriculum). We can raise questions and inquiry regarding the shapes produced by robotic arms with different joints and introduce motivating discussion for concepts in motion planning such as C-space.

There have been learning guides [2] to use robotics and introduce children to programming. Kumar's [2] manual uses the *Myro* robot to introduce python programming. We allude to Kumar's [2] manual because it shows how natural it is to present the programming of "papers, scissors, rock" with a uniform random choice. It is also interesting that although uniform random choice is the only mixed strategy that is a Nash equilibrium, and in that sense the best, Kumar's [2] manual also shows that if we know biases of the opponent's selection (and humans are creatures full of biases), we can take advantage of such preferences.

We have observed a constructivist revision in most students concept map regarding the notion of *machine*. The definitions and the popular culture are that once properly set up, machines perform an intended action deterministically. Now we are introducing machines we cannot predict what they will do (recall, the champion robot of "papers, scissors, rock" must be unpredictable).

4 Second Challenge

Our second challenge is simply stated.

Construct a program that travels the same circle (same trajectory) but at two different speeds. Explain (justify) why it works correctly.

This challenge favours PBL over IBL, because the emphasis shifts slightly from inquiry skills to problem-solving skills, and in particular reasoning strategies. But our intention is to fulfil Bell's first principle of *Connectedness* [5]. We expect that each new idea is linked to an old one enabling students to progress to a common understanding. We aim to guide them to produce a kinematic model for the differential drive. This is a challenging task. Depending on several assumption regarding the nature of the motors, it sits at the frontier of some current investigations (the derivation of the classical kinematic model for a differential robot is sometimes explained with not much notation [32, pp. 61–62] and rapidly [27]). However, sometimes, much more detailed explanations are provided [30, Sect. 2.4]. Kinematics models only some aspects of the situation. Thus, the nature of the potential open avenues to the question, or the opportunity for the pupils to inquiry about the assumptions for an acceptable solution (typical assumptions are suitable for kinematic modelling: the robot is on a rigid surface and not or sand or on a slippery surface, and learners adopted them the context of the room where we carry out the activities).

The analogies with physical situations (role-playing situation of pushing with only one motor, or both at the same speed) and producing drawings should provide *structure and context* (as per Bell's second principle [5]). But as before, we let the learners explore. If they are able to discuss with others we have found that the typical solution that emerges is one where the differential robot has one motor with no power, and the other motor uses the input value to set the speed. This is a correct solution, but it is a special case. Some student will produce some other extreme special cases. For example, one motor with the inverse (negative) speed as the other motor and using the input to regulate the magnitude of the speed. Under this conditions, the robot will spin on its centre, and at different speeds, and some students will argue that this is a very small circle. The exploration by the students take several directions, it is unpredictable. But this indeed how mathematics is constructed. Originally, the kinematic model of a differential drive was not found in a book, it was constructed. Building a program so that a differential robot travels the same circle but at different speed is relatively easy if one is familiar or has a description of the kinematic model for such robot. The challenge is to keep the radius constant although the speeds in the left and right wheels are different. This concept already can establish links to some mathematical thinking, in particular, to the notion of *curvature*.

We do not claim all students from Year 6 to Year 12 reach all the steps, of the intend reasoning path, but we have observer Year 6 pupils discover the entire argument, and we have observed Year 12 learners failing to complete it, but at least all have gained an appreciating for the reasoning behind the mathematical development. This is precisely what PBL proponents suggest [24]: to identify sub-problems, activating prior knowledge, build models, specify and elaborate on knowledge, and derive justifiable conclusions.

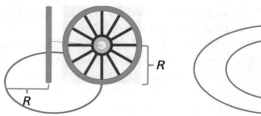 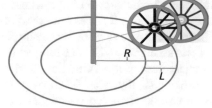

(a) Radius of wheel is same as radius of circular trajectory

(b) If two wheels complete laps in the same time, the outer spins faster

Fig. 2. Figures that assist the argument to build a kinematic model of a differential-drive.

Some of the explorations by students may lead them to the first observation: as long as the speed of both motors does not change (and is somewhat not too large), then the differential drive result is some circle. They probably will discover that the difference between the speed of the right wheel and the left wheel is a significant factor in regulating the diameter of the circle. Playing with different program usually inspires them to notice that very high speeds may cause undesirable effects, like the robot skidding and producing what seems random (or at least hard to predict) trajectories. Some of these examples are special solution that correspond to Polya's principle to explore special cases [26]. We can usually guide them to the discovery that the magnitude of the difference in speed contributes to the radius of the circle's trajectory. Moreover, we find again another special case, when the magnitude and direction of the speed of both motors is the same, the robot travels in a straight line. This allows us to suggest mathematical abstractions of geometry where the straight line is a special circle with its centre at infinity (note the link with the notion of *limit*).

As an aside, because the schedule for the activity does not allow, we consider in the booklet (and in presentations to high-school teachers of the activities to build lessons) some other examples to motivate exploring kinematics (as mathematics of motion without considering the forces that affect the motion). At least we hope teachers can suggest to learners that there is much to contribute in solving problems in robotics considering other wheel arrangements, familiar to students, such as a bicycle (equivalently a motorbike or scooter). We find that some concepts are within the intuition of students, for example, the notion of instantaneous centre of curvature.

But, for the allocated time, our activity plan motivates learners to follow the experimentation with working on reasoning with drawings. Producing drawings is also part of the problem-solving strategies by Polya [26] who suggest to draw pictures. But such drawings already consist of an abstraction, that removes what the problem-solver considers not essential. Those sketches are now models. A crucial model that emerges in some students is derived of the special case of considering just one wheel (like a mono-cycle). What trajectory is performed by a mono-cycle if the centre of the wheel is attached by a rigid string to a post, and the string slides (it does not wrap around the post)?

We inspire learners with some simple cases (Polya's principles once more): what if the string's length is as long as the radius of the wheel (refer to Fig. 2a)? This special case represents the wheel rotating on another wheel and can be illustrated physically with LEGO's gears. How many rotations will the wheel do over a circular trajectory that has double the radius? What if the trajectory is a circle with k-times the radius? We are aiming at learners discovering the meaning behind a formula that relates the linear speed of the wheel with the angular speed around the trajectory of the circle. Recall that our challenge is to identify how to command with software the wheels of a differential-drive robot around a circle with fixed radius (but to regulate the speed).

Typically students discover by experimentation that if they double (or multiply by the same constant) the magnitude of the speeds for both motors, the robot travels the same circle. What raises the bar about this exercise is the justification. The aim here is to reach relational understanding of mathematics as proposed by Skemp [34]. Relational understanding is equivalent, in our view, to understanding why an algorithm works. Instrumental understanding is equivalent to being able to execute the algorithm. Naturally, to inspire students into studying algorithms, we feel the relational understanding [34] of mathematics and computer science is of primary importance.

If another wheel rotated around the post further away (on a larger circle), such second wheel would spin faster (Fig. 2b), why? But as long as they go around the post at constant speed, they each rotate at their individual constant speed. Does the separation of the wheels play a role? Bringing the recently build knowledge assist with this. The physical robot is hard to modify to explore enlarging/reducing the separation L between the wheels, but the graphical model should enable student to discover that the further apart the second wheel is, the faster it must spin to circumvent the post. Again, most of the debate should be prompted encouraging and fostering reflection and critical thinking: alternating from *Why?* to *What if?* and attempting to re-formulate the problem into other problems. As part of our IBL approach, and since Polya [26], some of the elements of problem-solving consist of reformulating the question.

The next step guides learners to the observation that there are three circles, when the robot *travels at constant speed over a circle*. There is the circular trajectory of the left wheel, there is the circular trajectory of the right wheel and there is the circular trajectory of the centre of the axel that connects the wheels (and that we may chose that point as the origin for our frame of reference for the robot). When the robot travels at *constant speed* the angular velocity around the instantaneous centre of curvature is the same for these three circles. We may not use the term angular velocity, (but from the guided analysis, and in particular Polya's principle of *working in the one understanding the question* as a problem-solving strategy), the challenge statement is translated into going around in a circle at a controllable number of laps per unit of time.

We can proceed to some mathematical encoding, and move from the geometric model to the mathematical model. Again, we have found that the ability from students to reach these stages is independent of their schooling year classification. Although some familiarity with algebra and in manipulating linear equations does enable some students to build the argument themselves. If the radius of the circular trajectory is R and the separation between the wheels (the length of the axle in the robot is L, then fact that the rigid robot is travelling at constant speed means that the angular velocity of the wheels in the environment is the same. Depending on the frame of reference, whether the robot is drawn travelling clockwise or counter-clockwise (for our argument refer to Fig. 2b), we may get some symmetrical argument, but for the circles of the wheels:

$$\text{angular velocity} = V_r / \left(R + \frac{L}{2} \right) \qquad \text{and} \qquad \text{angular velocity} = V_l / \left(R - \frac{L}{2} \right)$$

where V_r is the linear velocity of the right wheel, and V_l is the linear velocity of the left wheel. Some learners could encode the situation as

$$V_r = \text{angular velocity} \times \left(R + \frac{L}{2} \right) \tag{1}$$

$$V_l = \text{angular velocity} \times \left(R - \frac{L}{2} \right) \tag{2}$$

already because of the earlier exploration of the single wheel on a circular trajectory which provides a formula of the form linear velocity $= R \times$ angular velocity where R is the radius of the trajectory.

Subtracting Eq. (2) from Eq. (1) gives $V_r - V_l = L \times$ angular velocity or equivalently

$$\text{angular velocity} = (V_r - V_l)/L. \tag{3}$$

This equation should be a mathematical model that represent all the initial hands-on exploration with the robot. Namely, it does represent that when both linear speeds are the same there is no angular velocity, the robot moves on a straight line! Moreover, if we make V_r, V_l of equal magnitude and opposite sign, then the robot spins on itself, with angular velocity proportional to the magnitude $|V_r| = |V_l|$.

This first part of the kinematic model describes how the orientation of the robot (what is it looking at straight ahead, or it bearing) changes as the software controls each motor speed (typically power and direction, already suggesting the notion of vector as magnitude and orientation). However, the challenge is to set the radius R. However, now adding Eqs. (2) and (1), and substituting the angular velocity given by Eq. (3) gives $R = \frac{L}{2} \frac{(V_l + V_r)}{(V_l - V_r)}$. Now is clear that if the new speeds are $V_l' = kV_l$ and $V_l' = kV_r$ the radius is the same, and Eq. (3) shows the angular velocity is not the same.

The previous argument is also the kinematic justification of why a robot with a differential-drive performs circles. While many kinematics concepts have traditionally been regarded as topics suitable for advanced undergraduates and graduate students [10,32], we argue here that the hands-on activity enables even Year-6 learners to gain significant understanding of the techniques and an appreciation for the usefulness of mathematical thinking.

5   Third Challenge

We introduce this challenge reviewing effects of the industrial revolution, the birth of engineering and wide-spread automation. We discuss the video *Humans Need Not Apply* (https://www.youtube.com/watch?v=7Pq-S557XQU) as it is thought-provoking because it suggest automation will move not only from replacing people for physical tasks, but also for intellectual tasks.

Our third challenge is an introduction to feedback-control systems. Although our pupils rarely have any exposure to calculus, they can be exposed to the notion of a feedback-loop by requesting their robot to follow a line or to maintain a distance to another moving robot. Actually, it is quite intuitive that the software must do something every time it detects an error in such regularisation challenges. We have no space to describe the plan for the activity, but suffice to say that rather than using the approach of control theory (that studies the proportional-derivative-integral (PID) controller using the tools from calculus and maybe then the tools of transforms to build the discrete PID), we motivate directly discrete PIDs. This turns our to be quite natural, since discrete PIDs are the only type that can be used with the software and the robots.

6   Conclusion

We have also described the activities to 3 groups of high-school teachers as part of two professional-development conferences to teachers. We performed a survey with the school teachers [13]. The survey consisted of 8 question in a 5-point Likert scale and was completed anonymously by 38 participants. The preferred response to all questions is highly indicative that the activities could be incorporated in high-school lessons and would foster the interest in learners for problem-solving using mathematical tools. For all questions, at least 84% of the responses were the extreme positive response. Moreover, never an educator chose the extremely negative response. For all questions, never more than 2 answers were not positive.

References

1. Åström, K.J., Murray, R.M.: Feedback Systems – An Introduction for Scientists and Engineers, 2nd edn. Princeton University Press, Princeton (2018)
2. Axelrod, B., Balch, T., Blank, D., Eilbert, N., Gavin, A., Gupta, G., Gupta, M., Guzdial, M., Jackson, J., Johnson, B., Kumar, D., Muhammad, M.N., O'Hara, K., Prashad, S., Roberts, R., Summet, J., Sweat, M., Tansley, S., Walker, D.: Learning computing with robots. IPRE Institute For Personal Robots in Education (2008). www.roboteducation.org

3. Barrington, F., Evans, M.: Year-12 mathematics participation in Australia. Australian Mathematical Sciences Institute (2017)
4. Beddoes, Z.M. (ed.): Special Report on Artificial Intelligence, 5, vol. 4. The economist, The address of the publisher (2016)
5. Bell, A.: Principles for the design of teaching. Educ. Stud. Math. **24**(1), 5–34 (1993)
6. Christofi, N., Talevi, M., Holt, J., Wormnes, K., Paraskevas, I.S., Papadopoulos, E.G.: Orbital robotics: a new frontier in education. In: 6th International Conference on Robotics in Education, pp. 20–26. Roboptics (2015)
7. Chung, C.J., Cartwright, C., Cole, M.: Assessing the impact of an autonomous robotics competition for STEM education. J. STEM Educ. **15**(2), 24–34 (2014)
8. Dessimoz, J.: International robotics competitions as excellent training grounds for technical education and student exchanges. In: 6th International Conference on Robotics in Education, pp. 86–93. Roboptics (2015)
9. Dessimoz, J.: Natural emotions as evidence of continuous assessment of values, threats and opportunities in humans, and implementation of these processes in robots and other machines. In: Bhatt, M., Lieto, A. (eds.) 1st International Workshop on Cognition and Artificial Intelligence for Human-Centred Design 2017 co-located with IJCAI, CEUR Workshop Proceedings, vol. 2099, pp. 47–55. CEUR-WS.org (2018)
10. Dudek, G., Jenkin, M.: Computational Principles of Mobile Robotics, 2nd edn. Cambridge University Press, New York (2010)
11. Eguchi, A.: Robotics as a learning tool for educational transformation. In: 4th International Workshop Teaching Robotics, Teaching with Robotics & 5th International Conference on Robotics in Education, pp. 27–34 (2014)
12. Eguchi, A.: RoboCupJunior for promoting STEM education, 21st century skills, and technological advancement through robotics competition. Robot. Auton. Syst. **75**, 692–699 (2016). https://doi.org/10.1016/j.robot.2015.05.013
13. Estivill-Castro, V.: Linking mathematics curriculum and concepts to robotics activities. In: INTED2019 Proceedings, 13th International Technology, Education and Development Conference, pp. 8012–8017. International Academy of Technology, Education and Development, IATED (2019)
14. Fagin, B., Merkle, L.: Measuring the effectiveness of robots in teaching computer science. SIGCSE Bull. **35**(1), 307–311 (2003). https://doi.org/10.1145/792548.611994
15. Flexner, A.: The Usefulness of Useless Knowledge. Princeton University Press, Princeton (2017)
16. Kandlhofer, M., Steinbauer, G., Sundström, P., Weiss, A.: Evaluating the long-term impact of RoboCupJunior: a first investigation. In: Obdržálek, D. (ed.) 3rd International Conference on Robotics in Education, pp. 87–94. MATFYZPRESS (2012)
17. Lammer, L., Lepuschitz, W., Kynigos, C., Giuliano, A., Girvan, C.: ER4STEM educational robotics for science, technology, engineering and mathematics. In: Robotics in Education, Advances in Intelligent Systems and Computing, vol. 457, pp. 95–101. Springer, Cham (2017). https://doi.org/10.1007/978-3-319-42975-5_9
18. Mayerová, K., Veselovská, M.: How we did introductory lessons about robot. In: 4th International Workshop Teaching Robotics, Teaching with Robotics & 5th International Conference on Robotics in Education, pp. 127–134 (2014)
19. Mohr-Schroeder, M.J., Jackson, C., Miller, M., Walcott, B., Little, D.L., Speler, L., Schooler, W., Schroeder, D.C.: Developing middle school students' interests in STEM via summer learning experiences: see blue STEM camp. Sch. Sci. Math. **114**(6), 291–301 (2014). https://doi.org/10.1111/ssm.12079

20. Mosley, P., Ardito, G., Scollins, L.: Robotic cooperative learning promotes student STEM interest. Am. J. Eng. Educ. **7**(2), 117–128 (2016)
21. Nugent, G., Barker, B., Grandgenett, N.: The effect of 4-H robotics and geospatial technologies on science, technology, engineering, and mathematics learning and attitudes. In: Luca, J., Weippl, E.R. (eds.) EdMedia + Innovate Learning 2008, Vienna, Austria, pp. 447–452. Association for the Advancement of Computing in Education (AACE) (2008)
22. Nugent, G., Barker, B., Grandgenett, N., Welch, G.: Robotics camps, clubs, and competitions: results from a U.S. robotics project. In: 4th International Workshop Teaching Robotics, Teaching with Robotics & 5th International Conference on Robotics in Education, pp. 11–18 (2014)
23. Nugent, G.C., Barker, B., Grandgenett, N.: The impact of educational robotics on student STEM learning, attitudes, and workplace skills. In: Robotics: Concepts, Methodologies, Tools, and Applications, pp. 1442–1459. Information Resources Management Association, IGI Global, Hershey (2012). https://doi.org/10.4018/978-1-4666-4607-0.ch070
24. Oguz-Unver, A., Arabacioglu, S.: A comparison of inquiry-based learning (IBL), problem-based learning (PBL) and project-based learning (PJBL) in science education. Acad. J. Educ. Res. **2**(7), 120–128 (2014). https://doi.org/10.15413/ajer.2014.0129
25. Pittí, K., Curto, B., Moreno, V., Ontiyuelo, R.: CITA: promoting technological talent through robotics. In: Obdržálek, D. (ed.) 3rd International Conference on Robotics in Education, pp. 113–120. MATFYZPRESS (2012)
26. Polya, G.: How to Solve It: A New Aspect of Mathematical Method, 2nd edn. Princeton University Press, Princeton (1957)
27. Robins, M., Somashekhar, S.H.: Trajectory tracking and control of differential drive robot for predefined regular geometrical path. Procedia Technol. **25**, 1273 – 1280 (2016). https://doi.org/10.1016/j.protcy.2016.08.221. 1st Global Colloquium on Recent Advancements and Effectual Researches in Engineering, Science and Technology - RAEREST 2016 on April 22nd & 23rd April 2016
28. Rubenzer, S., Richter, G., Hoffmann, A.: Development of a multi-grade curriculum: project "READY". In: 6th International Conference on Robotics in Education, pp. 60–65. Roboptics (2015)
29. Rursch, J.A., Luse, A., Jacobson, D.: IT-adventures: a program to spark IT interest in high school students using inquiry-based learning with cyber defense, game design, and robotics. IEEE Trans. Educ. **53**(1), 71–79 (2010). https://doi.org/10.1109/TE.2009.2024080
30. Salem, F.A.: Dynamic and kinematic models and control for differential drive mobile robots. Int. J. Curr. Eng. Technol. **3**(2), 253–263 (2013)
31. Samuels, P., Poppa, S.: Developing extended real and virtual robotics enhancement classes with years 10–13. In: Robotics in Education, Advances in Intelligent Systems and Computing, vol. 457, pp. 69–81. Springer, Cham (2017). https://doi.org/10.1007/978-3-319-42975-5_7
32. Siegwart, R., Nourbakhsh, I.R., Scaramuzza, D.: Introduction to Autonomous Mobile Robots, 2nd edn. The MIT Press, Cambridge (2011)
33. Sierra Rativa, A.: How can we teach educational robotics to foster 21st learning skills through PBL, Arduino and S4A? In: Robotics in Education, pp. 149–161. Springer, Cham (2018). https://doi.org/10.1007/978-3-319-97085-1_15
34. Skemp, R.: Relational understanding and instrumental understanding. Math. Teach. **77**, 20–26 (1976)

35. Stanley, K.: Learning with LEGO® education: how robotics can meet the NSW technology mandatory 7-8 syllabus outcomes. LEGO Education
36. Swenson, J.: Examining the experiences of upper level college students in 'introduction to robotics'. In: 6th International Conference on Robotics in Education, pp. 72–77. Roboptics (2015)
37. Turkle, S. (ed.): Reclaiming Conversation: The Power of Talk in a Digital Age. Penguin Press, New York (2015)
38. Veselovská, M., Mayerová, K.: LEGO WeDo curriculum for lower secondary school. In: Robotics in Education, pp. 53–64. Springer, Cham (2018). https://doi.org/10.1007/978-3-319-62875-2_5
39. Whitman, L.E., Witherspoon, T.L.: Using LEGOS to interest high school students and improve K12 STEM education. In: 33rd Annual Frontiers in Education, FIE 2003, vol. 2, pp. F3A_6–F3A_10 (2003). https://doi.org/10.1109/FIE.2003.1264721
40. Wu, H.: Basic skills versus conceptual understanding: a bogus dichotomy in mathematics education. Am. Educ. **23**(3), 1–7 (1999)

Integrating Mathematics and Educational Robotics: Simple Motion Planning

Ronald I. Greenberg[✉], George K. Thiruvathukal, and Sara T. Greenberg

Loyola University Chicago, Chicago, IL 60645, USA
rig@cs.luc.edu, gkt@cs.luc.edu, sgreenberg1@luc.edu

Abstract. This paper shows how students can be guided to integrate elementary mathematical analyses with motion planning for typical educational robots. Rather than using calculus as in comprehensive works on motion planning, we show students can achieve interesting results using just simple linear regression tools and trigonometric analyses. Experiments with one robotics platform show that use of these tools can lead to passable navigation through dead reckoning even if students have limited experience with use of sensors, programming, and mathematics.

Keywords: Computer science education · Trigonometry · Algebra · Linear regression · Robotics · Mathematics

1 Introduction

Providing robotics experiences has become a popular and successful mechanism for broadening participation in computing and STEM more generally, retaining more students in these fields, and improving their learning. For example, one study of student responses to brief computing outreach visits found that the most popular component of such visits was viewing of robotics videos [5]. The typical skill development focus in robotics programs is on logical thinking, computer programming, and/or engineering design. Mentors may be aware of the potential for integration with mathematics, but there is limited availability of level-appropriate materials focused on such integration for middle school and especially high school students. This paper focuses on motion planning built on the algebra and precalculus background typical for high school students.

We will work with a robot drawn as shown below that must navigate through a two-dimensional field. With two wheels that may be driven independently (forwards or backwards up to some maximum speed) and a third balance point such as a caster wheel or track ball, such a robot is generally referred to as a differential-drive robot, and this model is a good fit for most educational robots. Some of these platforms have a fixed distance w between the wheels, or *track*

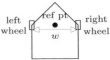

R. I. Greenberg—Supported in part by United States National Science Foundation grants CNS-1738691, CNS-1543217, and CNS-1542971.

M. Merdan et al. (Eds.): RiE 2019, AISC 1023, pp. 262–269, 2020.
https://doi.org/10.1007/978-3-030-26945-6_23

width, but this parameter can be a point of flexibility in many LEGO®-based designs. Intuitively, a small *w* makes the robot able to move more nimbly, assuming one avoids values that are so small as to lead to problems with rollover.

We focus on dead reckoning through a known terrain without substantial sensor use. For experimental work, we do use a robot with a built-in functionality of testing the amount of rotation of the motors controlling the robot wheels. This makes the results more robust in the face of possibly varying levels of battery charge, but passable results might also be achievable using only timing delays as long as the battery is frequently recharged. Experienced teams in robotics competitions do tend to make use of other sensors that are typically available, for example a range finder to judge closeness of approach to a landmark or a reflectance sensor to detect black/white boundaries on the driving surface. While such sensors can provide still more robust results in the face of such phenomena as imperfections in the driving surface, we show here that students can use accessible mathematics to achieve passable results without advanced sensors and to compare different navigation primitives and strategies. In addition to providing an opportunity to integrate mathematical and statistical or data-driven reasoning, the results can be valuable to beginning and advanced teams. Beginning teams may not have yet mastered the use of sensors and the requisite programming, while advanced teams may desire a fail-safe in case of failure of a sensor, its mount, or its electrical connection. In all cases, dead reckoning is likely to be helpful, even if only for initial rough positioning before using sensors, since this often can be done at higher speed and lower power consumption.

Within this dead reckoning context, this paper uses an intermediate level of mathematics (algebra and trigonometry) to analyze and experiment with intuitive alternatives for basic navigation tasks. This contrasts with both very simple exercises suitable for students as young as elementary grades (e.g. [6]), and with studies using calculus or other heavy mathematics (e.g., inertial navigation, overviewed in [2, pp. 77–80] for example, or proofs regarding optimal path types, using various constraints and criteria, for arbitrary changes in position and orientation of a differential-drive robot and even more complex car-like robots as in [1,3,7] and references therein). While Ben-Ari and Mondada [2] provide a rare example of explaining robotics concepts at primarily an intermediate level, this paper adds depth to their elementary discussions of odometry while stopping short of calculus-based discussions. (For a similarly intermediate-level discussion of integrating mathematics and data analysis into robot building, see [4].)

2   Navigation Framework and Primitives

We will think about moving a reference point on the robot from a start position at coordinates $(0,0)$ to a target position (x,y) but keeping in mind that the movement is constrained by the use of our two wheels at separation w. We also assume the initial orientation, or heading, is $0°$, and the analyses in this paper relate to bringing the robot to a final orientation θ that is also $0°$; working with other ending orientations would not add a great deal of complexity. As another

```
/* doticks is called by other procedures to move wheels correct number of ticks */
void doticks(float leftlim,float rightlim){int lticks,rticks; /* to track left/right wheels */
  cmpc(LEFT); cmpc(RIGHT); /* cmpc & gmpc clear & get motor position counter */
  do {lticks = abs(gmpc(LEFT)); rticks = abs(gmpc(RIGHT));
      if (lticks>=leftlim) motor(LEFT,0); if (rticks>=rightlim) motor(RIGHT,0);
      } while (lticks<leftlim || rticks<rightlim);}
void gostraight(float dist){ /* dist in millimeters */ int ticklim = straightfit(dist);
  motor(LEFT,.9*speed); motor(RIGHT,speed); doticks(ticklim,ticklim);}
  /* .9 chosen empirically; straightens our robot using a primitive that sets motor speeds */
void rotate(float theta){ /* theta in positive/negative radians for right/left rotations */
  float ticklim; float dist=theta*w/2; /* rotation radius w/2 */
  if (dist>=0) {motor(LEFT,speed); motor(RIGHT,-speed); ticklim=rrotfit(dist);}
  else {motor(LEFT,-speed); motor(RIGHT,speed); ticklim=lrotfit(dist);}
  doticks (ticklim,ticklim);}
void swing(float theta){ /* theta in positive radians for right swings; negative for left */
  float dist=theta*w; /* swing radius w */
  if (dist>=0) {motor(LEFT,speed); motor(RIGHT,0); doticks(rswingfit(dist),0);}
  else {motor(LEFT,0); motor(RIGHT,speed); doticks(0,lswingfit(dist));}}
```

Fig. 1. Example code for straight, rotation, and swing movements. LEFT and RIGHT give port numbers controlling corresponding motors. straightfit, rrotfit, lrotfit rswingfit, and lswingfit incorporate linear regression results described in Sect. 5.

simplification, and to focus on the most common navigational tasks, we assume $y \geq w$, and we focus within this paper on $x \geq 0$, since negative values of x just lead to a mirror image. (Much of the complication of more advanced works is devoted to analyzing the much broader range of possibilities for x, y, and θ.)

We presume students will be most interested in comparing time from different navigation strategies rather than total amount of wheel rotation or some other criterion. To keep the math simple, we assume, as in some other works, that the robot can accelerate and decelerate instantaneously as long as a bound on velocity is respected. In practice, students may need to experimentally analyze effects of inertia and compose overall movements from a sequence of navigational primitives separated by brief delays that allow the robot to come to rest.

Advanced works regarding optimal trajectories of differential-drive robots show that optimal paths always contain only a limited number of segments of straight-line movements, rotations (about the reference point as one wheel moves forward and the other backward), and swings (with one wheel moving so that the robot pivots around the fixed wheel). These also are natural primitives for students to program.

To account for inertial effects, we recommend teaching students to gather data points experimentally and perform linear regression to obtain formulas for performing straight line motions of specified distance and rotations and swings of specified angle. While linear regression sounds a little advanced, it is actually an easy task using ubiquitous spreadsheet programs, but is still a meaningful way of integrating mathematical understanding into the motion planning task.

As an example, we gathered data for a LEGO-based robot built from parts provided in the Botball educational robotics program and wrote functions for straight, rotation, and swing movements parameterized by the distance to travel or the angle to turn through as in Fig. 1.

Depending on the reliability of the routines to rotate or swing to a specified angle, students may also want to write routines that specifically turn 90° left and 90° right. In any case, we will begin by considering simplified navigation schemes in which all the straight-line motions are horizontal or vertical and the turns are all by 90°. Then we will consider more general navigation.

3   Horizontal and Vertical Navigation

With all navigation along horizontal and vertical lines with turns of 90°, the navigation primitives for turns can be simplified. Instead of gathering data and doing linear regression to write routines to swing and rotate through an arbitrary angle, one could just determine what is needed to turn right and left by 90°. In this context, an interesting first exercise for students (in addition to the testing and programming to create the navigation primitives) is to compare the effects on overall navigation from using rotations versus swings.

Figures 2(a) and (b) show the paths using rotations and swings when moving the reference point from $(0, 0)$ to (x, y) on a "middling" path through the terrain that is likely to avoid obstacles in a typical educational robotics setting. (If there is actually some more particular need to avoid an obstacle, the point where the first turn is taken can be easily adjusted without affecting the navigation time.)

Under the rotation approach of Fig. 2(a), both wheels are always in motion, so we can compute the time as being proportional to the distance traveled by either wheel, i.e., $x + y + \pi w/2$. Under the swing approach of Fig. 2(b), we can again consider either wheel, but when that wheel is stationary, we must account for the distance traveled by the other wheel; thus the time is proportional to $(x - w) + (y - w) + \pi w = x + y + (\pi - 2)w$. The swing approach is therefore superior at a time savings proportional to $2w - \pi w/2 \approx .43w$.

In practice, as previously noted, one may need a brief delay before and after each turn to allow for inertial effects where the theoretical path calls for a discontinuity in velocity, but we will have the same number of discontinuities, four (not counting start and end), whether using rotations or swings.

4   Generalized Navigation

While the navigational approach of the prior section is simple, we would expect to be able to navigate more quickly by proceeding on a path closer to a straight line. In addition, we can define paths with fewer points of velocity discontinuity at which we insert delays for inertial effects. As long as we have gathered enough data to calibrate the rotations and swings according to the angle desired, we just need to do some trigonometric calculations.

Figures 2(c) and (d) show the paths under the rotation and swing approaches when moving the reference point from $(0, 0)$ to (x, y) assuming we use the rotations or swings just to line us up for straight-line navigation. Both of these paths have just two (internal) points of velocity discontinuity.

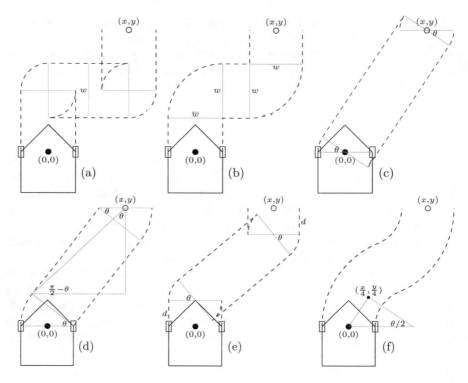

Fig. 2. The paths of the robot wheels for the routing methods considered in this paper. (a) and (b) for horizontal and vertical navigation, using rotations and swings, respectively. (c) and (d) for general navigation, using rotations and swings, respectively, and (e) and (f) for two other path types designed to avoid hitting a competition field boundary as may occur in (c). In (e), straight segments are added at the beginning and end of the path; these lengths are exaggerated for visual effect. In (f), we follow two mirror-image circular arcs.

Under the rotation approach of Fig. 2(c), both wheels are always in motion, so we can compute the time as being proportional to the distance traveled by either wheel, i.e., $\sqrt{x^2 + y^2} + w\theta$ with θ in radians and $\tan\theta = x/y$. Under the swing approach of Fig. 2(d), we can again consider either wheel, but when that wheel is stationary, we must account for the distance traveled by the other wheel; thus the time is proportional to $\sqrt{(x - w + w\cos\theta)^2 + (y - w\sin\theta)^2} + 2w\theta$ with θ in radians and $\tan\theta = (x - w + w\cos\theta)/(y - w\sin\theta)$. We have not found an analytical solution for this θ, but writing $\tan\theta$ as $\sin\theta/\cos\theta$, cross-multiplying and using the identity $\sin^2\theta + \cos^2\theta = 1$, we can obtain a somewhat simplified function of θ for which we seek to find a zero: $y\sin\theta - (x - w)\cos\theta - w$. Students can be taught to write a simple routine using the bisection method to find a root in the range $[-\pi/2, \pi/2]$ Without an analytical solution for θ, it is difficult to make a full comparison of the times for rotations and swings, but we can at least say immediately that as x or y or both get large, the values of θ in these two cases approach the same value, and the swing approach takes longer by an amount of time proportional to θw.

With the observations just above, students can see that rotations seem to be better than swings for general navigation, in contrast to what was observed when using horizontal and vertical navigation. In addition, the math is simpler for rotations, but this navigational approach may often be impractical in robotics competitions, because the robot may often butt right against a boundary of the competition field at the start or end of the path and therefore be unable to do a pure rotation at that point.

There are two more simple navigation strategies that students might like to analyze to cope with the potential problem of boundary walls abutting the start and end positions as shown in Fig. 2(e, f).

In Fig. 2(e), straight segments of length d are added at the beginning and end of the path, while otherwise using the basic rotation approach. A sufficient value for d to avoid hitting a boundary running horizontally just below the robot would be the difference of the distance from the reference point to the lower right corner of the robot and the distance from the right wheel center to the lower right corner of the robot. The resulting angle of rotation is such that $\tan \theta = x/(y - 2d)$, and the time spent on wheel rotation is $\sqrt{x^2 + (y - 2d)^2} + 2d + w\theta$.

Typically, the displacement d should be very small compared to other distances so that it should have little effect on the angle and wheel rotation time. But another difference is that we are adding two more points of discontinuous velocity where we will insert delays to handle inertial effects; this brings us back to the same number of internal discontinuities, four, as when we restricted movements to be horizontal and vertical.

In Fig. 2(f), we consider an approach that reduces the number of velocity discontinuities within the path to just 1. For $x \neq 0$, we route the reference point along a circular arc to $(\frac{x}{2}, \frac{y}{2})$ and then along a mirror image arc to the final destination. From the right triangle drawn for reference in the figure, we can see that the angle of rotation for each of the two arcs is given by $\theta/2 = \arctan(x/y)$, and the radius of the arc traced by the reference point is $r = (x^2 + y^2)/(4x)$. The left wheel initially rotates on an arc of radius $r + w/2$ and the right wheel on an arc of radius $r - w/2$, and then the roles reverse. These arcs can be realized by running the outer wheel at full power and the inner wheel at a relative power of $(r - w/2)/(r + w/2)$; in practice, an adjustment may be needed for wheels that do not behave entirely identically, similar to the adjustment in relative powers that may be needed just to make the robot drive straight. Aside from such an adjustment, the total time spent on wheel rotation is proportional to the lengths of the two wider arcs, which is $(r + w/2)2\theta = ([x^2 + y^2]/x + 2w)\arctan(x/y)$. For x of large magnitude, the path traced by the wheels is longer than under previous methods but with just one internal delay for switching wheel powers.

5   Empirical Results

To perform empirical tests, we began by writing routines as in Fig. 1 for straight, rotation, and swing movements parameterized by the distance to travel or the

angle to turn through; these routines loop until reaching a specified value from
a built-in gauge measuring the amount of rotation (number of ticks) of the
motors controlling the wheel(s). We initially coded `straightfit`, `rswingfit`,
`lswingfit`, `rrotfit`, and `lrotfit` to just return the argument given, and we
gathered data for millimeters traveled or degrees turned when running the routines of Fig. 1 with various limits. For turns, we separately handled positive (right
turn) and negative (left turn) arguments. Then we fitted a line to each data set
using a spreadsheet program; we also verified with a short R program. After
fitting the lines, a simple algebraic manipulation allowed us to rewrite the corresponding navigation primitives to actually travel a specified number of millimeters or turn a specified number of radians. Students can exercise their trigonometry and algebra knowledge by solving for *ticklim* after replacing x with *ticklim* in
the regression line from `gostraight` ($y = .188x + 31.086$), *ticklim/w* for `swing`,
and *ticklim/(w/2)* for `rotate`. They also must replace y in the regression lines by
dist for `gostraight` and by $\theta\frac{360}{2\pi}$ for `swing` and `rotate` for radian/degree conversion. Finally, for `swing` and `rotate`, we also replace θ by *dist/w* or *dist/(w/2)*,
respectively. In all cases, we complete simple algebraic manipulations to express
ticklim in terms of *distance*. The numerical results obtained appear in the code
squeezed into the white space in the graph below.

It is hard to show all the data from our experiments compactly, but if we
perform the above transformations to convert measured turning angles to distances (with $w = 150\,\text{mm}$ for our robot), and look at absolute values of tick
limits, then the data all falls very close to the regression line for the `gostraight`
data as shown here, and we added code as shown based on the regression results
and algebraic manipulations:

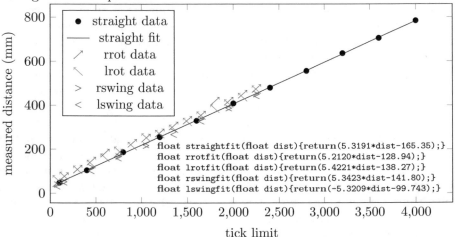

Then we ran some tests of navigating along paths as defined in Figs. 2(a–
d) using general procedures we wrote, incorporating delays as short as 100ms
at velocity discontinuities. We tested with (x, y) as $(100, 400)$, $(300, 600)$, and
$(600, 300)$. Almost all runs resulted in orientation and positions within a few
percent of the target, but $(100, 400)$ did not work very well for general naviga-

tion with rotations, because very small rotations were not generated reliably. In general, however, these results show promise for students to be able to use the techniques in this paper to achieve handy navigation to specified coordinates.

6 Conclusion

We have overviewed several ways in which students in the middle school to high school range can use algebra, data analysis, and trigonometry to compare motion planning strategies for a typical educational robot and program general routines to navigate by dead reckoning. Some qualitative observations are that routings considered in Sect. 4 are shorter and/or have fewer velocity discontinuities, but the horizontal and vertical routings considered in Sect. 3 are simpler to implement. Further, navigation is faster with swings than with rotations when routing horizontally and vertically, though rotations are generally a better basis when routing along more of a straight-line path. The results also show the expected result that navigation time generally increases with the track width.

A valuable next step would be to create worksheets directed towards students to help them work through the types of analyses in this paper. It would also will be nice to incorporate consideration of the case in which the orientation (angle) of the robot is to be changed in the final position relative to the start position.

References

1. Balkcom, D.J., Mason, M.T.: Time optimal trajectories for bounded velocity differential drive vehicles. Int. J. Robot. Res. **21**(3), 199–217 (2002)
2. Ben-Ari, M., Mondada, F.: Elements of Robotics. Springer, Cham (2018)
3. Chitsaz, H., La Valle, S.M., Balkcom, D.J., Mason, M.T.: Minimum wheel-rotation paths for differential-drive mobile robots. Int. J. Robot. Res. **28**(1), 66–80 (2009). https://doi.org/10.1177/0278364908096750
4. Greenberg, R.I.: Pythagorean approximations for LEGO: Merging educational robot construction with programming and data analysis. In: Proceedings of the 8th International Conference on Robotics in Education, RiE 2017. Advances in Intelligent Systems and Computing, vol. 630, pp. 65–76. Springer, Cham (2017)
5. McGee, S., Greenberg, R.I., Reed, D.F., Duck, J.: Evaluation of the IMPACTS computer science presentations. J. Comput. Teach. 26–40 (2013). International Society for Technology in Education. http://www.iste.org/resources/product?id=2853
6. Thiruvathukal, G.K., Greenberg, R.I., Garcia, D.: Understanding turning radius and driving in convex polygon paths in introductory robotics (2018). http://ecommons.luc.edu/cs_facpubs/202
7. Wang, H., Chen, Y., Souères, P.: A geometric algorithm to compute time-optimal trajectories for a bidirectional steered robot. IEEE Trans. Robot. **25**(2), 399–413 (2009)

Learning Symmetry with Tangible Robots

Wafa Johal[1,2(✉)], Sonia Andersen[1,2,3], Morgane Chevalier[2,3], Ayberk Ozgur[1],
Francesco Mondada[2], and Pierre Dillenbourg[1]

[1] CHILI Lab, EPFL, Lausanne, Switzerland
wafa.johal@epfl.ch
[2] MOBOTS, BIOROB Lab, EPFL, Lausanne, Switzerland
[3] HEP Vaud, Lausanne, Switzerland

Abstract. Robots bring a new potential for embodied learning in class-rooms. With our project, we aim to ease the task for teachers and to show the worth of tangible manipulation of robots in educational contexts. In this article, we present the design and the evaluation of two pedagogical activities prepared for a primary school teacher and targeting common misconceptions when learning reflective symmetry. The evaluation consisted of a comparison of remedial actions using haptic-enabled tangible robots with using regular geometrical tools in practical sessions. Sixteen 10 y.o. students participated in a between-subject experiment in a public school. We show that this training with the tangible robots helped the remediation of parallelism and perpendicularity related mistakes commonly made by students. Our findings also suggest that the haptic modality of interaction is well suited to promote children's abstraction of geometrical concepts from spatial representations.

Keywords: Robot for learning · Tangible · Haptic · Education · Child-robot interaction

1 Introduction

Children acquire knowledge in their everyday life and build mental representations through experiences, and the building of prior knowledge is particularly true for the domain of geometry. Didactisists in geometry differentiate *spatial knowledge*, linked with perceptible space, and *geometrical knowledge*, conceptualized and abstracted [1,2]. Hence, solving a geometrical problem goes beyond visual recognition of spatial relations between shapes or ability to use geometrical tools; it requires the learner to analyze these spatial relations and the properties of the shapes using mathematics and logic. In a way, geometrical knowledge provides a model of spatial knowledge. As pointed by Oberdorf et al. [3], practitioners in early childhood mathematics have observed that many children have numerous misconceptions in geometry. To avoid misconceptions when teaching about geometry, Laborde et al. [4] recommends: (1) the distinction between spacial graphical relations and geometrical relations, (2) the use of movement

© Springer Nature Switzerland AG 2020
M. Merdan et al. (Eds.): RiE 2019, AISC 1023, pp. 270–283, 2020.
https://doi.org/10.1007/978-3-030-26945-6_24

between spacial representations, (3) the recognition of geometrical relations and (4) finally the ability to imagine possible geometrical relation. In order to provide teaching tools to follow these recommendations, research in computer-assisted learning have investigated geometrical tools since decades. Several solutions have been proposed to use computer simulation to render geometrical transformations so that learners could understand properties of isometries [5,6]. Yet, computers are underused in geometry teaching, partially due to logistic issues. But, besides the logistical difficulties but also because several studies showed that tangible manipulations and manual interactions were important for development and can benefit learning [7,8].

Nowadays, robots are being used in broader contexts, among which of course is education. Recent research in Human-Robot Interaction reports multiple topics taught using robots [9–11]. In this article, we describe the design of two symmetry activities using tangible robots and a study that aimed to compare our activities to traditional practical activities in geometry. Our activities use the haptic-enabled Cellulo robots [12] that let the children feel geometrical shapes without seeing them on paper, providing embodied interaction with a simulation.

2   Related Work

2.1   Robots in Education

In education, the logo turtle entered schools nearly 40 years ago. However, if the field of interest is not new, the robots are. Indeed, they have changed a lot over time; sequentially or event-based programmable, they also integrate a wide spectrum of sensors and actuators in an increasingly minimal package. More recently, new robotic systems are designed and built to be brought into schools to teach subjects other than STEM in the curricula. Social robots (e.g. Nao and DragonBot) have been used in previous studies to teach languages [9], handwriting [13] and nutrition [11]. We also find new robotics tools that aim to be used broadly in schools. For instance, Ozgur et al. proposed bringing tangible robot interaction experience to learners [12,14].

In the present article, we propose to include teachers in a participatory design process and to specifically investigate the use of manipulation and haptic feedback that could benefit learners. In our approach, a school teacher was involved at every stage of the design of the learning tasks, starting from the design of concepts to the execution of the tests.

2.2   Tangibles and Haptics for Learning

The use of Tangible User Interfaces (TUIs) for education has been widely researched in order to foster learning [15–18]. However, these tangibles for the most part are passive, e.g. they are tokens moved manually by the user. The use of active tangibles in education have found recent developments with the Cellulo robots [12]. In their study [14], authors have started to explore the use of such active tangibles in a learning activity on wind. They specifically showed that certain types of haptic feedback (repelling forces) were less perceived than others (attracting forces). They also showed that the underlying graphics of a map of Europe could disturb the abstraction of concepts such as atmospheric pressure, leading children to focus on spatial references (e.g. cities, mountains) rather than what they were feeling with the robotic device. This leads us to understand that the use of haptics and graphics should be well balanced in the interaction design in order to allow the learner to build efficient mental representations.

An interesting dimension in learning with tangible devices is the use of shared resources to trigger collaboration among learners. It is established in the literature that collaboration is a lever of learning [19] which amounts to saying that one learns better with several collaborators, rather than being all alone. Collaborative Learning usually takes the form of an instruction in which students work in groups towards a common academic goal. Collaborative setups have also the advantage to make learners explicitly express their reasoning in order to engage with their peers.

Zacharia et al. gives a literature review of touch sensory feedback and its effect on learning through experimentation [20]. They reported several studies in which haptic feedback was used for learners (forces and fields, gears, biochemistry); showing positive relations between haptics and learning sciences. This review however concludes that touch sensory feedback can be beneficial in some cases for learners but under the condition of existing prior knowledge; i.e. having already encountered the notion before, the learners could then build a multimodal representation of the concept, constituting a more solid ground for future learning. Another conclusion of Zacharia et al. is that the touch sensory channel seems mostly relevant for abstract concepts.

In this paper, we present a participatory design with a primary school teacher for using haptic-enabled robots in classrooms to teach geometry. We present results from learning outcomes in using the robots compared to the classical practical sessions.

3   System Description

The system is composed of tangible and haptic-enabled Cellulo robots, paper posters used for localisation of the robots and an Android tablet application used to manage the activity logic and for the user to select the activities.

Cellulo (see Fig. 3) is a small tangible robot that was designed to be used in education [12]. It is 7.5 cm in diameter with a white mouse-like appearance. It has a three wheel omni-directional drive system (3DOFs) allowing back-drivability - the robot can move and can be moved without being damaged [21]. The on-board localization system uses a dotted pattern printed on paper [22] to calculate the absolute position and orientation of each robot with sub-millimeter accuracy. A QtQuick API allows to develop PC and tablet compatible activities featuring many robots. This application logs the robots' position and touch sensor data.

Cellulo allows various types of interactions. It can display RGB colors through its 6 top LEDs. 6 touch capacitive sensors allow to determine if the robot is grasped. The robot has also been used to render various types of haptic feedback [23], including to render direction and intensities of forces, a point, a line or a closed shape in a 2D space. As this tangible robot works only on dotted sheet of paper, the implementation uses extensively spatial zones in the 2D space to trigger events. One can for instance define a zone on paper that will change the LEDs color of the robot or make it vibrate (in an oscillating motion).

4   Design of the Activities

In order to test how tangible robots could be used by the teachers, we decided to co-design a practical session with a mathematics school teacher.

A first meeting session with the teacher aimed to present the capabilities of the robots. We showed a demonstration of the platform and then discussed what difficulties she might have in her class. One of the difficulties pointed out by the teacher was that children cannot easily abstract geometrical concepts from spatial perception [1,2]. Pens leaving a mark on paper when solving the symmetrical problem can lead children to think only in terms of spatial relations and not in terms of geometrical relations [4]. Children tend to stay at the perceptive level and this lack of abstraction to the geometrical concepts leads to misconceptions.

After several meetings discussing the abilities of the robots and potential needs of children in the teacher's classroom, we decided to focus on isometries; and in particular, on reflective symmetry (also known as orthogonal or axial symmetry). Several weeks of developments with a regular contact with the teacher enabled us to develop a suitable activity for the teacher.

A first pilot study was conducted with a classroom of twenty 8 y.o. children in Switzerland. The goal of this pilot was to test the understanding and usability of the system with robots by children, as well as to train the teacher to use the system. By the end of the pilot study, the teacher was able to run all activities by herself without the help of the researchers.

In practical geometry teaching, paper, pen and ruler are usually used. Children experience paper on the symmetrical axis and draw reflected objects accordingly to it. This drawing allows to keep the folding spatial reference of the parts of the object that is being transformed. However, axial or reflectional symmetries can be achieved through geometrical transformations (similarly to rotation and translation) but this transformation can be hard to notice or understand when the original shape remains in place.

Drawing on paper does not allow the rendering of the transformation as a process of isometries. Quite often, because of this limitation in the rendering properties of symmetries, children form misconceptions when taught symmetry. These misconceptions often only disappear much later on during the child's academic years [24]. Previous works in educational sciences presented that some of these misconceptions come from prior knowledge or are formed when manipulating paper for symmetry understanding [4,25].

In our work, we chose to focus on the following two misconceptions:
(1) Parallelism and orientation. This misconception comes from the fact that children tend to do a translation of the object when asked to perform an axial

symmetry. We can often observe that the child draws the symmetry with parallel edges to the original shape. As for an example of a mistake due to this misconception, the child would performed a translation instead of the symmetry. Paper with grid could enforce the parallelism misconception making the learner to follow the vertical or horizontal lines of the grid to draw the reflective shape [4].

(2) Perpendicularity of axis. This misconception is linked with the fact that children are often presented with vertical or horizontal axis of symmetry. Because of this, they tend to treat symmetries as left-right of up-down transformations and fail when the axis is oblique. An example of this type of mistake due to this misconception would be that the child treats the symmetry as if the axis was vertical.

4.1   Paper Posters

We prepared with the school teacher a poster sheet for each activity with the robots (800 × 600 mm) and several copies of the same sheets for the traditional tools (A3). The poster map in Fig. 1 is used in a first activity that aims to address the misconception 1 on parallelism and orientation. This activity is meant to focus on the reflective symmetry of a blue triangle with regard to a vertical axis. In addition to the vertical axis, we added a grid to help the learners to measure the distances from the vertices of the shape to the symmetry axis.

For the second misconception on perpendicularity, we designed a sheet for the reflective symmetry of a line segment (see Fig. 2). In the robot condition, the two vertices of the line segment are represented by two robots. The vertices of the symmetrical axis can be manipulated through two robots (the orange and blue robots).

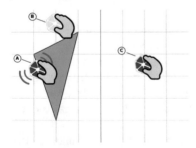

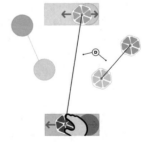

Fig. 1. Interaction design for Activity 1. (A): Border rendering as a vibration and LEDs lighted. (B): Rendering of empty space with LEDs white. (C): Rendering of inner of the symmetric shape with still LEDs blue color.

Fig. 2. Interaction design for Activity 2. (D) Black elastic string representing the segments.

4.2 Interaction Design

For the first activity, the border of the shape was rendered with haptic feedback (vibration) and the interior by lighting the robot in blue (see Fig. 1). Each pair of children would start by scanning the left side of the poster sheet on which the original shape was drawn. If the child was scanning outside of the shape the robot's LEDs would turn white. The reflective shape would similarly be rendered but only if the teacher activated this option through the tablet. The option of feeling the reflected shape was at first disabled, allowing time for children to make a hypothesis before they could check by scanning the sheet of paper.

Before starting the second activity, a 3D printed colored "hat" was attached to each robot to be able to differentiate them (see Fig. 2). These colored hats also had a nib with an elastic thread symbolizing the segment between two robot points. This way, the orange and blue robots were attached and symbolized the symmetry axis, and the green robots the shape to be reflected. Two areas printed on paper determined where the orange and the blue robot could be. By sliding the blue or orange robot in their respective areas, the symmetry segment symbolized by the green robots would adjust their position in real-time to keep the symmetry correct.

5 Research Hypothesis

In the following study, we propose to evaluate the influence of tangible manipulation of robots on correction of two types of common misconceptions in reflective symmetry compared to the use of classical geometric tools.

Practical sessions using paper to teach about orthogonal symmetry can lead to misconceptions [25, 26]. Without the drawing of the symmetry on paper, children were forced to build a mental representation of the geometry. We believe that this process will favor the acquisition of the skill and correct the misconceptions in the haptic conditions. We believe that pairs of learners that easily used the system would be able to better collaborate and perform the tasks in a more efficient way.

As the haptic modality provides the ability to render edges of a shape without imprint on paper, we believe that it would help children to abstract the geometric properties. We first hypothesize that our system is usable and that children can solve the task with the robot as they would with their regular geometry tools. Because of the above statements, we also expect more children in the experimental condition to correct their misconceptions.

6 Case Study

The experiment was conducted by a school teacher with her classroom following the five steps below:

Introduction: This first session with children aimed to clarify the vocabulary. It was done as a short intervention two days before the experiment. The teacher would show natural pictures of axial symmetry (a butterfly, a snowflake and the reflection of a tree on a lake). The teacher would ask children about particularities of the 3 images and ask to place one (or more) symmetry axes on the images. Then the teacher drew the axes and explain the terms "axial symmetry" and "axis of symmetry" to the whole class.

Pre-test: The pretest was taken by all the children in order to evaluate the presence of misconceptions and was taken with the regular pen and ruler tools.[1] Another school teacher of the same grade was in charge of the grading which consisted simply in denoting the presence or absence of the two misconceptions we focused on.

Pairing and Grouping: Children were then be paired according to their misconceptions. The goal was to have as many homogeneous pairs as possible with children sharing the same misconceptions. The pairs where then assigned to the control or experimental conditions. The control group would perform the same tasks but manipulating only pen, paper and ruler, whereas, the experimental group would use the robots on a printed map.

Problem Solving Activities: About one week after the pre-test, each pair had to solve two problems in a semi-guided practical activity (Sect. 6.2 describes the two activities). The teacher could intervene to ask the children to question their proposed solution. The interventions were scripted and were the same for the two conditions (with robots vs with paper).

Post-test: A post-test two weeks after the experiment concluded the experiment. The post-test material was a variation of the pretest (e.g. original image on the other side of the axis).

6.1   Participants and Apparatus

A classroom of 16 children of ages between 11 and 13 ($M = 12$, $SD = 1$), participated to a between subjects experiment in a school of Switzerland. The pre-test and post-test were individual but the practical sessions were performed in pairs (in order to force children to express their reasoning processes in solving the geometrical problem). We recorded the interaction with two video cameras.

6.2   Activities

A1: Finding the Symmetrical Figure. In this task, the figure used for the symmetry is a blue triangle. None of the edges of the triangle follow the grid underneath. The goal of this activity is for the children to hypothesize the position of the symmetrical figure and verify it through manipulation. This overall aim of this task is to correct the misconception 1 (parallelism and translation).

---

[1] We provide the material used in the experiment here: https://github.com/WafaJohal/Cellulo-Symmetry-Material.

Fig. 3. Activity 1 - Control condition. With pen and ruler as tools.

Fig. 4. Activity 1 - Experimental condition. Using a robot as a sensor with haptic feedback.

The control group (see Fig. 3) used regular tools such as a ruler and a pen. They drew the symmetry and then could check their hypothesis by folding the paper on the symmetry axis and placing the paper on the windows. Using the paper transparency, they could see if their hypothesis was correct or not. They could then refine their drawing and verify again until they were convinced that their answer was correct.

The experimental group used a robot (see Fig. 4). First, the children were invited to familiarize themselves with the interaction with the robot. The teacher asked them to grab a robot and to move it on the table. Children would then be asked to observe the behavior of the robot: When the robot was on the border of the geometrical shape, it would vibrate, giving a haptic rendering of the border; and when the robot would be inside the figure, the top LEDs of the robot would replicate the color of the geometrical figure (i.e. blue). The child would thus explore the blue triangle and be asked to notice the haptic and visual feedback. They would then be asked to hypothesize on the position of the symmetry by pointing at the potential place of the symmetrical triangle on the grid. In order to check their hypothesis, they would use the robot to feel the symmetric shape. As for the control group, they could refine their hypothesis freely until they were satisfied with their answer.

A2: Manipulation of the Orientation of the Symmetry Axis. In the second activity, the goal was to correct the misconception of children that the image of the axial symmetry is always either vertical or horizontal. In the activity, they were facing an oblique axis and had to estimate or observe the symmetry of a line segment. Children were asked to hypothesize on the symmetry and to check their answer by observing the symmetry. In the control condition, the children had 2 paper sheets, one with a vertical axis, one with an oblique axis. They had to draw the symmetric segment (Fig. 5a).

This implementation of the experimental condition made use of 4 robots. The robots were mounted with 3D printed colored tops that were used to visually group the robots into segments. Two robots (blue and orange) where linked with an elastic rope and acted as the vertices of the axis of symmetry. The two other

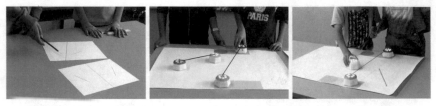

(a) Control condition. Two sheets of papers, with the same original shape being a segment. On one sheet the axis is vertical and on the other the axis is oblique

(b) Experimental condition without haptics. The blue and orange robots are linked with a rope and act as vertexes of the symmetrical axis. They can be moved within the blue and orange area respectively. The green robots act as vertexes of the symmetrical segment. As the axis changes orientation and move; they move accordingly to keep the symmetry correct.

(c) Bonus Activity Experimental condition with haptics. The blue and orange robots are linked with a rope and act as the vertexes of the symmetrical axis. They can be moved within the blue and orange area respectively. The third robot is used as an haptic device to feel the symmetrical segment.

Fig. 5. Setup of Activity 2 on manipulation of the axis from vertical to oblique orientation

robots, mounted with green tops, where also linked and displayed the position of the symmetric segment. The activity will start with the two green robots being in the original position of the segment, and the axis in a vertical position. The children were then asked to point at the position of the symmetrical segment. The teacher then announced that the symmetry would occur. The two green robots would then move to their symmetric positions. The children could then change the slope of the axis by moving the blue and orange robots like sliders within their area of action (respectively the blue and orange boxes).

7   Results

Both the control and the experimental groups took the same pretest and posttest. These tests aimed to measure the disappearance or persistence of misconceptions after manipulating either traditional geometry tools such as paper and ruler (control group) or the robots (experimental group).

Remediation of Misconception 1. The first misconception was concerning mistakes dealing with Parallelism and Orientation of the shape (i.e. doing a translation instead of a reflection). For the pretest, two pairs in the experimental

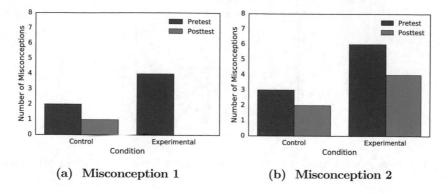

(a) **Misconception 1**   (b) **Misconception 2**

Fig. 6. Sum of misconceptions in each group for the pretest and post-test

condition and one pair in the control condition presented this misconception. After the practical session, the misconception was maintained for one member of the pair in control, and disappeared for all the students in the experimental condition. Even-though the remediation (number of disappearing misconceptions of Type 1) seems better in the robot condition, a Pearson Chi2 test shows that it is not significantly different compared to the traditional one (χ-*squared* = 1.56, *df* = 1, *p-value* = 0.21) (Fig. 6).

Remediation of Misconception 2. The misconception disappeared for some learners between the pretest and the post-test but several students in both control and experimental group still had the misconception after the practical activities. We could explain the persistence of the second misconceptions for certain children who associated spatial relations rather than geometrical abstractions. A Pearson Chi2 test did not show significant difference between traditional and robot activities for the correction of the second misconception (χ-*squared* = 1.1, *df* = 2, *p-value* = 0.58). These results tend to demonstrate that the robot activity is as effective as the teacher's traditional method to correct common misconceptions.

7.1   Assessment of Collaboration

In order to evaluate the quality of the interaction within the group, we proposed to the teacher who participated to the design to annotate video recordings of the practical sessions. We selected two stereotypical groups: one for which the activity seemed to work in correcting the misconceptions, one for which it didn't work as well. For these two groups we asked the teacher to comment the videos of interaction according to two dimensions: (1) nature of collaboration and (2) topic of the exchange. In the nature of collaboration we distinguished: conflicts (expressing disagreement) and arguments (arguing for a thesis). The topic of exchange could be either on the use of the robot ("you should mover it like that") or on the notion ("see, the axis is here") These assessments on collaboration for

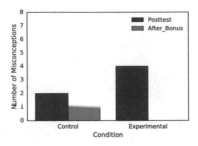

Fig. 7. Sum of misconceptions in each condition for the post-test and the after bonus test.

these two stereotypical pairs showed more arguments and conflicts in the pair that was able to repair successfully their misconception. These results are in line with the literature in collaborative learning stating that conflicts are positively correlated with learning gain in collaborative tasks. About the topics of intragroup communications, both groups were referring more to the notion than to the usage of the robotic tool. This argues in favor of our hypothesis claiming the ease of use of the device, even though it was novel for students.

7.2 Bonus Activity

As the haptic modality to render the shape seemed to give good results for the activity 1, we decided with the teacher to implement another instance of the activity 2 using haptic feedback. Children still had to manipulate the symmetry axis by moving the blue and orange robots (see Fig. 5c) and to observe its effect on the symmetric segment. But this time the symmetric segment was rendered with haptic and light feedbacks (similar to the blue triangle in the first activity). The pairs of children performed this bonus activity three weeks after the post-test and another test (a variation of the two first) was administrated two days after the bonus practical activity. The control group also took the bonus test but without any practical activity in between. Figure 7 shows that the bonus activity remediated for the misconception 2 for all students in the experimental condition. Concerning the control group, we observe that the retention was good. Only one student kept the misconception.

8 Discussion

The primary goal of this experiment was to evaluate our design in an ecological scenario involving the teacher. The results from the user study show that overall the robots led to similar solving of misconceptions compared to the traditional tools. Yet the haptic modality seemed to play an important role in the abstraction and solving of the tasks.

Children had high expectations after the teacher first mentioned an activity with robots. Some children felt that the robots were precious and fragile and hence were holding back from manipulating them freely. We believe that this novelty effect could disappear after some time and would let children adopt this new tool.

Another major contribution of our work is the co-design of ecologically tested tangible robots activities for geometry. The teacher was able to use the system alone in her classroom. Very few technical issues occurred during the experiment and most of them were due to Bluetooth connection (the teacher was able to fix and handle them without our intervention). The school teacher reported that she enjoyed this experience and proposed to continue our collaboration to develop new applications for other mathematical tasks.

Some limitations should be stated. First of all, we chose to keep the scenario identical for all the pairs of learner, starting with activity 1 then 2. The post-test were taken by the students only after the two activities and there could have been an order effect influencing the learning between activity. Further investigation with another experimental design and with more students could evaluate such effect. While the Bonus activity seems to repair the misunderstanding for all the students, it could have also been the case for a traditional practical session using the pen and ruler. Here again, our hypothesis concerning the impact of haptics could be tested by comparing haptic and non-haptic interactions in the same task, in a counterbalanced experimental design.

9   Conclusions and Future Works

This article's main contributions are two-fold: First, we present the participatory design and implementation of two learning activities using tangible haptic-enabled robotic devices. Our design process was driven by the teacher's constraints and needs in terms of misconceptions of her students and led to the implementation of a standalone tablet application with activities on reflective symmetry. Second, we present results from an ecological study that showed the feasibility for teachers to use robots in pedagogical scenarios. Children were able to experience various types of interaction with the same robotics device and within the same conceptual tasks; e.g. used as a scan device, tangible points. They contained very different sets of interaction affordances utilizing both active and passive robot behaviors, and were well received by the children. The fact that the symmetrical shape was not drawn on paper also forced children to develop mental representations. In this particular case, our results are in line with some previous works [27,28] showing that 2D shapes could be perceived with haptic feedback. However, further studies should be run to confirm these results that were obtained on a relatively small sample size.

Our future work will focus in studying the use of haptics in curricular activity in more depth, in order to better understand how this new learning modality with robots can help learners acquire abstract concepts. We also plan to investigate several configurations of collaborative setups (N robots and M children), in order to explore the effect of shared resources on collaboration. This investigation will aim to design smart behaviors for the robots to reallocate themselves to enhance collaboration among learners. For these future experiments, we plan to include more participants in our study to guarantee soundness of the results.

Acknowledgements. We would like to thank the Swiss National Science Foundation for supporting this project through the National Centre of Competence in Research Robotics.

References

1. Berthelot, R., Salin, M.-H.: L'enseignement de la géométrie à l'école primaire. Grand N **53**, 39–56 (1993)
2. Douaire, J., Emprin, F.: Teaching geometry to students (from five to eight years old). In: CERME 9-Ninth, pp. 529–535 (2015)
3. Oberdorf, C.D., Taylor-Cox, J.: Shape up!. Teach. Child. Math. **5**(6), 340 (1999)
4. Laborde, C., Kynigos, C., Hollebrands, K., Strässer, R.: Teaching and learning geometry with technology. In: Handbook of Research on the Psychology of Mathematics Education: Past, Present and Future, pp. 275–304 (2006)
5. Akkaya, A., Tatar, E., Kağızmanlı, T.B.: Using dynamic software in teaching of the symmetry in analytic geometry: the case of geogebra. Procedia-Soc. Behav. Sci. **15**, 2540–2544 (2011)
6. Laborde, C.: Integration of technology in the design of geometry tasks with cabri-geometry. Int. J. Comput. Math. Learn. **6**, 283–317 (2002)
7. Abrahamson, D.: Embodied design: constructing means for constructing meaning. Educ. Stud. Math. **70**(1), 27–47 (2009)
8. Smith, C.P., King, B., Hoyte, J.: Learning angles through movement: critical actions for developing understanding in an embodied activity. J. Math. Behav. **36**, 95–108 (2014)
9. Westlund, J.K., Gordon, G., Spaulding, S., Lee, J.J., Plummer, L., Martinez, M., Das, M., Breazeal, C.: Learning a second language with a socially assistive robot. In: Conference Proceedings New Friends (2015)
10. Yadollahi, E., Johal, W., Paiva, A., Dillenbourg, P.: When deictic gestures in a robot can harm child-robot collaboration. In: IDC 2017, pp. 195–206 (2018)
11. Short, E., Swift-Spong, K., Greczek, J., Ramachandran, A., Litoiu, A., Grigore, E.C., Feil-Seifer, D., Shuster, S., Lee, J.J., Huang, S., et al.: How to train your DragonBot: socially assistive robots for teaching children about nutrition through play. In: RO-MAN International Symposium, pp. 924–929 (2014)
12. Özgür, A., Lemaignan, S., Johal, W., Beltran, M., Briod, M., Pereyre, L., Mondada, F., Dillenbourg, P.: Cellulo: versatile handheld robots for education. In: International Conference HRI 2017, pp. 119–127 (2017)
13. Lemaignan, S., Jacq, A., Hood, D., Garcia, F., Paiva, A., Dillenbourg, P.: Learning by teaching a robot: the case of handwriting. IEEE Robot. Autom. Mag. **23**, 56–66 (2016)
14. Özgür, A., Johal, W., Mondada, F., Dillenbourg, P.: Windfield: learning wind meteorology with handheld haptic robots. In: International Conference HRI 2017, pp. 156–165 (2017)
15. Celentano, A., Dubois, E.: Metaphors, analogies, symbols: in search of naturalness in tangible user interfaces. In: Intelligent Human Computer Interaction, IHCI 2014, vol. 39, pp. 99–106 (2014)

16. Zuckerman, O., Arida, S., Resnick, M.: Extending tangible interfaces for education: digital Montessori-inspired manipulatives. In: Proceedings of CHI Conference, pp. 859–868 (2005)
17. Marshall, P.: Do tangible interfaces enhance learning? In: TEI 2007, pp. 163–170 (2007)
18. Cuendet, S., Bonnard, Q., Do-Lenh, S., Dillenbourg, P.: Designing augmented reality for the classroom. Comput. Educ. **68**, 557–569 (2013)
19. Dillenbourg, P.: What do you mean by collaborative learning? (1999)
20. Zacharia, Z.C.: Examining whether touch sensory feedback is necessary for science learning through experimentation: a literature review of two different lines of research across k-16. Educ. Res. Rev. **16**, 116–137 (2015)
21. Özgür, A., Johal, W., Dillenbourg, P.: Permanent magnet-assisted omnidirectional ball drive. In: IROS International Conference, pp. 1061–1066 (2016)
22. Hostettler, L., Özgür, A., Lemaignan, S., Dillenbourg, P., Mondada, F.: Real-time high-accuracy 2D localization with structured patterns. In: 2016 IEEE International Conference on Robotics and Automation (ICRA), pp. 4536–4543. IEEE (2016)
23. Özgür, A., Johal, W., Mondada, F., Dillenbourg, P.: Haptic-enabled handheld mobile robots: design and analysis. In: CHI Conference, pp. 2449–2461 (2017)
24. Yetiş, S., Ludwig, M.: Plane geometry: diagnostic and individual support of children through guided interviews-a preliminary study on the case of line symmetry and axial reflection. In: CERME 8th (2013)
25. Grenier, D.: Construction et étude du fonctionnement d'un processus d'enseignement sur la symétrie orthogonale en sixième. PhD thesis, Université Grenoble Alpes (1988)
26. Chesnais, A., Mathé, A.-C.: Articulation between students' and teacher's activity during sessions about line symmetry. In: CERME 9-Ninth, pp. 522–528 (2015)
27. Ballesteros, S., Millar, S., Reales, J.M.: Symmetry in haptic and in visual shape perception. Percept. Psychophys. **60**(3), 389–404 (1998)
28. Kalenine, S., Pinet, L., Gentaz, E.: The visual and visuo-haptic exploration of geometrical shapes increases their recognition in preschoolers. Int. J. Behav. Dev. **35**(1), 18–26 (2011)

Lessons from Delivering a STEM Workshop Using Educational Robots Given Language Limitations

Daniel Carrillo-Zapata[1,2,3], Chanelle Lee[1,2,3]([✉]),
Krishna Manaswi Digumarti[1,2,3], Sabine Hauert[1,3], and Corra Boushel[2]

[1] University of Bristol, Bristol, UK
{daniel.carrillozapata,c.l.lee,km.digumarti}@bristol.ac.uk
[2] University of the West of England, Bristol, UK
[3] Bristol Robotics Laboratory, Bristol, UK

Abstract. Educational robots are increasingly being used in schools as learning tools to support the development of skills such as computational thinking because of the growing number of technology-related jobs. Using robots as a tool inside the classroom has been proved to increase motivation, participation and inclination towards STEM subjects at both primary and secondary levels; however, language has usually not been considered as a mitigating factor. This paper reports our experience delivering nine workshops in English, using Thymio robots, to over two hundred students aged 9–12 across a week in the French cities of Nancy and Metz. Our goal was to test whether students would still have fun, learn something new and gain an interest in STEM even when the workshop was conducted in a foreign language. Our results indicate that using language that is easy to understand, although foreign, has a strong direct correlation ($p \sim 10^{-3}$) with having fun and that the latter positively affects learning and increased interest in STEM.

Keywords: Educational robots · Bio-inspired robots · Biomimicry · Foreign-language workshop · Thymio

1 Introduction

Coding is steadily gaining importance in school classrooms as a result of society's demand for skills related to Information and Communication Technologies (ICT) [1]. A report by the European Schoolnet [2] (a network of 34 European Ministries of Education) noted that countries such as Belgium, Estonia, France, Israel and Spain have been offering coding as an optional subject in primary education since 2015, while others include it in their mandatory curriculum, as is the case in Finland, Slovakia and state maintained schools in UK. As part of this, educational

---

D. Carrillo-Zapata, C. Lee, K. M. Digumarti — These authors contributed equally to the work.

M. Merdan et al. (Eds.): RiE 2019, AISC 1023, pp. 284–295, 2020.
https://doi.org/10.1007/978-3-030-26945-6_25

robots are increasingly being used in schools as learning tools to support the development of skills such as computational thinking [3]. Many believe the technology improves the learning experience of students [4,5] and is a useful aid for improving STEM interest at both primary and secondary levels [6].

Several educational robots with different functionalities, purposes and costs have been proposed and used. Some examples are LEGO Mindstorms [7], Edison [8], Boe-Bot and Scribbler [9], e-puck [10], Finch [11], mBot [12], Nao and Pepper robots [13], and Thymio [14]. Benitti [15] has found LEGO Mindstorms to be most commonly used in research studies; however, despite its impressive functionality, its high price point limits its suitability as an educational robot for use in schools [16]. Some robots are even more expensive, such as the e-pucks, whose target market is mainly higher education institutes, costing in excess of £600 per unit [14]. Others have been primarily designed for use as peers, tutors or socially assistive robots, e.g. Softbank Robotics' Pepper and Nao. Among the low-cost alternatives, Thymio (version II) has the widest range of sensors and actuation capabilities making it a good choice when the trade-off between functionality and cost is taken into consideration [14]. It has been demonstrated in practical workshops that Thymio robots appeal to children and adults of both genders, are a suitable tool for different activities and coding skill levels and make the user feel that they have learnt something new [17].

Educational robots have been proved to increase motivation, participation and attitude towards STEM subjects [18]; however, language has never been considered as a mitigating factor, i.e, workshops are often in the participants' native language. Do educational robots keep their motivational value even if the workshop is in a foreign language? To the best of the authors' knowledge, there is currently no work reporting the educational outcomes of a robotic workshop where participants are non-fluent in the language used.

This paper reports our experience delivering nine robotic workshops in English, using Thymio II robots, to over two hundred students aged 9–12 across a week in the French cities Nancy and Metz, under the umbrella of the Science in Schools British Council international outreach program. Our main goal was to test whether students would still have fun, learn something new and gain an interest in STEM subjects even when the workshop is delivered in a foreign language using educational robots, i.e. whether the benefits and gain of using educational robots would be language invariant. Moreover, does language and the ease of use of the robots contribute to these three outcomes? From a qualitative perspective, the other goals of the workshop were to help students practice a foreign language, introduce them to coding and robotics and expand their understanding of related research. Our approach follows the philosophy of the "Robots vs Animals" project which trained early career researchers in developing workshops on bio-inspired robotics aimed towards children [19].

2   Methods

Although educational robots are believed to have an impact on students' learning experience and attitudes towards STEM, Benitti notes in [15] that factors other than the use of robots also play an important role. Many important features found in successful robotic workshops, as identified in the literature and summarised by Benitti, were incorporated into our workshop design, with particular attention paid to the language limitations of the students.[1] These features and a small description of their implementation are given as follows:

- **Teachers play a big part in stimulating students** [20,21]. Three facilitators were in charge of delivering each workshop and assisting students, but most importantly, students were also assisted by their teachers.
- **Students should have a big space to play with the robots and explore different solutions** [21]. Workshops took place in the schools' computer rooms, which allowed the students to use the robot on either the tables or the floor, as seen in Fig. 1.
- **Each robot should be used by a maximum of two to three students** [21]. In most workshops, students shared one Thymio between two. In a few cases, it was necessary for a single robot to be shared by three.
- **The workshop should address content relevant to the students** [21]. As the workshop focused on bio-mimicry, the content was related to ecology and animals' behaviour, relevant topics in their science subjects.
- **Students perform better if they are familiarised with the robots before facing problem-solving activities with them** [22]. School teachers were informed a few weeks in advance of the type of robots used in the workshop. Most of them introduced the robots to the students or had used them in their science classes prior to the workshops. In addition, students were tasked with exploring the pre-programmed behaviours of the Thymios before attempting the coding challenges of the workshop.
- **Middle-school students should be guided to help them make the link between the robotic activities and the science/engineering behind** [23]. For this reason, our workshop was split in to three different sections: (1) Presentation, where we presented our research work on bio-inspired robotics and how bio-mimicry works; (2) robot interaction, where students faced coding challenges to mimic animal behaviour, and (3) questions and answers (Q&A) session, where students had the chance to ask the facilitators questions about their lives and research.
- **High-level activities as opposed to low-level ones are believed to be more successful** [24]. Coding challenges were kept simple and high-level through the use of animal behaviour examples. For example, rather than asking the students to use conditional statements, we instead asked them to consider how a chameleon changes colour in response to different stimulus and whether they could replicate this on the robots.

In the next subsections, a more detailed description of the workshop and materials is given.

---

[1] Material used in the workshops including the presentations, worksheets, questionnaire participant responses are accessible at
https://caidin.brl.ac.uk/k2-digumarti/data-for-rie2019-carrillo-zapata.

Fig. 1. Example of one session.

2.1 Thymio Specifications

The Thymio robot [17] was developed as a low-cost, robust and open hardware educational tool with the purpose of introducing children to software and robotics. The multitude of sensors on the robot (distance, touch and audio) and the actions that it can perform (move, change colour and play sounds) make it a suitable choice to demonstrate several bio-mimetic behaviours (Fig. 2). These robots can be programmed and interacted with, in a desktop setting next to a computer or as part of larger group where they can perform collective behaviours. The Thymios were chosen as they have been proved to be well-suited for educational activities and come with a suite of pre-programmed behaviours (friendly, explorer, fearful, investigator, obedient, and attentive [25]) that significantly reduced our preparation workload. This also means that the workshop is easily reproducible on other Thymios, even with little to no coding expertise. In addition, these robots have been deliberately designed to appeal to children of all genders across diverse age groups, to be easy and quick to use and to promote creativity [14].

2.2 Description of Workshop

The workshop was repeated over a period of five days with a maximum of two sessions per day, with a total of 219 primary and secondary school students taking part. Their demographics are presented in Table 1. The workshop was delivered entirely in English, with occasional translation into French by teachers at the schools. Care was taken to avoid scientific jargon and ideas were conveyed pictorially, as this is suggested as an effective way to communicate to children [27]. Each session lasted for three hours (including approximately 30 min in breaks) and comprised of three segments (Table 2). First, the students were introduced to the concept of bio-inspired design of robots through numerous examples presented using both textual and visual media. This was followed by a short activity demonstrating the emergence of order from chaos,

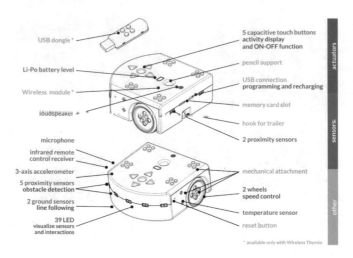

Fig. 2. Thymio hardware specifications [26]

Table 1. Participant demographics from each of the nine sessions in five different locations. A1 and B1 refer to language proficiency as defined by the common European framework of reference for languages (CEFRL).

Geographical location	Bar-le-Duc		Metz		Uxegney	Saint Mihel		Nancy	
Session	1	2	3	4	5	6	7	8	9
Age-group	10–11	10	10–11	10	9–10	11–12	10	10–11	10
No. of students	26	27	22	22	19	23	20	30	30
English proficiency	A1	A1	A1	A1	A1	A1	A1	B1	B1

a common concept in bio-inspired swarms [28]. Participants were challenged to synchronise their claps, starting from random clapping speeds. Eventually, they were expected to self-organise into a single clap. This particular activity—inspired by swarms of fireflies synchronising theirs flashes—was chosen because it requires no additional resources and presents an interesting and observable natural behaviour. Any other activity with similar results could be chosen.

In the second segment of the workshop, participants interacted with the robots. For this segment, two worksheets were handed out. The first one helped students familiarise themselves with the robots through a task (activity 2 in Table 2) aimed at identifying the built-in behaviours of the robots (Sect. 2.1). Each behaviour is uniquely associated with a colour. The activity required the participants to play with the robots, observe them in each colour mode to discover their behaviours and fill in the worksheet. The worksheet consisted of a matching task where the students had to match each colour with the description of the behaviour and the corresponding adjective (or name) in English.

The second worksheet presented the programming challenges. After being given an introduction to the Visual Programming Language (VPL) and a

Table 2. Outline of the workshop.

Introduction (45 min)	Researchers and workshop introduction
	Introducing bio-inspired robotics with examples
	Activity 1 - Synchronised clapping - Emergence of order from chaos
Interaction with the robots (1h 30min)	Introduction to Thymios
	Activity 2 - What behaviours can you see?
	Introduction to Aseba visual programming interface
	Challenge 1 - Chameleon colours
	Challenge 2 - Fast and slow
	Challenge 3 - What other animals can you copy? (Open-ended)
Q&A (45min)	Q&A with the researchers

Table 3. Programming challenges for participants to implement on the Thymio robots using the Aseba Visual Programming Language.

Challenge 1:	Chameleons can change their colour to match their environment. It helps them to hide from predators. They also change their colours to communicate with other chameleons.
Tasks:	Can you program your robot to change colours when you touch different buttons?
	Can you use this to communicate with another robot in the room?
Challenge 2:	Tortoises move slowly to save energy. Cheetahs move very fast to catch prey.
Tasks:	Can you program your robot to be slow like a tortoise?
	Now, can you program your robot to move fast like a cheetah?
Challenge 3:	Natural inspiration
Task:	What other animals can you program your robot to copy?

reference sheet, students had to solve three programming challenges (Table 3) to put the concept of bio-mimicry into practice, as well as learning how to program the robots using VPL. For the first two challenges, a short informative sentence about the animal that is being used as inspiration for a behaviour was presented in English followed by a set of programming tasks to mimic this behaviour on the robots, as shown in Table 3. The third challenge was open-ended but confined to the bio-mimicry aspect of the workshop and was included with the varying programming abilities among participants in mind. Those new to programming could apply skills learnt from the previous challenges whereas more experienced students could apply their knowledge to the context of bio-mimicry.

Table 4. The five categorical (Q1–Q5) and multiple-choice questions (MCQ) asked on the questionnaire.

Q1:	Was the workshop fun?
Q2:	Was the language easy to understand?
Q3:	Were the robots easy to use?
Q4:	Did you learn anything new today?
Q5:	Are you now more interested in Science, Technology, Engineering and Mathematics?
MCQ:	Which activity was most fun?

Finally, the third segment of the workshop, a Q&A session, allowed students to ask the facilitators any questions they had relating to the workshop, robotics or a research career.

2.3 Questionnaire

At the end of the workshop, participants were handed out a questionnaire to assess their opinion of the activities. It consisted of five categorical questions and one multiple-choice question (Table 4). This questionnaire was designed to assess the outcome of the workshop in terms of participants having fun, learning something new and being more interested in STEM subjects, even if delivered in a foreign language. To overcome the language barrier, the questions included translations in French provided by the teachers at the schools. Additionally, the words in English were deliberately chosen from common parlance and with an unambiguous equivalent in French. Each of these questions had five options (from 1 to 5) to express their level of satisfaction, ranging from extremely negative to extremely positive. Images with facial expressions (emoji) with same colour were used instead of words to label the options. In addition to the categorical questions, there was a single multiple-choice question which asked participants to select the parts of the activity that were the most fun.

3 Results

A statistical analysis of the responses to the questionnaire was performed to study correlations between answers to questions and to understand which part was more engaging. Analysis of the data was performed using MATLAB and R software. A histogram of responses was calculated to understand the spread of responses. This is shown in Fig. 3. The mean response was then calculated for each question, found to be 4.78, 3.61, 4.59, 4.57 and 4.45 respectively. We then performed a chi-squared test of independence with one degree of freedom between each pair of categorical questions to test for statistical correlation between the responses to the questions. Such responses were split into binary groups of YES and NO to ensure that each group had sufficient number of respondents to be

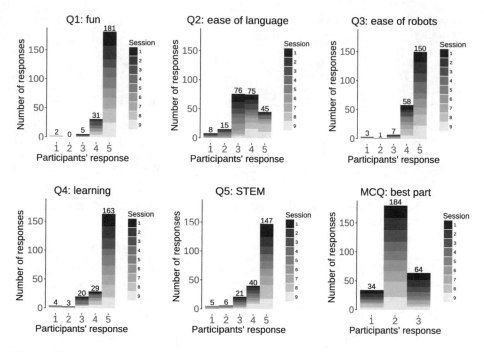

Fig. 3. Histogram of responses from all 9 sessions of the activity to questions 1–5 and multiple-choice question. (See Table 4 for the questions.)

amenable to a chi-squared test. The splitting of participants into groups was based on the mean response to avoid using a hard-coded threshold. In practice, this meant that every answer above the mean for such question was categorised as YES, and as NO for the opposite. The contingency tables for each pair of questions are shown in Table 5.

The null hypothesis in each test was that responses on the pair of questions were independent, while the alternative hypothesis was that there was a relation between responses to the questions. The corresponding p-values resulting from the chi-squared test are shown in Table 6a. Finally, the odds ratio was used as a statistical measure to quantify the strength and direction of association between every pair of questions. The corresponding values are shown in Table 6b.

4 Discussion

Results from the p-values from Table 6a strongly support that participants having fun is the most important aspect of a workshop if the desired outcome is for them to learn something new and become more interested in STEM, as also stated in other works [18]. This is also supported by the higher values of odds ratios (in particular higher than one) for comparison between these responses, indicating a strong positive association. This implies that students who had fun were also more likely to report that they had learnt something new and were

Table 5. Contingency tables between each pair of questions. (See Table 4 for the questions.)

	Q2			Q3			Q4			Q5			Q3	
	No	Yes		No	Yes		No	Yes		No	Yes		No	Yes
Q1 No	26	12	Q1 No	19	19	Q1 No	20	18	Q1 No	24	14	Q2 No	38	61
Q1 Yes	73	108	Q1 Yes	50	131	Q1 Yes	36	145	Q1 Yes	48	133	Q2 Yes	31	89

	Q4			Q5			Q4			Q5			Q5	
	No	Yes		No	Yes		No	Yes		No	Yes		No	Yes
Q2 No	30	69	Q2 No	42	57	Q3 No	18	51	Q3 No	29	40	Q4 No	27	29
Q2 Yes	26	94	Q2 Yes	30	90	Q3 Yes	38	112	Q3 Yes	43	107	Q4 Yes	45	118

Table 6. Results of the chi-square test of independence for each pair of questions. (See Table 4 for the questions.)

	Q2	Q3	Q4	Q5
Q1	2.8×10^{-3}	1.2×10^{-2}	6.3×10^{-5}	2.9×10^{-5}
Q2	-	6.5×10^{-2}	1.9×10^{-1}	9.7×10^{-3}
Q3	-	-	1	7.2×10^{-2}
Q4	-	-	-	7.6×10^{-3}

a) p-values

	Q2	Q3	Q4	Q5
Q1	3.21	2.62	4.48	4.75
Q2	-	1.79	1.57	2.21
Q3	-	-	1.04	1.80
Q4	-	-	-	2.44

b) Odds ratios

more interested in STEM. Conversely, the ones who gave a negative response to fun were more likely to give a negative one to the other two.

The chi-square results also strongly suggest that easily understandable language has a direct correlation with having fun and therefore, it is an important consideration when designing a workshop, especially if there are limited language capabilities. Although students were not fluent in English, the first histogram of Fig. 3 shows that most of them reported having fun, implying that our methodology was successful over the nine workshops. The role of the teachers was also very important in overcoming language barriers, especially when helping students understand a difficult concept, e.g., the difference between the verbs describing the pre-programmed behaviours "explore" and "investigate", which happen to be very similar in French. We found that is it important to constantly check whether students understood an important concept/task, and ask them to explain it to their classmates. This was particularly useful in cases of students with mixed English language abilities, as often the more able students would explain things to their peers using English words they were more familiar with. Furthermore, we found that the language barrier could be significantly lessened by using other means of explaining concepts, e.g., visually using images or physical movements. Thymios and the VPL were particularly good for this, due to their use of symbols and being able to physically demonstrate the robot.

Although the interaction with robots was chosen by most students to be the best part of the workshop (last histogram in Fig. 3), we believe the whole experience is important for success, as described at the beginning of Sect. 2. For example, the Q&A section was also highly rated, and we feel that giving the students the opportunity to freely question the experts facilitating the workshop helps them to link their robotic experience with the bigger science/engineering picture, as noted by Nugent et al. [23]. It is important to leave enough time for wider questions beyond the scope of the workshop. Based on our experience, the first few questions might be about concepts covered in the workshop, but they later can begin to start asking deeper questions, e.g. applications of robotics, possible robots that they could create or about our research and ourselves.

Finally, the survey carried out by Riedo et al. [17] with Thymios robots suggests that the accessible design of the robot made participants feel they had fun; however, surprisingly we found insufficient evidence to support an association between ease of use of the robots and either gaining interested towards STEM or having fun (due to a total of ten multiple comparisons, p-value for an overall 5% probability of error should be 5×10^{-3} applying Bonferroni correction). This could be attributed to the students having little to no experience with any type of robot and thus having no baseline to measure the ease of use against. A possible follow up study with a different robot and/or a control group might offer more concrete conclusions.

5   Conclusion and Future Work

This paper shares the results from a bio-inspired robotics workshop in English using the educational robot Thymio for a total of 219 students from French schools over nine sessions. The novelty of our work is that it investigated whether the fact that workshops were delivered in a language in which students were beginners affected their perception of having fun, learning something new and being more interested in science, technology, engineering or maths (STEM) subjects, as well as whether the ease of use of robots had a correlation with those. Our study suggests that there is a strong, direct correlation between having fun in the workshop and both learning something new and becoming more interested in STEM, as stated in the literature of educational robotics. Most importantly, our study also suggests that there is a strong, direct correlation between the language being easily understood and the ability to have fun, and therefore learning and engaging in STEM subjects.

The main contribution of this work is to demonstrate that a fun, highly-interactive and motivating robotics workshop in a foreign language can still be successful even if language is somewhat of a barrier for users, provided that factors identified in the literature such as teachers' participation, low ratio of students per robot, large room space, relevance of the topic, pre-familiarisation with the robots, guidance during the process and high-level activities are taken into account. We believe our workshop uses those features to be motivating, despite the students' lack of fluency in the delivered language. In particular, the role of teachers before, during and possibly after the workshop is crucial to overcoming language limitations. Moreover, the advantage of our methodology is that it can be easily adapted to different ages/expertise by changing the scientific depth of explanations (talking more/less about the scientific and technical concepts) or how challenging the coding tasks are, while ensuring the general structure is maintained. It could also be used as an introduction to robotics in the first lessons of a technology class, or as a tool to explain another topic apart from biology in a science class. If used this way, training for teachers would be key for robots acceptance and subsequent success [15]. Moreover, it could be used as an integration and teaching tool in environments where language is a barrier, e.g., a coding club for refugees.

Finally, we have identified that having a questions and answers session with the experts facilitating the workshop supports overall student satisfaction. We firmly believe that educational robots are an excellent tool to improve the learning journey, but other aspects, such as having a human conversation, also play a big role in making the experience more complete.

Acknowledgement. DCZ, CL and KMD would like to thank Mireia Bes-Garcia, Chloe Anderson, Ellie Cripps, Katie Winkle, Carole Hemard, Irene Daumur, Elisabeth Pirlot and all the schools staff involved for their help during the project. This work was supported by the EPSRC Centre for Doctoral Training in Future Autonomous and Robotic Systems (FARSCOPE, grant EP/L015293/1) at the Bristol Robotics Laboratory where DCZ, CL and KMD are PhD students. Financial help was provided by the University of Bristol Public Engagement Seed Fund. The activity was conducted for the Science in Schools program of the Bristish Council in France and the rectorat de l'académie de Nancy-Metz. The authors would like to disclose that they received a discount on the purchase of Thymios from Mobysa.

References

1. Coding - the 21st century skill. https://ec.europa.eu/digital-single-market/coding-21st-century-skill. Accessed 13 Dec 2018
2. Balanskat, A., Engelhardt, K.: Computing Our Future: Computer Programming and Coding-Priorities, School Curricula and Initiatives Across Europe. European Schoolnet (2015)
3. Mubin, O., Stevens, C.J., Shahid, S., Al Mahmud, A., Dong, J.-J.: A review of the applicability of robots in education. J. Technol. Educ. Learn. **1**(209–0015), 13 (2013)
4. Johnson, J.: Children, robotics, and education. Artif. Life Robot. **7**(1–2), 16–21 (2003)
5. Papert, S.: The children's machine: rethinking school in the age of the computer. In: ERIC (1993)
6. Grover, S.: Robotics and engineering for middle and high school students to develop computational thinking. In: Annual Meeting of the American Educational Research Association, New Orleans, LA (2011)
7. Lego Mindstorms robot. www.lego.com/en-gb/mindstorms. Accessed 14 Dec 2018
8. Edison robot. https://meetedison.com/. Accessed 14 Dec 2018
9. Boe-bot and scribble 3 robots. https://www.parallax.com. Accessed 14 Dec 2018
10. Mondada, F., Bonani, M., Raemy, X., Pugh, J., Cianci, C., Klaptocz, A., Magnenat, S., Zufferey, J.-C., Floreano, D., Martinoli, A.: The e-puck, a robot designed for education in engineering. In: Proceedings of the 9th Conference on Autonomous Robot Systems and Competitions, vol. 1, pp. 59–65. IPCB: Instituto Politécnico de Castelo Branco (2009). http://www.e-puck.org/. Accessed 14 Dec 2018

11. Lauwers, T., Nourbakhsh, I.: Designing the finch: creating a robot aligned to computer science concepts. In: AAAI Symposium on Educational Advances in Artificial Intelligence, vol. 88 (2010). http://www.finchrobot.com/. Accessed 14 Dec 2018
12. mBot robot. https://www.makeblock.com/steam-kits/mbot. Accessed 14 Dec 2018
13. Nao and pepper robots. https://www.softbankrobotics.com/. Accessed 14 Dec 2018
14. Mondada, F., Bonani, M., Riedo, F., Briod, M., Pereyre, L., Rétornaz, P., Magnenat, S.: The Thymio open-source hardware robot. IEEE Robot. Autom. Mag. **1070**(9932/17), 2 (2017)
15. Benitti, F.B.V.: Exploring the educational potential of robotics in schools: a systematic review. Comput. Educ. **58**(3), 978–988 (2012)
16. Kradolfer, S., Dubois, S., Riedo, F., Mondada, F., Fassa, F.: A sociological contribution to understanding the use of robots in schools: the Thymio robot. In: International Conference on Social Robotics, pp. 217–228. Springer (2014)
17. Riedo, F., Chevalier, M., Magnenat, S., Mondada, F.: Thymio II, a robot that grows wiser with children. In: 2013 IEEE Workshop on Advanced Robotics and Its Social Impacts (ARSO), pp. 187–193. Eidgenössische Technische Hochschule Zürich, Autonomous System Lab (2013)
18. Karim, M.E., Lemaignan, S., Mondada, F.: A review: can robots reshape K-12 stem education? In: 2015 IEEE International Workshop on Advanced Robotics and Its Social Impacts (ARSO), pp. 1–8. IEEE (2015)
19. Fogg-Rogers, L., Sardo, M., Boushel, C.: Robots vs animals: establishing a culture of public engagement and female role modeling in engineering higher education. Sci. Commun. **39**(2), 195–220 (2017)
20. Hussain, S., Lindh, J., Shukur, G.: The effect of lego training on pupils' school performance in mathematics, problem solving ability and attitude: Swedish data. J. Educ. Technol. Soc. **9**(3), 182–194 (2006)
21. Lindh, J., Holgersson, T.: Does lego training stimulate pupils' ability to solve logical problems? Comput. Educ. **49**(4), 1097–1111 (2007)
22. Williams, D.C., Ma, Y., Prejean, L., Ford, M.J., Lai, G.: Acquisition of physics content knowledge and scientific inquiry skills in a robotics summer camp. J. Res. Technol. Educ. **40**(2), 201–216 (2007)
23. Nugent, G., Barker, B., Grandgenett, N.: The effect of 4-H robotics and geospatial technologies on science, technology, engineering, and mathematics learning and attitudes. In: EdMedia: World Conference on Educational Media and Technology, pp. 447–452. AACE (2008)
24. Mitnik, R., Nussbaum, M., Soto, A.: An autonomous educational mobile robot mediator. Auton. Robots **25**(4), 367–382 (2008)
25. Getting started: begin with Thymio: pre-programmed behaviours. https://www.thymio.org/en:thymiostarting. Accessed 03 Dec 2018
26. Thymio specifications: what is Thymio composed of. CC BY-SA 3.0. https://www.thymio.org/en:thymiospecifications. Accessed 03 Dec 2018
27. Magnenat, S., Riedo, F., Bonani, M., Mondada, F.: A programming workshop using the robot "Thymio II": the effect on the understanding by children. In: 2012 IEEE Workshop on Advanced Robotics and Its Social Impacts (ARSO), pp. 24–29. IEEE (2012)
28. Şahin, E.: Swarm robotics: from sources of inspiration to domains of application. In: International Workshop on Swarm Robotics, pp. 10–20. Springer (2004)

Using Robots as an Educational Tool in Native Language Lesson

Ariana Milašinčić, Bruna Anđelić, Liljana Pushkar,
and Ana Sović Krzić[✉]

Faculty of Electrical Engineering and Computing, University of Zagreb,
Zagreb, Croatia
{liljana.pushkar,ana.sovic.krzic}@fer.hr

Abstract. In this paper, a pilot study of using robots in teaching native language is presented. The goal of the study is to find activities with educational robots that can be used in everyday classes. Robots were used as a tool to check the understanding of the read stories in two activities: as actors in a role play and as vehicles in a race. The technical performance analysis and the questionnaire showed that the robot race can be used in everyday classes since: (a) it checks the knowledge about the stories, (b) promotes team work and team spirit, (c) the activity is short with no technical problems, (d) it could be performed by one teacher, and (e) all pupils gave positive grades and comments about the race.

Keywords: Lego mindstorms EV3 · Book reports · Robot race · Educational robotics · Robot role play

1 Introduction

Technology and its fast-paced development changed society, and as society changes it calls for its key elements to adapt to this change. In recent years, the focus has been on educational robotics as a springboard to the STEM fields. Benitti in his systematic review found that 80% of the non-robotic fields, where robots were used as educational tools, pertained to mathematics and physics [1]. A research of case studies from three different science teachers, showed that using robotics in science classes where students had limited English proficiency, helped pique interest in the studied field, and prompted students' development of English language vocabulary [2]. Also, educational robots are used to teach a second language. Most of the attention in recent research has been on using humanoid robots as tutors [3]. Different learning scenarios where showed in [4] for primary school second language lessons, where it was shown that educational robots as tools showed advantages over other instructional tools. In our previous work we studied how to teach the basic university-level programming concepts to the first graders [5], how to use Lego Mindstorms robots in science education [6], how to foster creativity [7] and how to promote creative thinking and problem solving [8]. However, there is almost no research about using robots in learning native language. In this paper, we describe our pilot study for using robots during a book report lesson in Croatian language class. The goal of the study is to find activities with educational robots that can be used in everyday lessons in teaching native language.

© Springer Nature Switzerland AG 2020
M. Merdan et al. (Eds.): RiE 2019, AISC 1023, pp. 296–301, 2020.
https://doi.org/10.1007/978-3-030-26945-6_26

2 Methodology

In this study, the LEGO Mindstorms EV3 robots are used in the third grade of primary school for a lesson in Croatian language. 23 pupils participated in the class.

The lesson was focused on two stories from the author Božidar Prosenjak: "How Marko bought happiness" (in Croatian "Kako je Marko kupio sreću") and "In bed by nine" (in Croatian "U devet u krevet") [9]. The pupils had an assignment to read the stories at home before the class, and the goals of the lesson were to detect did the pupils understand the stories and help them to understand them correctly. The lesson plan was as follows: (1) Introduction; (2) Pre-test; (3) Role play activity; (4) Robot race activity; (5) Post-test; and (6) Questionnaire.

The pre- and post-tests checked the students' knowledge in a traditional way using paper and pencil. Three yes/no questions about each story were asked. The questions differed slightly in the two tests (Fig. 1). At the end of the lesson, the pupils answered a questionnaire about their satisfaction using the robots in the class.

The robots and the programs for the activities were prepared in advance. Each robot was a vehicle with an appearance of a cartoon character. Two big motors and an ultrasonic or infrared sensor were used for each robot (Fig. 2). For the role play activity three robots were connected using the *Daisy Chain Mode*, and for the robot race activity the robots were controlled using infrared remote controls.

The first robot activity was a role play: robots' movement was programmed in advance, and the students had to record the voices of the robots. Small scripts about both stories were prepared beforehand. The pupils were split into six groups and each group had an adult assistant. After reading practice (Fig. 3), students recorded their roles, and watched all role plays played by the robots.

The second activity was a robot race, i.e. robot-quiz: pupils answered yes/no questions using a remote control (Fig. 4.). The class was divided in 4 groups with 5 or 6 pupils and an assistant. The teacher read three test questions, and then 18 about the stories (at least three per student). As a demonstration, the first test question was answered by the assistants. After them, the students answered the two remaining test questions as a trial attempt. To avoid confusion or pressing of wrong buttons, all left buttons were for the 'Yes' answer, and all right buttons for the 'No' answer. After a button was pressed, the robot moved forward if the answer was correct, and backward if it was wrong. After the movement, the remote control was blocked for 15 s, so the teacher had time to read the next question and the students had time to swap places.

Fig. 1. Pre-test and post-test.

Fig. 2. Robots.

Fig. 3. Recording voices for the robots with help of the assistant.

Fig. 4. Robot race by answering questions about the stories.

3   Results

Analysis of the lesson can be divided into two parts: (1) technical performance, and (2) data collected from pupils. Both parts helped us to determine which activities can be used in everyday classes.

For the role-play activity, most of the pupils were very excited that they could show off their acting skills and that the robots would have their voices. However, recording the voices required a silent surrounding which was almost impossible to achieve in the classroom while the other groups were still practicing the reading. Therefore, the recording was moved to the hallway. Finding a quiet place prolonged the activity and the pupils were getting bored. Another problem occurred with the *Daisy Chain Mode* of the robots: motors on the connected robots were occasionally blocked, so it was not possible to execute the preprogrammed action. Repairing the hardware of the robots further prolonged the lesson. Lastly, the recorded voices were not loud enough, so some voices were hard to hear or understand even though the pupils were very quiet. Based on the technical performance, this type of activity is not appropriate for the classroom because (a) it requires a big number of assistants and (b) ideal conditions for the recording, (c) possible technical problems with robots can occurred, and (d) the activity needs a long-time frame.

The robot race activity is an alternative to the review at the end of the lesson that lasts approximately 10 min, including the preparation for the race. In addition to testing the knowledge about the stories, the activity promotes team work and team spirit through sharing one remote control. The students cheered for one another and encouraged the pupils that gave wrong answers. At the end, almost all questions were answered correctly. Although, we had an assistant for every group, one teacher could manage it by her/himself: she/he can explain the usage of the remote control to all the pupils in the same time, and she/he can control the pupils and the development of the race during the pauses between the questions. In this activity we did not have any technical problems with the robots. Therefore, based on the technical performance, this activity is highly appropriate for implementation in everyday classes.

The data were collected in three parts during the class: pre-test, post-test and questionnaire. All 23 pupils were tested individually, but one was excluded from the analysis because he did not finish the post-test. The questions in pre- and post-test were of two complexity levels: (a) answers could be derived directly from the test, and (b) answers required logical reasoning. The results the students achieved did not show a big difference in total scoring (improvement 8 pupils, deterioration 7 pupils, no improvement or deterioration 5 pupils, and 2 pupils with 100% in both tests), thus we cannot conclude the class impacted their knowledge substantially. Deeper analysis showed that all questions that were explicitly from the stories were improved in the post-test, while questions with logical reasoning kept the similar number of correct answers (Table 1). Results could be improved by upgrading the robot-activities.

Table 1. Comparison of results for each question in pre- and post-test. One sentence is given if it was the same in both tests, and a slash separates differences in sentences between the tests. Percentages show correct answers.

	Pre-test	Post-test
In bed by nine		
In this story, the mum jokes with the daughter Comment: Mum was not the only one who joked with the daughter. In our robot-play activity we omitted the mum as a role to make the play shorter	32%	27%
At the end of the story the girl turned off the computer/the TV	95%	100%
The girl reacted instantly to the jokes of the adults Comment: Grandfather, grandmother and mum joked with the daughter through the whole story, and the daughter reacted at the end	55%	46%
How Marko bought happiness		
In the beginning of the story Marko is scared/sad	86%	100%
Marko started playing the guitar/going to football practices Comment: Different pupils gave wrong answers	91%	86%
Marko's friend is called Jakov	100%	100%

The questionnaire consisted of two parts: (1) choosing one emoji face (happy/like, indifferent/neutral or sad/dislike) that the pupils associated with the activity, and (2) three open-ended questions which required writing their opinion. Questions were related to all activities (Table 2). Results show that the overall level of satisfaction is high, while the activity that showed the lowest satisfaction level was practice reading. This activity was, at the same time, the only *traditional* activity during the lesson. Table 3 shows some of the most common, interesting and useful answers to open-ended questions. Those answers can be helpful for finding the activities that can be used in everyday classes. Based on the pupils' answers to all questions in the questionnaire, the robot race was the most liked part of the class which is consistent with the previous technical performance analysis. This result is especially fascinating because this activity required thinking about the read stories, and it is used as a recapitulation of everything learned, both in class and home.

Finally, we asked the teacher for her comment. She expressed her overall satisfaction and approval of this type of robot implementation, claiming that she clearly sees the benefits the pupils have.

Table 2. Pupils' results of the first part of the questionnaire: choosing one emoji face (happy/like, indifferent/neutral or sad/dislike) that is associated with each activity.

How did you like	Like	Neutral	Dislike
The homework – reading of the story	19	3	0
The practice reading of the story	15	7	0
Recording the story	21	0	1
The role play with the robots	21	1	0
The robot race	22	0	0
The class	22	0	0

Table 3. Most common and interesting answers on the open-ended questions from the questionnaire.

What did you learn today?
Today I learned that book reports can be wonderful and fun. (3 similar)
Today I learned how robots copy (reproduce) our voices and how our voices can be transferred to the computer. (4 similar comments)
I did not like robotics before, but I learned to love it and learned a lot about robots. (7 similar comments)
Today I learned how to work with a computer and to be thankful and generous. (4 similar comments)
Today we learned that there is no talking about training.
I learned that Marko is kind-hearted.
Today I learned a lot. + Today I have not learned anything.
What did you especially like?
I particularly liked the competition with robots. (12 similar comments)
I liked that we acted with the robots. (5 similar comments)
I especially liked how we controlled robots. (2 similar comments)
I liked the robot race, voice recording and the introduction of robots.
I like it when the teacher laughs. + one comment that was unreadable
What would you do different?
I would not change anything, a perfect class and the best book report ever. (10 similar comments)
I would make the robots louder and a bit bigger. (6 similar comments)
I would make the acting part longer.
I would add more characters to story „In bed by nine".
Blackboard, homework, chairs and tables.
I would make a water park, and anybody who answered a question correctly could slide down the water slide. + I would make different games.
I would write neatly today.

4   Conclusion

In this paper two robot-based activities for a native language class are presented: a role-play and a robot-race. We conducted two analyses to find activities that are usable in everyday classes. Our study showed that a necessary condition is that the activity could be performed by a teacher and without additional assistants for managing the robots. Additionally, the activities should be easy to implement, preparation for them and possible technical problems during the activity should be minimal. The activities should also follow the curriculum and give an additional knowledge to the pupils. The bonus is if the pupils like the activity.

In our experiment, the robot role play showed many technical problems, needed more assistants and lasted too long. Consequently, pupils became bored and the activity did not meet the expectations. At the same time, the robot race represents an example of a good activity for including robots in everyday classes, since it satisfies all the conditions.

Acknowledgment. This work has been supported in part by Croatian Science Foundation under the project UIP-2017-05-5917 HRZZ-TRES and by Croatian Science Foundation under ESF DOK-01-2018 project "Projekt razvoja karijera mladih istraživača – izobrazba novih doktora znanosti".

References

1. Benitti, F.B.V.: Exploring the educational potential of robotics in schools: a systematic review. Comput. Educ. **58**(3), 978–988 (2012)
2. Robinson, M.: Robotics-driven activities: can they improve middle school science learning? Bull. Sci. Technol. Soc. **25**(1), 73–84 (2005)
3. Han, J.-H., et al.: Comparative study on the educational use of home robots for children. J. Inf. Process. Syst. **4**(4), 159–168 (2008)
4. Chang, C.-W., et al.: Exploring the possibility of using humanoid robots as instructional tools for teaching a second language in primary school. J. Educ. Technol. Soc. **13**(2), 13–24 (2010)
5. Sović, A., Jagušt, T., Seršić, D.: How to teach the basic university-level programming concepts to the first graders? In: IEEE Integrated STEM Education Conference. Princeton, USA, pp. 1–6 (2014)
6. Sović, A., Jagušt, T., Seršić, D.: Using lego mindstorms robots in science education. In: Proceedings of the International Science Education Conference, Singapore, pp. 1656–1684 (2014)
7. Jagušt, T., Cvetković-Lay, J., Sović Kržić, A., Sersic, D.: Using robotics to foster creativity in early gifted education. In: Robotics in Education. Advances in Intelligent Systems and Computing, Sofia, Bulgaria, vol. 630, pp. 126–131 (2017)
8. Sović Kržić, A., Jagušt, T., Pushkar, L.: Promoting creative thinking and problem solving through robotic summer camp. In: International Conference Educational Robotics EDUROBOTICS, Rome, Italy, pp. 1–4 (2018)
9. Prosenjak, B.: Sijač sreće. Alfa d.d. Zagreb (2016)

Robotics Competitions

Educational Robotics Competitions and Involved Methodological Aspects

Eftychios G. Christoforou[1]([✉]), Panicos Masouras[2,7],
Pericles Cheng[3,7], Sotiris Avgousti[2], Nikolaos V. Tsekos[4],
Andreas S. Panayides[5], and George K. Georgiou[6]

[1] Department of Mechanical and Manufacturing Engineering,
University of Cyprus, Nicosia, Cyprus
e.christoforou@ucy.ac.cy
[2] Department of Nursing, School of Health Sciences,
Cyprus University of Technology, Limassol, Cyprus
[3] Department of Computer Science and Engineering,
European University Cyprus, Nicosia, Cyprus
[4] Department of Computer Science, University of Houston, Houston, TX, USA
[5] Department of Computer Science, University of Cyprus, Nicosia, Cyprus
[6] Novatex Solutions Ltd, Nicosia, Cyprus
[7] Cyprus Computer Society, Nicosia, Cyprus

Abstract. The present article provides perspectives on educational robotics competitions based on teaching experiences with university students as well as the organization of *"Robotex Cyprus"*, a widely-attended annual event. The focus is on the involved methodological aspects both from an educational and a technological/engineering viewpoint. Common approaches and good practices relevant to robot design and programming are reviewed, while selected popular competition challenges are revisited to exemplify the concepts.

Keywords: Educational robotics · Robot competitions · STEM education

1 Introduction

Robotics impacts many aspects of human activity. Despite being an exciting scientific field, robotics also provides a platform for STEM (Science, Technology, Engineering, and Mathematics) education [1]. Relevant courses are commonly integrated into the curricula of schools and universities, while robotics competitions are often embraced by educational systems [2, 3], as an added stimulus (see Table 1).

In preparation for a competition, it is important for the team to follow a structured approach and this is the main topic of this paper, which is organized as follows. Section 2 highlights experiences from *Robotex Cyprus 2018*, a robotics competition held in Cyprus. This is based on the well-established *Robotex International* competition organized annually in Tallinn, Estonia. Specific competition challenges are revisited in Sect. 3 to exemplify the ideas that will be presented. In Sect. 4, is explained how the

© Springer Nature Switzerland AG 2020
M. Merdan et al. (Eds.): RiE 2019, AISC 1023, pp. 305–312, 2020.
https://doi.org/10.1007/978-3-030-26945-6_27

Table 1. Benefits from educational robotics competitions

Inspire interest in STEM	Foster creativity and innovation
Help understand science and technology	Cultivate technical skills
Develop computer programming skills	Bridge gap between theory and practice
Sharpen problem-solving capability	Practice teamwork and social skills
Develop design and integration skills	Improve presentation skills

general engineering design approach can be used for robot design, while Sect. 5 is dedicated to practical robot design issues. The use of an algorithmic procedure towards solving a problem and its flowchart representations are discussed and demonstrated with basic examples in Sect. 6. The conclusions are given in Sect. 7.

2   Robotex Cyprus 2018 Competition

Robotex Cyprus 2018 engaged four different robotics platforms: (1) Lego Mindstorms, (2) Engino, (3) Edison, and (4) Arduino-based. Some challenges were platform-specific while others were more open to participation. Furthermore, for each challenge the teams were allocated into an age group reflecting a corresponding skill level. The challenges and the number of participants per age group are collected in Table 2. The basic challenges will be discussed in Sect. 3. A newly introduced challenge was the "Color picking", which required robots to collect colored cubes from within the field, considering that red ones gave negative points. In the case of sumo challenges, the main difference between Lego Sumo and Mini Sumo was the maximum size and weight specifications for the robots in each category. "Lego WeDo" and "Engino Mini" were themed competitions aimed at younger primary school children. The goal was to create a programmed robot to perform a mobile activity relevant to the given subject: *"Robots in Cyprus in 100 Years"*.

The number of participants per challenge (Table 2) reflects both their popularity and their degree of difficulty. Apparently, the most popular was the line following challenge followed by sumo. Interestingly, in the line following challenge even though most teams were pursuing similar algorithmic approaches the recorded performances varied considerably. This emphasizes the importance of the refinements to the robot hardware and software. In sumo competitions the defense/attack strategies and the robot's rapid reactions proved equally important. More involved challenges such as the enhanced line following and the maze solving attracted comparatively less participation.

Another observation is that most of the teams participated with robots built using commercial robotics sets rather than off-the-shelf components. The latter involve more advanced skills. Most of the teams were from the "High School 1–3 Grade Students" age group, followed by the "Elementary School 4–6 Grade Students". This fact can be partly attributed to the encouragement from their educational institutions and the alignment of these extracurricular activities to their programs of study.

Table 2. Number of teams at each *Robotex Cyprus 2018* challenge per age group.

Challenge/Age group	Elem. school 1–3 Grade	Elem. school 4–6 Grade	High school 1–3 Grade	High school 4–7 Grade	University	Special (other adults)	Total
Line Following Lego	N/A	27	41	15	N/A	N/A	83
Line Following Engino	N/A	17	6	1	N/A	N/A	24
Line Following Arduino & Edison	N/A	7	15	10	4	8	44
Enhanced Line Following (Arduino, Engino, Lego)	N/A	1	8	3	2	1	15
Color Picking Engino	N/A	N/A	0	0	0	7	7
Mini Sumo (Lego)	N/A	2	5	3	0	7	17
Lego Sumo	N/A	30	36	13	N/A	N/A	79
Maze Solving Lego	N/A	6	10	6	N/A	N/A	22
Maze Solving (Arduino, Edison, Engino)	N/A	N/A	6	4	2	9	21
Lego WeDo	32	6	N/A	N/A	N/A	N/A	38
Engino Mini	13	3	N/A	N/A	N/A	N/A	16
Total	45	99	127	55	8	32	366

Fig. 1. Robotics competition challenges from *Robotex Cyprus 2018*. (a) Color picking; (b) Line following; (c) Maze solving; (d) Mini sumo.

3   Common Robotics Competition Challenges

Robot competitions include challenges that are sometimes themed (e.g., "Recycling", "Mini sumo", "Search and rescue", "Firefighting") or they focus on a generic task (e.g., "Line following", "Maze solving", "Sorting"). Sometimes a robot competes against a robot opponent (e.g., "Mini sumo") or a team of robots competes against other teams (e.g., "Robot soccer"). In the latter case, robots are required to communicate between them to coordinate and collaborate as robot swarms, towards a common goal. The aim of a challenge can either be solving a problem or scoring points. To better illustrate the ideas three of the most popular challenges in robotics competitions are briefly reviewed, namely: line following, maze solving, and mini sumo. We discuss their standard versions, as well as the ones adopted in *Robotex Cyprus 2018* (see Fig. 1).

3.1   Line Following

The basic task for line following robots is to autonomously drive through a track marked with a black line over a white background, as fast as possible. In *Robotex Cyprus* participants were expected to develop a robot and a generic code that can adapt and perform successfully on any field within the given specifications.

A more complex version is the "Enhanced line following" in which further to the primary line following task the robot has to overcome certain obstacles along its way. The obstacles constitute extra levels of complexity that have to be addressed as part of the system design and programming. In the case of *Robotex Cyprus* obstacles included line breaks on track sections, physical obstacles on the track (swing, mountain), expansion/constriction of the line width, and loops next to the track.

3.2   Maze Solving

Maze solving involves a robot finding its way through a 2-D maze from the start point to the finish, without any prior knowledge of the maze. A variation in *Robotex Cyprus* involved autonomous robots driving through the maze from a specified corner to its center, in the shortest possible time. The structure of the maze was such that a "wall-hugging mouse" (see Sect. 6) would not be able to reach the target. This version of the problem has further complexity and requires a navigation system and map-building capabilities, as for example in [4].

3.3   Mini Sumo

This challenge also draws its popularity from the fact that it is exciting and spectacular. It involves two robots attempting to push each other out of a ring platform ("Dohyo") and earn as many effective points ("Yuko points") as possible. The robots are often installed with an angled blade (movable or fixed shovel) whose purpose is to scoop the opponent up. A critical capability is to be able to locate the opponent quickly and attack first. Equally important to the attack strategy is to have an effective defense strategy. A distinct characteristic of this challenge is that the robot does not have any a priori knowledge of the opponent's strategy and is required to deal with this uncertainty.

4 Application of the Engineering Design Approach

A starting point in preparation for a robotics competition is the formulation of a
"strategy", which will be the high-level plan for all subsequent actions including the
robot design. A robot may be created using a commercial educational robotics platform
or through integration of off-the-shelf parts. In either case, a useful methodology is the
general engineering design approach which can be customized, as shown in Fig. 2.

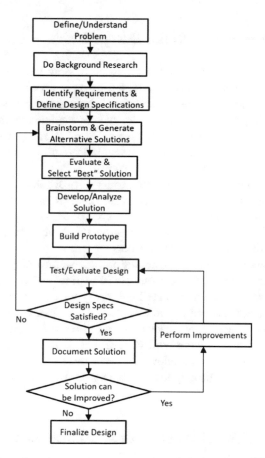

Fig. 2. The general engineering design approach can be applied to the design of a robotic
system for educational robotics competitions.

The first step is to understand the problem and the rules. This is followed by an
extensive background research to collect relevant information, discover ideas and avoid
"reinventing the wheel". At this stage, the design specifications must be defined.
Beyond the functional and performance specifications the list may include other issues
like cost, compactness, weight, and aesthetics. Typically, engineering design problems
have several different solutions. Brainstorming can yield alternative feasible solutions.

The solutions are systematically evaluated and "engineering tradeoff" considerations are involved while selecting the "best" one. A detailed implementation plan is then generated, a functional prototype is constructed and the software is developed. The prototype system is then systematically tested and evaluated. It is often required to return to a previous step and reiterate through the process until the test results are satisfactory. At this stage the final solution is documented. Lastly, further testing of the system is required to refine both the hardware and the software components. Documentation of the work that is carried out at each step is essential.

5 Practical Competition Robot Design Issues

When teams design a robot they need to be aware of how a robotic system operates: input is collected from sensors providing information about its environment (and perhaps its own state as well), the information is processed at the processing/control unit on the basis of a control algorithm/logic, and appropriate commands are issued to the actuators. The system itself may consist of various subsystems, which can be developed and analyzed separately before integration (e.g., locomotion system, collision avoidance system, and manipulation system).

Structurally, it is important for the robot to be compact and firmly assembled. The most common form of actuators in educational robotics has been the direct current (DC) motors, which are compact and easy to control. A variety of sensors (contact/contactless) provides versatility in implementing the control logic including touch, sound, light, ultrasound, and position encoders.

The most common approach to locomotion has been to use wheels. A common configuration involves two actuated wheels with a third castor wheel to serve as a follower and keep the vehicle's balance. Different types of motions/turns are possible by controlling the relative speed of the two motors. Another commonly used method for locomotion involves continuous tracks (treads) with drive wheels. The weight of the vehicle is distributed over a large contact area and this allows it to transverse soft ground. Tracked locomotion also allows to drive over rough terrain, obstacles, and ditches. Steering involves adjusting the relative speed of each track, in analogy to the wheeled case.

6 Algorithmic Control Procedures and Flowchart Representations

Depending on the challenge, the robot is required to operate based on a systematic control procedure (a control algorithm). An effective way to represent an algorithm is through a flowchart, which is also an intermediate step prior to translating it into programming code. Below, two popular algorithmic approaches that pertain to the line following and the maze solving challenges, respectively, are presented. In these examples the flowcharts only show the core part of the algorithm. An example that refers to the mini sumo challenge can be found in [5].

6.1 Basic Algorithms for Line Following

A basic line following approach involves a downwards-looking light sensor attached to the front of the robot. As long as the sensor detects the dark line, the robot moves forwards and turns slightly to the right. If it detects the white surface, then it moves forward and turns to the left. The result is to track the right edge of the line. This approach is represented in Fig. 3a. A more involved approach uses two light sensors at the front of the robot, which tries to keep them both over the dark line while moving forwards. In case one of them detects the white surface, the robot reacts by turning towards the opposite direction, as depicted in Fig. 3b. The threshold for the light intensity used to distinguish between black and white floor areas is a critical parameter. Robots are often programmed to perform a self-calibration procedure at the start line.

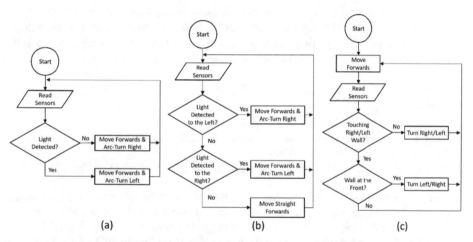

Fig. 3. Flowchart representations of basic algorithmic methods: (a) Line following robot with a single light sensor. (b) Line following robot with two light sensors over the dark line. (c) Maze solving robot using the "wall-follower" method.

6.2 Basic Algorithms for Standard Maze Solving

There exist various algorithms of different complexity and efficiency referring to the challenge of finding an exit to the maze. A simple one is the "random mouse" approach, which is about following a corridor and turning randomly when a junction is encountered. If the maze is solvable the robot will eventually find its way out, but the process can be extremely slow. Another popular approach is the "wall-follower" (or "mouse hugging" approach), which involves moving forwards and turning while making sure that the "right/left hand always touches the wall" (e.g., the implementation in [6]). This approach is also called the right-hand-rule or the left-hand-rule depending on the side that is selected for the implementation. The approach works if all the walls of the maze are "simply connected" (i.e., all walls are connected together or to the maze's outer boundary) but the route that will be followed may not be the shortest. The basic idea is represented with the flowchart in Fig. 3c. Implementation may involve touch sensors or noncontact distance sensors (e.g., ultrasound) for detecting the walls.

The most used algorithm for robotics competitions where the target is known (e.g., the center of the maze in *Robotex Cyprus*) is the *flood-fill* algorithm, which allows building a shortest path solution towards the target cell. This algorithm is more efficient than the previously mentioned algorithms [7, 8].

7 Conclusions

Educational robotics is becoming an integral part of teaching STEM courses at both secondary and tertiary education. Robotics competitions are complimentary to these courses and they have been gaining popularity among students. Competitions provide participants with opportunities to expand their knowledge, gain and share experiences, and network while developing new skills. Both from the organizers', the educators', and the participants' perspective it is important to promote systematic approaches to the design of competitive robotic systems, as it was described herein.

References

1. Afari, E., Khine, M.S.: Robotics as an educational tool: impact of lego mindstorms. Int. J. Inf. Educ. Technol. **7**(6), 437–442 (2017)
2. Murphy, R.R.: "Competing" for a robotics education. IEEE Robot. Autom. Mag. **8**(2), 44–55 (2001)
3. Akagi, T., Fujimoto, S., Kuno, H., Araki, K., Yamada, S., Dohta, S.: Systematic educational program for robotics and mechatronics engineering in OUS using robot competition. In: IEEE International Symposium on Robotics and Intelligent Sensors, vol. 76, pp. 2–8 (2015)
4. Chauhan, A.G., Suthar, H.A.: Evaluation of modified flood fill algorithm for shortest path navigation in robotics. Int. J. Electron. Commun. Technol. **3**(2), 59–63 (2012)
5. Ciarnoscki, P.F., Hoffmann, K., Scortegagna, R.G.: Improvement of a mobile autonomous robot to participate in sumo competitions. In: IEEE 24th International Symposium on Industrial Electronics, pp. 633–637 (2015)
6. Saman, A.B., Abdramane, I.: Solving a reconfigurable maze using hybrid wall follower algorithm. Int. J. Comput. Appl. **82**(3), 22–26 (2013)
7. Mishra, S., Bande, P.: Maze solving algorithms for micro mouse. In: IEEE International Conference on Signal Image Technology and Internet Based Systems, pp. 86–93 (2008)
8. Dang, H., Song, J., Guo, Q.: An efficient algorithm for robot maze-solving. In: IEEE 2nd International Conference on Intelligent Human-Machine Systems and Cybernetics, vol. 2, pp. 79–82 (2010)

Participants' Perceptions About Their Learning with FIRST LEGO® League Competition – a Gender Study

Despoina Schina[✉] ⓘ, Mireia Usart ⓘ,
and Vanessa Esteve-Gonzalez ⓘ

Universitat Rovira i Virgili, Tarragona, Spain
{despoina.schina,mireia.usart,vanessa.esteve}@urv.cat

Abstract. Robotics Competitions as FIRST LEGO® League (FLL) Competition are gaining more and more popularity, however, what are the participants learning? The present study investigates participants' perceptions regarding their learning with FLL Competition. In particular, it explores participants' perceptions about their learning about the world, learning to solve problems, learning to engage, learning to apply knowledge, learning to communicate, learning to apply the technology cycle, studies participants' perceptions on the competition activities, but, more importantly, it explores gender differences in students' perceptions about their learning in the competition. A quantitative methodology is used, a questionnaire collecting data on participants' perceptions was first validated by a team of experts and then completed by 84 participants of the Finals of FLL Competition 2018 in Greece. Results show that participants' perceptions on their learning are very positive: (a) they report that they enjoy taking part in the competition as they are engaged in activities of their preference, (b) participants consider the competition as a great opportunity for learning about real word problems and for the acquisition of skills in STEM areas of studies, (c) participants view that they acquire social, collaborative and communication skills and (d) regarding gender, females tend to be more engaged, enthusiastic, creative, eager to experiment and more likely to adopt collaborative strategies than males.

Keywords: Educational Robotics · FIRST® LEGO® League competition · Gender studies

1 Introduction

There is a growing interest in studies and professional careers related to STEM disciplines (Science Technology, Engineering and Mathematics) internationally. The total number of STEM graduates increased from around 755,000 in 2007 to 910,000 in 2012 in the EU countries, corresponding to an average annual 3.8% growth rate and an overall 20% increase over the period [1]. Similarly, over this period, the percentage of tertiary STEM graduates in Greece has increased around 4% [1]. However, despite the numerous gender specific actions taken by the European Commission and the local governments, female participation rates in STEM studies are considerably lower than

© Springer Nature Switzerland AG 2020
M. Merdan et al. (Eds.): RiE 2019, AISC 1023, pp. 313–324, 2020.
https://doi.org/10.1007/978-3-030-26945-6_28

male participation in the disciplines of computing and engineering [1]. STEM labor market is expanding and so needs to expand female presence in the fields of computing and engineering. Main objective of the European Commission is to achieve a better gender balance in STEM studies [2]. A wider female involvement in computing and engineering studies would encourage their future employability.

Student learning and acquisition of skills in the areas of computing, engineering and STEM disciplines in general can be facilitated through classroom use of Educational Robotics (ER). Research in the field has proven that Educational Robotics (ER) can effectively be used to equip students with skills in the area of STEM education [3, 4] and promote students' interests in STEM subjects [5]. ER is a teaching tool that engages both male and female students and encourages their affinity to STEM disciplines. ER offers immense opportunities for teaching and learning; due to its cross-curricular nature, it can be implemented in a variety of disciplines [4, 6, 7]. Students through ER project-based activities can work on real-world problems and challenges [8]. Educational robotics is an engaging learning tool that through hands-on activities makes learning fun and motivates students in learning [8]. Regarding gender, ER is attractive to both girls and boys, however, activities engaging to both males and females need to be designed [3] in particular, activities that maintain females' interest in robotics during their teens. Robotics competitions may promote both genders' learning and acquisition of STEM skills and this will be further explained in the section below.

2 Theoretical Background

2.1 Educational Robotics Competitions' Impact on Students' Learning

Educational robotics is not integrated into the school curriculum - most of the times, robotics activities are extracurricular [9]. ER activities often take place in informal educational contexts after-school clubs or summer camps that often lead to the participation in a competition. Educational robotics competitions provide a unique and vital learning opportunity for school age children [5]. Significant are the learning outcomes that participants reap of ER competitions in the area of STEM. In robotics competitions children independently identify and understand principles, concepts, and elements of practice that are fundamental to programming and engineering [10]. In addition, through robotics competitions students better understand principles and concepts that they had previously found challenging while trying to make the robot function [10].

Student participation in ER competitions brings additional benefits beyond the understanding of STEM principles and concepts. For example, it brings an increased student interest and motivation to take additional STEM courses in their studies [11]. Participants display a more positive attitude toward science and science related areas [12]. Robotics competitions have proven to be very motivating for children [10, 13]. Students' intrinsic and extrinsic motivation, self-determination and self-efficacy concerning robotics are maintained at a high level throughout robotics competitions [4]. As a result, participating in competitions is a motivating ER activity that positively affects students' affinity to STEM disciplines.

In this particular study, students' perceptions on their learning within competitions will be examined in the context of FIRST® LEGO® League (FLL) competition. FLL competition is an international robotics and research competition operated by the non-profit organization For Inspiration and Recognition of Science and Technology (FIRST) and involves students from 9 to 16 years old [14]. Each year a new themed challenge is released and the teams research, design, and present results of their work to the community and panels of judges [14]. During students' preparation for the competition and in the competition itself, the participants have gains in a variety of fields. First of all, students show high and positive attitudes and motivation towards ER both before and after the competition [4]. In addition, students develop a greater appreciation for and a more positive attitude regarding the importance of science and the social implications of science-related issues [15]. The competition also sharpens students' problem-solving skills as problem-solving is embedded in all parts of the competition - robot design, robot game, project and core values [16]. Through the competition, participants' confidence in their problem-solving skills is boosted [17]. More detailed, participants improve technological problem-solving styles, problem clarification, developing a design, evaluating a design solution, and overall technological problem-solving performance [17]. To sum up, according to the above-mentioned authors, robotics competitions offer considerable learning outcomes, however, are they popular with both boys and girls?

2.2 Gender Differences Observed in Educational Robotics Competitions

ER competitions do not seem to attract as many female participants as males. In particular, in a research carried out in the context of FIRST® LEGO® League competition in 2009, "the large majority of FLL participants were boys (approximately 70%) with the average team consisting of five and six boys and two to three girls" [18]. Despite the fact that both male and female participants' learning was positively influenced by their participation in the competition, a divergence was observed regarding their interests and self-perceived skills gained [18]. On the one hand, males were more likely to show interest in computers and technology and technology-related careers and report gains on skills in these fields [18]. On the other hand, females tended to show more interest in projects and in going to college and report gains on social skills and communications rather than on technology-related skills [18]. Females tend to display lower self-efficacy in robotics than males, despite the fact that they are as competent as males are [19].

To tackle female lower self-efficacy in educational robotics, it is suggested that females practice building and programming through hands-on activities [19]. The Roberta Initiative was also launched proposing themes and experiments that are more interesting for girls but do not exclude boys, aiming at increasing females' self-efficacy in educational robotics [20]. In a recent research, it has been observed that as females grow older, they are less involved in programming, meaning that their programming interest decreases as the years go by [21]. It is suggested that girls get involved in informal computing experiences to maintain their interest in pursuing STEM-related careers [21]. Interestingly, it was also observed that females showed more positive attitudes and motivation for learning robotics after their participation in FLL competition [4]. Apart from their attitudes and motivation, female students' self-efficacy was boosted as "their attitudes and motivation increased to the point that they felt that they were better than male students" [4]. All in all, noticeable is the positive impact of the competition on females' attitudes, motivation and self-efficacy regarding ER and STEM.

3 Methodology

Objective of this research project is to look into the participants' self-perceived learning within the competition and to further examine whether there are significant gender differences in the participants' perceptions regarding their learning with the competition. The variables of our study are students' learning about the world, learning to solve problems, learning to engage, learning to apply knowledge, learning to communicate, learning to apply the technology cycle and their opinions about the FLL competition. For the purpose of this study, an instrument measuring participants' self-perceived learning was validated by a team of experts.

3.1 Research Questions

– What are the participants' perceptions about their learning in FLL Competition?
– To what extent do females' perceptions about FLL Competition differ compared to males'?

3.2 Context and Sample

The sampling was conducted with self-selected students among the participants of the Final Competition of FLL in Greece held in March 2018 themed "Hydrodynamics". A total of 84 participants took part into the research. The sample consists of 45 males and 39 females representing a 54% and 46% of the sample respectively. The average age of the participants was 13.05, (SD = 1.58) while the age range varied between 10 and 16 years old. A significant majority of the participants (64% of the total sample) is between the age of 13 and 15 years old, while 33% of the participants is younger - between the age of 10 and 12 years old - and only a 3% of the total sample is over 15 years old. The participants completed the questionnaire on the second day of the competition right before the closing ceremony when all FLL challenges had been completed.

3.3 Research Instruments

The questionnaire distributed to the participants of the Final Competition of FIRST LEGO® League (FLL) is the translation into Greek of Chalmers' questionnaire used in the Australian Regional FLL Competition [22].

Questionnaire Content

The questionnaire focused on students' self-appraisal of their learning and they were asked to respond to statements about: learning about the world, learning to solve problems, learning to engage, learning to apply knowledge, learning to communicate, learning to apply the technology/engineering cycle, and specific questions relating to the FLL activities [22]. Precisely, regarding the content of the questionnaire, there are 4 items collecting demographic data (participants' gender/age/team) and the main body of the questionnaire includes 36 Likert scale statements ranging from Almost Always to Almost Never and 7 open-ended questions. The items are categorized into seven dimensions: 9 items about learning about the world, 7 about learning to solve problems, 6 about learning to engage, 4 about learning to apply knowledge, 5 about learning to communicate, 6 about learning to apply the technology cycle, 6 about the participants' opinions and together with the 4 demographic items, the total number of the questionnaire items rises to 48.

Questionnaire Validation

As a part of our research, the questionnaire was translated into Greek and validated by 15 experts at the field of educational technology and FLL Competition (7 males and 8 females), 8 of the experts were members of the judging committee in the FLL Competition 2018, while 7 of the experts were members of eduACT, the non-profit organization in charge of organizing FLL Competitions in Greece [23]. The experts validated the questionnaire and evaluated its items in terms of ambiguity, pertinence and importance by completing an online questionnaire on google forms[1].

First of all, experts' evaluation in terms of ambiguity will be presented. Based on the analysis of experts' answers, the questionnaire items were thought to be clear enough and easily understood by the children with the exception of the following four items. However, none of them reached an ambiguity rate higher than 50%. In particular, Q10 - "I got a better understanding of the world outside of school" and Q11 - "I learnt interesting things about the world outside of school" were considered to be ambiguous by 46% and 40% of the experts respectively. According to the feedback provided by the experts, Q10 was viewed as too obscure and misleading to the students. As far as item Q11 is concerned, the experts reported that the statement included the word 'things' and for this reason the item was considered to be unclear in the given context. Regarding Q25 - "I was always engaged", 47% of the experts pointed out the ambiguity of the item as they considered the word 'engaged' semantically difficult to the students. Last but not least, Q39 - "I was able to check my work" was considered ambiguous by 40% of the experts who underlined that the difference of this item when compared to the following item Q40 - "I was able to think about my work" is not clear enough and stated that a further explanation is required. The rest of the questionnaire items were considered unambiguous by the great majority of the experts and received positive feedback.

Furthermore, experts evaluated the questionnaire items regarding their pertinence to the research objectives on a scale from 1 to 4. The questionnaire items were considered to be pertinent to the research objectives as 87.5% of the items received a score of 3.5

---

[1] The online questionnaire in Greek is available at https://goo.gl/forms/AvU6nAA1fBhhSE3g1.

or higher. Interestingly, only item Q25 - "I was always engaged" received a low score (2.4) and was considered by the experts to be the least pertinent to the research.

Finally, the experts evaluated the items regarding their importance to the research objectives on a scale from 1 to 4. All items were thought to be important to the research objectives as 77% of them were evaluated with 3.5 or higher. Items Q1, Q22 and Q25 received lower scores equivalent to 3.0 or less. More precisely, Q1 - "Which is your team?" received a mean score of 3.0 and this implies that it was not considered very important to the research objectives by the experts. Q22 - "I did hands-on activities" also achieved 3.0 and the experts pointed out that an example of hands-on activities should have been provided to make the question more explicit. As in the previous sections, Q25 'I was always engaged' received a low score around 2.5 implying that this item was not very important according to the research objectives.

After expert validation, Cronbach's alpha test was run in order to measure the internal consistency or reliability of the validation questionnaire [24]. For the purpose of this research, the Cronbach's alpha test was once run for items of the validation questionnaire altogether and then it was run again separately per questionnaire part (ambiguity, pertinence, importance). The Cronbach Alpha for the total number of items of this questionnaire was 0.95 suggesting a high internal consistency. Regarding each scale, Ambiguity ($\alpha = 0.81$), Pertinence and Importance, ($\alpha = 0.94$) for both and this shows a high internal consistency of each scale.

3.4   Results and Discussion

The results of the research project will be analyzed per research question.

Research Question 1: What Are the Participants Perceptions About Their Learning in FLL Competition?

Participants in the FIRST LEGO® League (FLL) competition provided positive feedback. When analyzing the questionnaire answers on a five-point scale, the means per section (see Fig. 1) show that participants view their participation in the FLL competition as a very enriching experience as they learn about the world, they learn how to solve problems, to apply knowledge, to engage, to communicate and apply the technology cycle. Remarkable is the students' evaluation of FLL Competition activities achieving a mean rate around 4,6 out of 5 (see Fig. 1 - Section 7). Our findings indicate that competition activities are considered interesting, enjoyable, fun, challenging and

Fig. 1. Students' perceptions per section

well designed by most of the participants. Our findings are in line with previous research in the field [22].

More than 1/3 of the items are rated between 4.5 and 4.87 out of 5.0 (see Table 1), meaning that participants report that these practices were taking place very often or almost always in the competition. Furthermore, 58.3% of the statements were marked with a rate ranging from 4.0–4.49 out of 5.0, underlining that these practices were taking place often or very often. A limited amount of statements (4 out of 36) received a frequency rate ranging from 3.5–3.99. The limited number of items that received a less high evaluation, confirms that FLL participants view the competition as a very appealing and fruitful experience.

Table 1. Students' perceptions per item and gender - Means and Standard Deviation

Item	M(SD)			Item	M(SD)			Item	M(SD)		
	Males	Females	Total		Males	Females	Total		Males	Females	Total
Q1	4.36(.61)	4.62(.63)	4.49(.63)	Q13	4.39(.72)	4.41(.75)	4.40(.73)	Q25	4.26(.72)	4.68(.58)	4.47(.69)
Q2	4.43(.82)	4.56(.60)	4.49(.72)	Q14	4.22(.88)	4.36(.74)	4.29(.81)	Q26	4.31(.78)	4.49(.65)	4.39(.72)
Q3	4.34(.83)	4.56(.68)	4.45(.77)	Q15	4.52(.70)	4.65(.59)	4.58(.65)	Q27	4.44(.85)	4.51(.82)	4.48(.83)
Q4	4.09(.91)	4.28(.76)	4.18(.84)	Q16	4.26(.85)	4.24(.98)	4.25(.91)	Q28	4.20(.99)	4.05(.86)	4.13(.93)
Q5	4.40(.80)	4.38(.85)	4.39(.82)	Q17	4.24(.85)	4.62(.72)	4.43(.81)	Q29	4.60(.54)	4.72(.56)	4.66(.55)
Q6	4.18(.82)	4.44(.75)	4.31(.79)	Q18	4.55(.59)	4.59(.60)	4.57(.59)	Q30	4.35(.65)	4.56(.55)	4.45(.61)
Q7	4.55(.70)	4.67(.70)	4.61(.70)	Q19	3.76(.95)	3.97(1.01)	3.86(1.03)	Q31	4.30(.83)	4.33(.74)	4.32(.78)
Q8	4.34(.75)	4.62(.54)	4.48(.67)	Q20	3.98(.99)	3.81(.90)	3.90(.95)	Q32	4.81(.67)	4.79(.50)	4.80(.58)
Q9	4.36(.69)	4.49(.60)	4.42(.65)	Q21	3.93(.92)	3.76(1.01)	3.85(.96)	Q33	4.86(.47)	4.87(.34)	4.87(.41)
Q10	3.89(1.0)	4.23(.78)	4.06(.92)	Q22	4.16(.97)	4.24(1.00)	4.20(.96)	Q34	4.84(.43)	4.92(.35)	4.88(.40)
Q11	4.45(.77)	4.72(.56)	4.58(.68)	Q23	4.67(.65)	4.81(.57)	4.74(.61)	Q35	3.81(1.22)	4.05(.98)	3.93(.99)
Q12	4.73(.54)	4.77(.62)	4.75(.58)	Q24	4.29(.74)	4.27(.99)	4.28(.86)	Q36	4.45(.85)	4.64(.59)	4.54(.74)

The analysis of the students' perceptions leads us to summarize our findings:

- Participants report that they enjoy the competition as they are engaged in activities of their preference.

  Remarkable are the high mean frequencies in Q33 "FLL activities were enjoyable" (M = 4.87) and Q34 "FLL activities were fun" (M = 4.88), together with low standard deviation (SD = .41) and (SD = .40) respectively, highlighting that the activities were enjoyable and fun almost always for everyone (see Fig. 1). In addition, students consider interesting the activities held in FLL Competition as they rated the item Q32 - "FLL activities were interesting" on average with 4.8 (SD = .58). Our findings are line with the fact that educational robotics is an efficient learning tool that helps to create a fun and engaging learning environment and keeps students interested and engaged in learning [8].

- Participants view the FLL Competition as a great opportunity for learning.

  Students consider FLL competition as a great opportunity for learning about the world outside of school (Q1–Q8), implying a relation between competition with real word problems and STEM areas of studies. Results from previous research report that students who participated in robotics competition had a more positive attitude towards science and science related areas [12]. Additionally, in our research findings it was observed that participants clearly see the impact of the FLL activities on their ability to solve problems (Q9–Q13). The fact that FLL Competition cultivates students' problem-solving skills was also observed in [16]. From our findings it is inferred that participants have the chance to develop their Computational thinking skills and more specifically they practice and acquire skills in the area of "Computational Problem-Solving Practices" [25]. Interestingly, Q12 "I learnt that there can be more than one solution to a problem" was impressively highly rated (M = 4.75) (SD = .58) implying that through the competition the computational thinking practice "Using Computational Models to Find and Test Solutions" is developed [25]. Similarly, Q18 "I was finding new ways to improve what I was doing" could be linked to "Assessing Different Approaches/Solutions to a Problem" computational thinking practice (Kotsopoulos 2017). Quite high is the mean in Q22 - "I was able to apply my knowledge in technology to solve problems" (M = 4.2) implying that participants apply previously acquired practical knowledge in the competition often or almost always. Our findings are line with previous studies that reported that Educational robotics offers opportunities for hands-on and project-based activities that are closely connected to real-world problems and challenges [8].

- Participants view the FLL Competition as a great opportunity for collaboration and socialization.

  Students' answers in Q23–Q26, Q10 and Q14, show that FLL is a rich context for collaboration and socialization. Most students agree that they had the chance to get in touch with other students (Q23 - M = 4.74). Regarding collaboration, in Q10 "I learnt how others solved problems" and Q14 "I showed others how I solve problems", the mean of the students' answers indicate that these behaviors were adopted often or very often and this confirms that FLL Competition is a great opportunity for collaborative learning. The results of our study verify the view that FLL Competition helps to increase participants' social skills [18] and confirm previous research results according to which students after participating to the competition, felt that they had learnt how to work as part of a team and how to communicate effectively [26].

Research Question 2: To What Extent Do Females' Perceptions About FLL Competition Differ Compared to Males'?

Aim of the research is to study any potential differences between males' and females' perceptions about their learning with FLL Competition: the two groups' answers will be compared and contrasted separately per questionnaire section. From a quick look at Table 1, groups' means - between male and female students do not differ substantially, however, there is a tendency that female students demonstrate slightly more positive perceptions in the questionnaire than males. More positive attitudes of female participants after the FLL Competition were also observed in previous research [4]. However, it needs to be pointed out that the participants joined the competition voluntarily - females participating in the competition might not exemplify average females as they may be more confident with STEM subjects than other females. Males' and females' perceptions will be analyzed below per questionnaire section.

- Section 1 - Learning about the world
  In all questionnaire items in this section except for item Q5, females' perceptions on learning about the world are slightly better than the ones of males, however their differences are not significant. Females seem to be slightly more aware of the connection of the FLL Competition challenges to the real world and everyday problems. Informal STEM activities that are structured around real-world problems and providing help to others, are attractive to girls [27]. In FLL Competition female participants seem to be more enthusiastic about the learning outcomes of the competition, its usefulness and connection to the real world.
- Section 2 - Learning to solve problems
  In section two, non-significant differences were reported regarding participants' perceptions: females tended to provide more positive feedback regarding their learning to solve problems than males. However, in previous research it was observed that female FLL students perceived their overall technological problem-solving style no differently than males [17]. Females more often report that they are eager to learn from others e.g. Q10 - "I learnt how others solve problems" when compared to males. This could imply that females would better collaborate than males. Furthermore, females provide slightly better feedback regarding testing new and different approaches to find a solution (Q11 - "I experimented new ways to solve problems") and this could also imply that females tend to be more creative than males in the context of the competition.
- Section 3 - Learning to engage
  Females perceive their learning to engage slightly higher than males, however, the differences are not significant with the exception of item Q17 "I was trying new ideas". The T-test did not reveal statistically important differences, however, the Pearson correlation test, showed a significant positive correlation between gender and Q17 ($r = 0,238$, $p = 0.35$). Based on these findings, it could be inferred that when females learn within the context of the FLL, they seem to be more engaged, enthusiastic, creative and eager to experiment with new approaches than males. These findings are in line with [26], in which it was observed that females seemed to have a more positive learning experience within the FLL Competition than males and considered it more rewarding.
- Section 4 - Learning to apply knowledge
  In this section, non-significant gender differences were reported. Females perceived their learning to apply knowledge in science and mathematics within the competition slightly lower than males while higher in technology. These findings may suggest a higher self-efficacy in technology in females after the competition also noted in [4].
- Section 5 - Learning to communicate
  As far as the questionnaire section "learning to communicate" is concerned, females provided higher rated feedback compared to males. Significant gender differences were observed in Q25 – "I asked other students to explain their ideas". The T-test carried out did not reveal statistically important differences, however, the Pearson test, showed a significant positive correlation between gender and Q25 ($r = 0,305$, $p = 0,006$). These results show that it is more common for females than males to

learn by asking other students to explain their ideas in the context of the FFL. It has been previously observed that FLL helps increasing participants' social skills [18], collaborative and communication skills [26]. Research findings suggest that in our context females adopt collaborative strategies more often than male participants. It is more common for females to acquire knowledge by interacting with others while giving and receiving advice

- Section 6 – Learning to apply the technology cycle
  In this section, non-significant gender differences were reported. Females perceived their learning to apply the technology cycle in the FLL Competition slightly higher than males in all related items apart from Q28. These findings are in line with [4], suggesting females' higher self-efficacy in technology in this competition. Items in this section related with the ability to investigate, design, create and reflect on tasks are closely connected to computational thinking practices as defined by Weintrop [25]. Females seem to perceive their acquisition of computational skills higher than males, although differences are not significant.
- Section 7 - Opinions
  FLL Competition activities receive positive feedback from both males and females. Given the almost equally rated first three items of the section (Q32, Q33, Q34), the FLL activities seem equally interesting, enjoyable and fun for both males and females.

4   Conclusions

The present study consisted of the expert validation of a questionnaire on participants' learning within FLL Competition and the analysis of participants' perceptions. Data was collected from 84 participants of the FLL Finals in Greece. Results showed that: (a) FLL Competition participants report to enjoy taking part in the competition as they are engaged in activities of their preference, (b) participants consider the competition as a great opportunity for learning about real word problems and acquiring skills in STEM areas, and (c) participants view that they develop their social, collaborative and communication skills in the competition. Regarding gender, results show a generalized but not statistically significant tendency of female participants to perceive their learning more positively than males. Based on statistically significant results, females in the context of FLL Competition seem to be more engaged, enthusiastic, creative, eager to experiment and more likely to adopt collaborative strategies than male participants. Finally, given these findings, future research is needed to study greater samples and to draw a comparison between self-perceived learning/self-efficacy and gender in informal robotics activities. It is also recommended to study the actual process of learning with FLL through observation, following a qualitative approach. The relation between self-perceived learning and actual learning could additionally be explored in future research.

Acknowledgements. This project received funding from the European Union's Horizon 2020 research and innovation programme under the Marie Skłodowska-Curie grant agreement No. 713679 and from the Universitat Rovira i Virgili (URV). The authors thank eduACT - The Organization For Future Education for their support.

References

1. Shapiro, H., Østergård, S.F., Hougard, K.F.: Does the EU need more STEM graduates? Publications Office of the European Union, Luxembourg (2015)
2. Caprile, M., Palmén, R., Sanzé, P., Dente, G.: Encouraging STEM studies labour market situation and comparison of practices targeted at young people in different member states. Publications Office of the European Union, Luxembourg (2015)
3. Johnson, J.: Children, robotics, and education. Artif. Life Robot. 7(1–2), 16–21 (2003)
4. Kaloti-Hallak, F., Armoni, M., Ben-Ari, M.: Students' attitudes and motivation during robotics activities. In: Workshop in Primary and Secondary Computing Education, pp. 102–110 (2015)
5. Eguchi, A.: RoboCupJunior for promoting STEM education, 21st century skills, and technological advancement through robotics competition. Rob. Auton. Syst. 75, 692–699 (2016)
6. Ribeiro, C.R., Coutinho, C.P., Costa, M.F.M.: Robotics in child storytelling. In: Science for All, Quest for Excellence. 6th International Proceedings on Hands-on Science, Ahmedabad, vol. 9, pp. 198–205 (2009)
7. Benitti, F.B.V.: Exploring the educational potential of robotics in schools: a systematic review. Comput. Educ. 58(3), 978–988 (2012)
8. Eguchi, A.: Educational robotics for promoting 21st century skills. J. Autom. Mob. Robot. Intell. Syst. 8(1), 5–11 (2014)
9. Alimisis, D.: Educational robotics: open questions and new challenges. Themes Sci. Tecnol. Educ. 6(1), 63–71 (2013)
10. Petre, M., Price, B.: Using robotics to motivate 'back door' learning. Educ. Inf. Technol. 9(2), 147–158 (2004)
11. Hendricks, C.C., Alemdar, M., Ogletree, T.W.: The impact of participation in Vex robotics competition on middle and high school students' interest in pursuing STEM studies and STEM-related careers. In: 2012 ASEE Annual Conference & Exposition, Texas (2012)
12. Welch, A.G.: Using the TOSRA to assess high school students' attitudes toward science after competing in the first robotics competition: an exploratory study. Eurasia J. Math. Sci. Technol. Educ. 6(3), 187–197 (2010)
13. Theodoropoulos, A., Antoniou, A., Lepouras, G.: Teacher and student views on educational robotics: the Pan-Hellenic competition case. Appl. Theory Comput. Technol. 2(4), 1–23 (2017)
14. Rosen, J.H.: FIRST LEGO League participation: perceptions of minority student participants and their FLL coaches. In: 120th ASEE Annual Conference & Exposition (2013)
15. Welch, A.G.: The effect of robotics competition on high school students' attitudes toward science. Sch. Sci. Math. 111(8), 416–424 (2011)
16. Chen, X.: How does participation in FIRST LEGO League robotics competition impact children's problem-solving process? In: Lepuschitz, W., Merdan, M., Koppensteiner, G., Balogh, R., Obdržálek, D. (eds.) Robotics in Education. RiE 2018. Advances in Intelligent Systems and Computing, vol. 829. Springer, Cham (2018)
17. Varnado, T.E.: The effects of a technological problem solving activity on FIRST LEGO League participants' problem solving style and performance. Virginia Tech (2005)
18. Melchior A., Cutter T., Cohen F.: Evaluation of FIRST LEGO league. Center for Youth and Communities, Heller Graduate School, Brandeis University, Waltham (2004)
19. Milto, E., Rogers, C., Portsmore, M.: Gender differences in confidence levels, group interactions, and feelings about competition in an introductory robotics course. In: 32nd Annual Frontiers in Education. IEEE (2002)

20. Bredenfeld, A.H., Leimbach, T.: The roberta initiative. In: Proceedings of SIMPAR 2010 Workshops International Conference on Simulation, Modeling and Programming for Autonomous Robots, Darmstadt, pp. 558–567 (2010)
21. Witherspoon, E.B., Schunn, C.D., Higashi, R.M., Baehr, E.C.: Gender, interest, and prior experience shape opportunities to learn programming in robotics competitions. Int. J. STEM Educ. 3(16), 18 (2016)
22. Chalmers, C.: Learning with FIRST LEGO League. In: Society for Information Technology and Teacher Education (SITE) Conference, pp. 5118–5124. Association for the Advancement of Computing in Education (AACE), New Orleans (2013)
23. eduACT Homepage. https://eduact.org. Accessed 23 Jan 2019
24. Cohen, L., Manion, L., Morrison, K.R.: Research Methods in Education, 6th edn. Routledge Falmer, London (2007)
25. Weintrop, D., Beheshti, E., Horn, M., Orton, K., Jona, K., Trouille, L.: Defining computational thinking for mathematics and science classrooms. J. Sci. Educ. Technol. 25 (1), 127–147 (2016)
26. Ball, C., Moller, F., Pau, R.: The mindstorm effect: a gender analysis on the influence of LEGO mindstorms in computer science education. In: Proceedings of the 7th Workshop in Primary and Secondary Computing Education, pp. 141–142. ACM (2012)
27. Dasgupta, N., Stout, J.G.: Girls and women in science, technology, engineering and mathematics: STEMing the tide and broadening participation in STEM careers. Policy Insights Behav. Brain Sci. 1, 21–29 (2014)
28. Kotsopoulos, D., Floyd, L., Khan, S., Namukasa, I.K., Somanath, S., Weber, J., Yiu, C.: A pedagogical framework for computational thinking. Digit. Exp. Math. Educ. 3(2), 154–171 (2017)

Young Roboticists' Challenge - Future with Social Robots - World Robot Summit's Approach: Preliminary Investigation

Amy Eguchi[1][(✉)] and Hiroyuki Okada[2]

[1] Bloomfield College, Bloomfield, NJ, USA
amy_eguchi@bloomfield.edu
[2] Tamagawa University, Tokyo, Japan
h.okada@eng.tamagawa.ac.jp

Abstract. The World Robot Summit (WRS) Junior Robot Category's School Robot Challenge provides an opportunity for school-age children to learn to program the Pepper robot, a sociable humanoid robot offered by SoftBank Robotics, Inc. to realize their ideas for living with a social robot in school. This paper presents a preliminary investigation on the programming performance of the participating teams in the School Robot Challenge. The data for the pilot study was collected from the WRS 2018 Junior Robot Category's School Robot Challenge held in Tokyo, Japan in October 2018. Analysis of the programming performance of the participating teams at WRS 2018 Junior School Robot Challenge indicates that having access to Pepper, the standard platform robot, does not guarantee the success of the teams in terms of achieving higher rankings. Although it is not clear what contributed to the success of the teams, there are some indications that access to the standard platform robot before the competition and the programming experience have some influence on the teams' performances. Further investigation is needed to determine what contributes to the teams without access to the sociable robot becoming successful with its programming.

Keywords: Social robot · Human-Robot Interaction · Educational robotics · Robotics competition · Computer science education

1 Introduction

The World Robot Summit (WRS) is an initiative dedicated to creating a society where humans and robots coexist and collaborate in a way in which robots and humans augment each other in various aspects of our lives. Following the New Robot Strategy proposed by the Japan government, the Japan Ministry of Economy, Trade, and Industry (METI) and the New Energy and Industrial Technology Development Organization (NEDO) organized an executive committee involving experts and researchers from various robotics related fields for the preparations to host two WRS events (2018 and 2020). The WRS theme, *Robotics for Happiness*, highlights the goal to bring together innovators who work toward creating a society where humans and robots live and work side by side in harmony. The WRS aims to accelerate the research

M. Merdan et al. (Eds.): RiE 2019, AISC 1023, pp. 325–335, 2020.
https://doi.org/10.1007/978-3-030-26945-6_29

and development of robotics technologies and advancement of the performance of robots in manufacturing, search and rescue, and service sectors, as well as provide opportunities for young roboticists to acquire and develop skills and knowledge necessary to become successful citizens coexisting and collaborating with robots in the future. The WRS 2018 was held in October 2018 in Tokyo and the WRS 2020 is scheduled to be held in October 2020 in Aichi prefecture, Japan. The WRS consists of the World Robot Competition (WRC) and World Robot Expo (WRE). The WRC has four categories: Service Robotics, Industry Robotics, Disaster Robotics and Junior Robotics Categories (http://worldrobotsummit.org/en). The Junior Robotics Category Challenges are for students up to 19 years of age. The Junior Robotics Category consists of two challenges: School Robot Challenge and Home Robot Challenge.

2 Motivation

Recent years, social robots started to move out of laboratories and are becoming more and more accessible for regular consumers to use in everyday life. For example. every year at the CES, social robots are becoming one of the popular exhibitions. Following Jibo in 2017, a handful of social robots including CLOi, BUDDY, and Tapia among many others were introduced in 2018. Those are social home robots or partner robots designed to interact and carry out a conversation with humans. Sony's *aibo* came back in November 2017, after more than 10 years of silence [1, 2]. The new *aibo* robot features advanced mechatronics and AI, making smarter, more sophisticated, and natural performances resembling dogs possible. We are living in a transitional period when human-robot coexistence is rapidly becoming a reality, not only in factories and laboratories but also in our everyday lives. Similar to how personal computers and smartphones have become technologies that many of us cannot live without, robots in many different forms are becoming an integral part of our lives. It is happening whether we are ready or not.

Social robots have already started to be tested and integrated into children's lives, especially at school, as one of the educational/learning technologies. The roles that social robots have played in educational settings range from tutors [3, 4], teaching assistants [5], and fellow learners [6], to learning [7, 8] and therapeutic companions [9]. For example, NAO, a humanoid robot developed by Aldebaran Robotics has been used as a therapeutic robot with students with autistic spectrum disorder (i.e. [10, 11]).

The Junior Category Challenges are inspired by the WRS's aim to envision a society where humans and robots coexist and collaborate. Existing robotics competitions for school-age children usually have a focus on games (sports or a game to accomplish targeted tasks) or creating a solution to a theme set by the competition, which may not be authentic or relate to the daily lives of students. The WRS Junior Category Challenges focus on the idea of making robotics as partners that augment participating students' everyday lives. The challenges provide unique opportunities for students to imagine robots coexisting in their lives at home and school, and create and realize their dream collaboration with robots in settings which they are familiar with. Following the WRS's theme – *Robotics for Happiness*, the Junior Category Challenges ask students to create a scenario to achieve human-robot coexistence and cooperation at school or home.

In this paper, we introduce the World Robot Summit's Junior School Robot Challenge, where competitors are asked to program an interactive social robot in a school setting, imagining a society living with social robots interacting and collaborating together with humans. In addition, we discuss the pilot study examining the programming performance of the participating teams in WRS's Junior School Robot Challenge.

3   Junior School Robot Challenge

If you had a humanoid robot at school, what would you want it to do? What is the role of a humanoid at school?" [12]. Those are the questions to which the School Robot Challenge requires teams to respond. Teams participating in the School Robot Challenge are tasked to demonstrate their ideas for living and collaborating with social robots at school by programming a standard platform robot. The use of standard platform robot in competitions is not rare; however, the use of a sophisticated social robot with school-age children as a standard platform is quite unique. For example, the VEX IQ Challenge is a robotics competition for elementary and middle school students requiring teams to use VEX IQ kits (https://www.vexrobotics.com/vexiq/competition). First LEGO League is a robotics competition for students in grades 4–8 that uses LEGO Mindstorms kits as the standard platform (https://www.firstinspires.org/robotics/fll). The social robot used in the WRS is Pepper robot offered by SoftBank Robotics. Pepper is currently commercially available in a limited number of countries, including Japan, France, and some European countries. Pepper is "the world's first personal robot that reads emotions" (http://www.softbank.jp/en/robot/) developed for domestic use.

Participating teams are required to demonstrate their knowledge and understanding of Pepper and its unique functions, as well as their idea for living and cooperating with a social robot in a school environment. Pepper is a robust robot equipped with various functions. Pepper is an ideal standard platform for School Robot Challenge because of a variety of sensors which enable Pepper to *sense* its environment, react to objects, and interact with humans around it. In addition, Pepper's default perception and interaction capabilities include human detection and tracking, face and voice recognition, gaze detection, and speech capability through its proprietary NAOqi framework [13]. Pepper is capable of actively approaching people via sound, voice, and/or its movements. Those capabilities and characteristics make Pepper an ideal standard platform for the School Robot Challenge. Pepper can be programmed with Choregraphe, a visual programming language, and other programming languages including Python, C++, and Java. It can be also controlled by Android and SoftBank Robotics provides SDK for Android Studio.

Through their participation in the School Robot Challenge, future generations will be inspired to become *active users* of social robots. Participating students are provided with the opportunity to explore ideas for how to optimize the environment where students and robots coexist in school and engage in programming a sociable humanoid robot to demonstrate their idea.

4   The Study

The study considers the Junior School Robot Challenge as an intervention and investigates the effects that the competition might have on participating students. It consists of multiple focuses. This paper reports the investigation of students' programming performances as demonstrated at the WRS 2018 held in October 2018 at Tokyo Big Sight in Tokyo, Japan. The research questions we ask are:

- Could participating teams successfully learn to program Pepper to create their solutions to the competition tasks?
- If so, what contributed to the success?

4.1   The Competition

The Junior School Robot Challenges is designed to be a venue where teams demonstrate their skills and knowledge of robotics as well as a creative and innovative idea for living and collaborating with robots in a school environment. The WRS 2018 School Robot Challenge consists of a 3.5-day workshop and the competition. The workshop provides time and support for teams to prepare their solutions to given competition tasks. During the workshop, teams learn how to control Pepper using the programs that they developed before participating in the workshop since not all teams have access to Pepper prior to the competition. In School Robot Challenge, teams are judged in three areas: Skills Challenge, Open Demonstration, and Technical Interview.

The Skills Challenge entails a set of tasks that test a team's ability to control their robot (Fig. 1). It consists of four required tasks that teams have to solve. With the WRS 2018 School Robot Challenge, the Skills Challenge tasks were disclosed at the beginning of the workshop so that all participating teams would begin working on them at the same time.

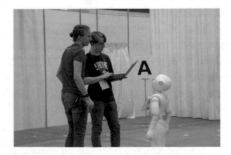

Fig. 1.   School Robot team members programming and preparing for their Skills Challenge

The Open Demonstration provides an opportunity for teams to demonstrate their idea(s) for human-robot coexistence and collaboration in a school environment. Each team has two chances to show their demonstration to showcase their solution to an identifiable challenge within a school-based context. The length of each demonstration is 5 min. During the demonstration, teams are asked to present a summary of the algorithms controlling the behavior or performance of the robot(s) and the problems that they encountered and overcame during the development process. A panel of judges evaluates their demonstration.

The Technical Interview is a face-to-face 15-min interview with a panel of judges. During the interview, each team explains their robotic performances and algorithms/programs, as well as how they worked together as a team. The focus of the interview is on students' understanding of the robotic technologies including sensors and functions, and algorithms/programs they have developed, and assessment of their work against technical criteria. Judges are very interested in understanding the processes students went through to arrive at their preferred solutions, and the authenticity and originality of their work.

4.2   The Workshop

The workshop at the WRS 2018 lasted 3.5 days providing time for teams to learn to control Pepper and work on the solutions to the competition tasks (Fig. 2). The workshop, first, provided the information focusing on how to handle the sophisticated social robot correctly, followed by the introduction to Choregraphe which includes the software interface and the important functions, such as how to wirelessly connect their laptop with Pepper, very important information for teams without access to Pepper prior to the competition. Those who had prior experience with the hardware and ready to start working on their programs could get right into programming. All teams were provided with supporting materials giving basic information about how to use Choregraphe when their team was selected. The lessons in the supporting materials were developed to cover the necessary functions and programming components that help teams to work on the Skills Challenges, as well as the Open Demonstration. Choregraphe can be downloaded free from the SoftBank Robotics' website. Thus, teams could work on their preparation before the competition even without the robot. Choregraphe comes with a simulator with which a program can be tested before running on the actual hardware. It also allows programmers to enhance the behaviors with their own Python code. In addition, Softbank Robotics provides online tutorials that help teams acquire more advanced knowledge and techniques on how to program Pepper (http://doc.aldebaran.com/2-5/software/choregraphe/index.html).

Fig. 2. Students programming and preparing for their presentation

4.3   Participants

There were 15 teams participated in the competition. Seven teams were from Japan (with one from a German School in Japan). Eight teams were from the Philippines, Malaysia, Australia, the Netherlands, Austria, Chile, and the U.S. (two teams). Two teams had all female members, and six teams had all male members.

Table 1.   Team demographic information and pepper access

Team ID	Residence	School	Pepper Access
N1	Philippines	High	No
N2	Chile	High	No
N3	Malaysia	High	No
N4	The Netherlands	High	No
N5	Austria	High	No
N6	USA	Middle	No
N7	USA	Middle	No
Y1	Japanese	High	Yes
Y2	Japanese	High	Yes
Y3	Japanese	Middle	Yes
Y4	Japanese	High	Yes (2 month+)
Y5	Japanese	High	Yes (2 month+)
Y6	Japanese	High	Yes (2 month+)
Y7	Australia	High	Yes (3 month+)
Y8	Japanese	High	Yes

There was a total of 57 students - 44 were male (77%) and 13 were female (23%) students. In addition, 39 students (68%) were in high school (15–19-year-old), while three were elementary school age students (5%, younger than 12-year-old).

Among the participating teams, eight teams had access to Pepper for some length of time prior to the competition. Most of those teams were Japanese except for one team from Australia that participated in the WRS Junior Category School Robot Challenge Trial event held in Summer 2017 (the Trial 2017) and was able to borrow Pepper for about three months to work on their open demonstration project. Among the seven Japanese teams, four teams had Pepper at their school and could use it for their practice. Three other Japanese teams obtained Pepper about two months prior to the competition. All other non-Japanese teams had no prior access to Pepper. One of those teams participated in the Trial 2017 where some of the team members worked with Pepper during the event. Table 1 shows the information about the country and school as well as Pepper access per team.

4.4 Programming Experience

The participating teams had various levels of programming experiences. Among the teams with Pepper access, four teams have text coding experience, mainly C language (C, C++), except for one team that had extensive experience in text coding with a variety of languages including Java, Javascript, Visual Basic, Python, HTML, C, CSS, as well as Mindstorms' visual programming language. One team whose school has Pepper for about one year had no prior programming experience but Choregraphe which they started to learn when they received Pepper. Another team without prior programming experience also learned to program with Choregraphe for 3 months. The other two teams have used LEGO Mindstorms which uses a visual programming language.

Among the seven teams without Pepper access, five teams had text coding experiences including C, C++, C# and/or Python, except for one team that also had programming experiences in JavaScript, SQL, HTML, and CSS. The other two teams have used Mindstorms' visual programming language.

4.5 Competition Performance

After the workshop, teams competed in the WRS' World Robot Competition which was held for five days. Despite different programming experiences and preparation experiences, all teams demonstrated their solutions for Skills Challenge and Open Demonstration during the competition. However, there were various factors contributed to the quality of the demonstrations. In this section, teams' performances are analyzed through their Skills Challenge tasks and Open Demonstration.

The level of difficulty of the Skills Challenge tasks in 2018 was much higher than those in the Trial 2017. The 2018 tasks required not only the understanding of the Choregraphe's basic functions but also some understanding of the advanced functions and the NAOqui APIs (http://doc.aldebaran.com/2-5/naoqi/index.html).

With the Skills Challenge tasks, as shown in Fig. 3, many teams did not score high. The highest average score of the four tasks was 85. Only three teams scored above 60. Six teams scored below 20. One team's score was zero.

The Skills Challenge result shows that having prior access to Pepper did not necessarily guarantee better ranking. Although among the top five teams, four teams had prior access to Pepper, the top team was the one that did not work with Pepper before the workshop. Focusing on the teams with Pepper access, five teams were among the top six teams. The length of the time that they had access to Pepper did not necessarily guarantee better ranking. The top two out of those five teams used Pepper for only two months. Four teams that had Pepper more than one year were below the two top teams. Comparing the top two teams with Pepper, one team had text coding experience; however, the other team had no prior programming experience. The third team among the five teams also did not have prior programming experience until they started working on Pepper with Choregraphe.

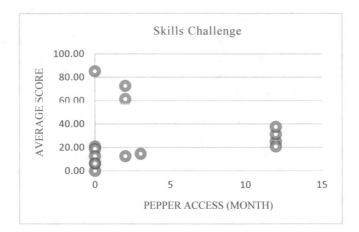

Fig. 3. Skills challenge scores by team

Focusing on the teams without Pepper, top three teams have text coding experience. The two teams who have experience only with a visual programming language (Mindstorms) had difficulties completing the tasks. A team with extensive text coding experience did not score well also. However, this is due to the lack of time that they could spend preparing for the Skills Challenge since they did not participate in the workshop.

With the Open Demonstration result, the teams that had Pepper prior to the workshop performed better (Fig. 4). However, there were teams without access to Pepper prior to the workshop outperformed some of the teams that had access to Pepper. Among the top three teams who had the Pepper access, top two teams were the ones with longer access to Pepper (over one year) with no prior programming experience or with a visual programming language (Mindstorms). The third team had Pepper for only two months but had text coding experience prior to participating in the competition. With the Open Demonstration, having longer access to Pepper seemed to have some advantage because teams could have a better preparation for the competition by creating the solution for the Open Demonstration before the competition. A team that had access to Pepper for two months but did not have prior programming experience scored lower than four teams without Pepper. It indicates that the longer access to Pepper had a stronger contribution to the result among the teams with Pepper access. Also, the result indicates that teams with text coding experience could be successful without a lengthy preparation period with Pepper.

Although the top three teams all had prior access to Pepper, among the top ten teams, four teams did not have Pepper prior to the workshop. The top team among the teams without Pepper had programming experience only with a visual programming language. Other three had text coding experiences. Since all of the teams without Pepper among the top teams had prior programming experience, either visual programming or text coding, programming experience might have helped the teams. The team with very little visual programming experience did struggle to successfully

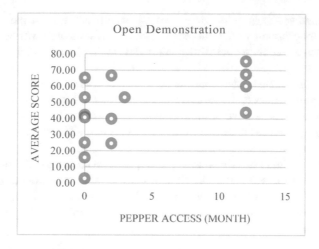

Fig. 4. Open demonstration scores by team

demonstrate the Open Demonstration solution. The team was also at the bottom with the skills challenge.

However, having extensive programming experiences with text coding did not necessarily help the teams either. Two teams with extensive programming experiences including multiple text coding programming languages did not score high. Their performance ranked in the middle group. One of the teams did not have Pepper, as well as did not participate in the workshop, while the program that they prepared was quite advanced. The other team did have Pepper for three months. Although their preparation looked promising, the complex program they created with taking some risks to complete challenging tasks might have made it difficult to show a perfect performance.

5   Conclusion and Next Step

Analysis of the programming performance of the participating teams at WRS 2018 Junior School Robot Challenge indicates that having access to Pepper, the standard platform robot, does not necessarily guarantee success in terms of achieving higher rankings. Looking at the top five teams of the Skills Challenge and Open Demonstration, though most of the teams had Pepper at their schools, the teams that did not have access to Pepper could also achieve scores high enough to rank within the top five.

However, when focusing on the teams with prior Pepper access, there seem to be two factors contributed to the success of the Open Demonstration performance, which are (1) length of the access to Pepper, and (2) prior text coding experience. If a team does not have access to Pepper, having programming experiences seem to give some advantage to the team.

With the Skills Challenge, five teams with Pepper access were among the top six teams. However, the length of the time that they had Pepper did not necessarily give the teams an advantage. Their prior programming experience did not have a clear indication of its contribution to the success either. With the teams without Pepper, text coding experience seems to help them with developing the solutions for the Skills Challenge tasks.

Although it is not clear what contributed to the success of the teams, there are some indications that the access to the standard platform robot before the competition and the programming experience have some influence on the teams' performances. Further investigation is needed to determine what contributes to the teams without access to the sociable robot becoming successful with its programming.

We plan to continue the study with participating teams at the WRS 2020 competition, where it is expected that up to 20 teams, from Japan and around the world, will participate. Using the data from the pilot study and future WRS events, we aim at providing the necessary support for the teams to successfully participate in the WRS 2020 competition.

References

1. Boyd, J.: Sony unleashes new Aibo robot dog. In: IEEE Spectrum. IEEE (2017)
2. Carman, A.: Sony's robot dog Aibo is headed to the US for a cool $2,899 (2018). https://www.theverge.com/2018/8/23/17773084/sony-aibo-dog-us-release-robot. Accessed 23 Dec 2018
3. Vogt, P., et al.: Child-robot interactions for second language tutoring to preschool children. Front. Hum. Neurosci. 11(73), 1–7 (2017)
4. Kennedy, J., et al.: Heart vs hard drive: children learn more from a human tutor than a social robot. In: The 11th ACM/IEEE International Conference on Human-Robot Interaction (HRI 2016). IEEE, Christchurch (2016)
5. Oh, K., Kim, M.: Social attributes of robotic products: observations of child-robot interactions in a school environment. Int. J. Des. 4(1), 45–55 (2010)
6. Hood, D., Lemaignan, S., Dillenbourg, P.: When children teach a robot to write: an autonomous teachable humanoid which uses simulated handwriting. In: The Tenth Annual ACM/IEEE International Conference on Human-Robot Interaction. ACM, Portland (2015)
7. Diyas, Y., et al.: Evaluating peer versus teacher robot within educational scenario of programming learning. In: The Eleventh ACM/IEEE International Conference on Human Robot Interaction. IEEE Press, Christchurch (2016)
8. Wong, C.J., et al.: Human-robot partnership: a study on collaborative storytelling. In: The Eleventh ACM/IEEE International Conference on Human Robot Interaction. IEEE Press, Christchurch (2016)
9. Arnold, L.: Emobie™: a robot companion for children with anxiety. In: The Eleventh ACM/IEEE International Conference on Human Robot Interaction. IEEE Press, Christchurch (2016)
10. Shamsuddin, S., et al.: Initial response of autistic children in human-robot interaction therapy with humanoid robot NAO. In: IEEE 8th International Colloquium on Signal Processing and Its Applications, Melaka (2012)
11. Tapus, A.A., et al.: Children with autism social engagement in interaction with Nao, an imitative robot: a series of single case experiments. Interact. Stud. 13(3), 315–347 (2012)

12. World Robot Summit - Junior Category Competition Committee. Junior Category - School Robot Challenge: Rules and Regulations (2018). https://worldrobotsummit.org/download/rulebook-en/rulebook-School_Robot_Challenge.pdf. Accessed 12 Dec 2018
13. de Jong, M., et al.: Towards a robust interactive and learning social robot. In: The 17th International Conference on Autonomous Agents and Multiagent Systems (AAMAS 2018). International Foundation for Autonomous Agents and Multiagent Systems, Stockholm (2018)

Evolution of Educational Robotics in Supplementary Education of Children

Anton Yudin[1,2]([✉]), Andrey Vlasov[1], Maria Salmina[3], and Marina Shalashova[2]

[1] Bauman Moscow State Technical University, Moscow, Russia
skycluster@gmail.com
[2] Moscow City University, Moscow, Russia
[3] Lomonosov Moscow State University, Moscow, Russia
http://www.bearobot.org

Abstract. The paper presents a retrospection of educational robotics and supporting technologies via the analysis of the evolution of a robotic competition, its rules and designs of participating robots. Corresponding evolution of tools for technical education is discussed in a context of digital fabrication and educational robotics focused on schoolchildren of 6 to 18 years old. Blended learning development is suggested to increase technical literacy among larger groups of children.

Keywords: Educational robotics · Technical literacy ·
Eurobot retrospection · Technology retrospection · Digital fabrication ·
Blended learning

1 Introduction

Mass education is weakly receptive to the achievements of science and technology. Years and sometimes decades pass before the "yesterday's" advanced technologies reach classrooms and begin to be used effectively there. Such a time lag will always be present in the "body" of human society, characterizing the rate of metabolic processes.

Industrial revolutions are triggers of fundamental changes in education systems. And in the absence of important inventions (such as, for example, the invention of the printing press) it is difficult to imagine the development of mass education in principle. The transformations never take place at once, but rather they have the nature of oscillation, ranging from impulses of the positive influence of enthusiastic innovators, who inspire with their exceptional educational results, to the massive introduction of seemingly best practices, but deployed in unsuitable conditions.

Successes and failures in the search for actual solutions to the problems facing the education system lead, over time, to a stable equilibrium, which should in the future ensure the next industrial revolution by training new generations of specialists, which in turn start the next cycle of conversion.

Below issues will be considered in the context of supplementary technical education (in Russia, voluntary after-school education for children of 6 to

© Springer Nature Switzerland AG 2020
M. Merdan et al. (Eds.): RiE 2019, AISC 1023, pp. 336–343, 2020.
https://doi.org/10.1007/978-3-030-26945-6_30

18 years old) related to the opportunities and limitations arising from the need for individualization and improving the quality of education of schoolchildren. A robotic competition retrospection is provided in an effort to show the possible mechanism of bringing educational content up-to-date and renewing the teaching process by raising questions actual for the industry and the society. The competition's evolution affects approaches and tools used in class for students to succeed in it. When exercised for several years such "competitional" education encourages adaptive behavior for both the teacher and the student resulting in a more capable graduate.

2   Project Activities in Engineering Education of Schoolchildren

In preparing future engineers, it is important, in addition to the transfer of direct scientific knowledge, to reproduce, if not always fully, the development process typical for such a specialist and the type of activity in general [1]. Since an engineer is engaged in project activities, their features need to be cultivated in class. Today, such attempts are made even at school, introducing project activities in the curricula, but according to educators working in this field [2] they face a number of problems and this happens inefficiently.

For supplementary technical education, project work is a natural form of transferring skills and knowledge to students. The authors have many years of successful experience of participating in the organization of supplementary education for schoolchildren in the field of educational robotics [3–6]. The purpose of training is to create a mobile robot in a certain period of time in a team and demonstrate results in competitions [7]. Such robot development cycles can be called educational. They have a pronounced project design character and contribute to the development of creative thinking and engineering skills of students while preserving well-known steps for the teacher.

Having gained experience in educational cycles, students can proceed to the development of individual projects that are distinguished by an increased level of autonomy in development. A number of such successful transitions makes it possible to say that the educational construction of robots results in an understanding of engineering activity, design work and provides a natural transition to self sufficiency for the student [8].

Unfortunately, the individualization of work leads to a sharp increase in the burden on the teacher. The load is expressed not only in the number increase of different projects, but also in the need for each student to look for solutions to problems, very often not similar to each other, inevitably arising in each project.

The complexity of individual engineering design work leads to the need of reducing the number of students in a group to preserve the quality of education. On the other hand, society requires teachers to ensure the mass and accessibility of education. There is a contradiction with which every teacher meets sooner or later. Some teachers in this case begin to work only with very few gifted children, while others are forced to limit the depth of educational material in order to preserve the size of groups.

Being known for more than 50 years, since the dawn of computers, blended learning seems to be a natural way of solving this contradiction of request and abilities. While some successful implementations exist there is still no common solution for an ordinary school. In the given context of technical education aiming development of creativity, abilities and knowledge it is even more important to find the proper ways of supporting individual student's progress development.

3   Eurobot Competition Retrospection

To be more precise on the aims and means of the discussed technical education it is important to address a real life example for tasks which students will likely face when they finish the proposed studies.

Figure 1 shows such an example in a progression of Eurobot mobile robot competition from its start in 1994 to the present time (the competition still exists).

First of all it is important to see that competition rules which affect the looks of the playing field and the design of competing robots changed drastically in a series of steps.

First step from 1994 to 1998 introduced competitions mostly for the university students. Environment requires rather simple navigation and the playing elements robots interact with are same if any at all. Designs of the robots are simple and parts are mostly done by hand or with a construction kit. The most capable teams start to work on vision systems and better autonomy for their robots.

Second step from 1998 to 2004 starts with accepting teams from other countries to the otherwise national competitions and brings common field dimensions and diversity of problem solving still relying on simple navigation for most robots. Possibility to win even with the smallest budget attracts hobbyists and enthusiasts with different backgrounds to the participation. The most capable teams start using navigation systems based on beacons. Vision systems are evolving but are most of the time a good research topic rather than a win necessity. Looking at the bulky computer hardware teams used to bring to their boxes at that time one can get the right feeling of the speed of technological changes our world is experiencing.

In 2004 the popularity of the competitions and increasing number of participants bring an international association to life. From 2004 to 2008 fields start to lose navigation lines and challenge all teams to develop beacon-oriented navigation systems. While most of the teams continue to build robots by hand top-rated teams rely on heavy and precise production machinery to achieve the needed robot part movement precision. Self-made electronics is a necessity to win.

In 2008 a special edition of the rules appears aimed at schoolchildren up to 18 years old. This starts Eurobot Junior edition and challenges educators from all over the world to prepare the young to build electromechanical devices. Since then Eurobot Junior follows the footsteps of its older "brother". This means also that technologies becoming common for Eurobot participating robots are adopted by Eurobot Junior with a lag of several years.

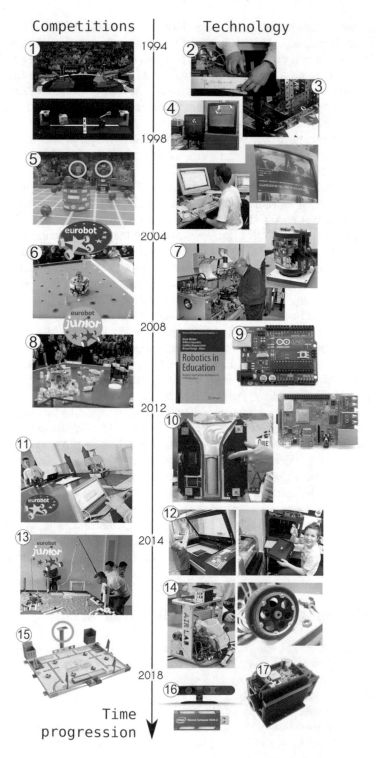

Fig. 1. Competition and technology retrospection.

From 2008 to 2012 playing field provides more and more tasks for the teams to solve simultaneously. Different playing elements to manipulate mean more mechanical design and diverse control. Strategies for robot operation also become more diverse. It is no longer sufficient to do more similar actions in a row to win, it is important to adapt in the process. In this period Arduino appears and becomes popular allowing beginner and smaller teams to compete better because of the electronics simplification. Some of the most capable teams start using 3D-printers and digital fabrication equipment which is still too expensive for massive usage. At this time the term "robotics in education" becomes official with the foundation of the conference.

From 2012 to 2017 rules are merging the previously introduced good practices and start giving even more freedom for teams in strategy planning. For example, navigation lines return on the playing field for beginner teams to be able to participate from scratch. In this step Eurobot Junior starts to copy the Eurobot rules, making it possible for the young to build similar mechanics and learn from the experienced older teams during the competitions. Eurobot in turn receives from Eurobot Junior the possibility to use 2 robots per team in a match.

With the appearance of affordable digital fabrication tools like a 3D-printer and a laser cutter it becomes possible to achieve precision while building mechanics for most of the teams. Precise milling machine is also providing more possibilities for custom electronics design implementation. Because of the technology shift teaching becomes possible in a digital fabrication lab for schoolchildren.

Since 2018 it is possible for a participating team to additionally use a standalone autonomous vision system to track the state of the field and playing elements. Further usage of affordable and compact vision systems by teams brings artificial intelligence tasks to the playing field. At the same time this step again brings new teams starting from scratch into play when compared to teams already participating for many years which means they have better solutions for common tasks on the field (i.e. navigation).

At this time the number of teams in Russia (156 teams in 2018, est. 175-200 teams in 2019) can be compared to the number of teams in France both countries giving almost 2/3 of the total number of teams in the world. This is seen by the authors as a result of a constant effort in educational systems of both countries orienting the students interested in engineering to the described competition practice.

4   Evolution of Tools for Technical Education

While visual changes of the rules are obvious it is important to note that they evolve in the direction of technical complexity and this could not be possible if the used technologies would stay the same.

Another important conclusion is addressing educational tools which need to change together with the tasks of the competition.

Figure 2 shows authors' experience within supplementary education groups of children from 6 to 18 years old. The picture refers to annual Eurobot competition practice involving about 200 students: tools' capabilities used in the educational process and corresponding student groups. Percentage showing average number of students. Acquaintance level renews each year with new students. Technical literacy level is composed of students staying at the lab and learning for 2–3 years. Preprofessional level is composed of students deeply involved in robot design for 4 and more years and aiming to enter a technical university upon school graduation.

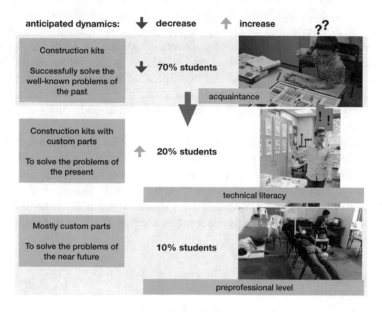

Fig. 2. Educational tools and resulting outcome.

Construction kits while solving the tasks of the past and aiding the teacher to start working with a group of students would never win the competition. It is possible to address the actual tasks if we start modifying the kit or develop own custom designs. With more and more affordable digital fabrication equipment it becomes real even for schoolchildren.

To be productive the educational environment of the near future should provide means of building such actual kits at class to increase the number of technically educated students. While blueprinting of the design could happen literally anywhere (provided there is electricity plug and Internet available) the actual development will still happen at class. This vision supports the idea of increasing efficiency when adding blended learning elements for such a context.

5   Conclusion

To be efficient and competitive in the new technological reality an engineer of the near future has to be inventive and creative. Such skills require practice which

Eurobot competitions constantly provide and prove to be effective in this matter. Top-edge technology now will be used by the most capable teams "tomorrow" and in two "days" it will be usual to see a solution on a common participating robot.

Digital form of robot's parts allows it to be stored on a computer and make it available within a network coverage. Blended education elements could be used to naturally document and deliver such past experience once it is properly organized and used to adapt the best solutions to the actual tasks and needs.

Current technological state shows a lot of potential in increasing the number of technically educated people by using blended education elements within supplementary education groups of children. It could be achieved by making educational process more individual and motivating [9]. While digital fabrication tools widely spread in networks of laboratories today allow design to be carried out any time and outside the actual place of studies, it is still a necessity to gather and share with the teacher and more experienced students to progress and improve.

In the future work, it is important to identify the factors hindering the development of blended learning to increase technical literacy among larger groups of children. Other educators are working in this field as well [10]. While using construction or robotic kits in education has its place in the educational system [11] it is important to find the means of quicker renewal of those technologies they utilize to cope with and fully exploit the increased speed of communication available today.

References

1. Shalashova, M., Shevchenko, N., Mahotin, D.: Development of functional literacy of school and university students, Espacio, T. 39, N. 30 (2018)
2. Yushkov, A.: Research and project activity in the secondary school: problems, stages of standards' cognition, educational results. Na putyah k novoy shkole, no. 3, pp. 37–42 (2014). ISSN 0869-690X
3. Yudin, A., Salmina, M., Sukhotskiy, V., Dessimoz, J.-D.: Mechatronics practice in education step by step, workshop on mobile robotics. In: 47th International Symposium on Robotics. ISR 2016, 590–597 (2016)
4. Sukhotskiy, D., Yudin, A.: Startup robotics course for elementary school. In: Communications in Computer and Information Science, CCIS, vol. 156, pp. 141–148 (2011). https://doi.org/10.1007/978-3-642-27272-1_13
5. Vlasov, A., Yudin, A.: Distributed control system in mobile robot application: general approach, realization and usage. In: Communications in Computer and Information Science, CCIS, vol. 156, pp. 180–192 (2011). https://doi.org/10.1007/978-3-642-27272-1_16
6. Yudin, A., Vlasov, A., Salmina, M., Sukhotskiy, V.: Challenging intensive project-based education: short-term class on mobile robotics with mechatronic elements. In: Advances in Intelligent Systems and Computing, vol. 829, pp. 79–84 (2019)
7. Yudin, A., Sukhotskiy, D., Salmina, M.: Practical mechatronics: training for mobile robot competition. In: 6th International Conference on Robotics in Education, RiE 2015, pp. 94–99 (2016). ISBN 978-2-9700629-5-0

8. Yudin, A., Kolesnikov, M., Vlasov, A., Salmina, M.: Project oriented approach in educational robotics: from robotic competition to practical appliance. In: Advances in Intelligent Systems and Computing, vol. 457, pp. 83–94 (2017). https://doi.org/10.1007/978-3-319-42975-5_8

9. Shakhnov, V., Vlasov, A., Zinchenko, L., Rezchikova, E.: Visual learning environment in electronic engineering education. In: International Conference on Interactive Collaborative Learning, ICL 2013, pp. 379–388 (2013)

10. Lepuschitz, W., Koppensteiner, G., Merdan, M.: Offering multiple entry-points into STEM for young people. In: Advances in Intelligent Systems and Computing, vol. 457, pp. 41–52 (2017). https://doi.org/10.1007/978-3-319-42975-5_4

11. Lammer, L., Vincze, M., Kandlhofer, M., Steinbauer, G.: The educational robotics landscape exploring common ground and contact points. In: Advances in Intelligent Systems and Computing, vol. 457, pp. 105–111 (2017). https://doi.org/10.1007/978-3-319-42975-5_10

Robot League – A Unique On-Line Robotics Competition

Richard Balogh[1]([✉]) and Pavel Petrovič[2]

[1] Slovak University of Technology in Bratislava, Bratislava, Slovakia
richard.balogh@stuba.sk
[2] Comenius University Bratislava, Bratislava, Slovakia
ppetrovic@acm.org

Abstract. We have successfully organized seven years of a competition in building and programming robots in Slovakia named Robotická liga (Robot League) with the motto 'The joy of learning through solving and sharing'. The activity is deeply based in the didactic theory of Constructionism, we beleive it is one of the most genuine examples of organized constructionist activities utilizing modern on-line technologies. A side result is a set of about 80 challenges with solutions, a useful learning resource for both learners and teachers.

The competition has a unique format allowing the teams to compete remotely from their home, school or club. In other conventional competitions, the participants pay too much attention to their own performance, and the sharing and mutual learning aspect tends to be neglected.

Traditional competitions require a long preparation that culminates at a tournament, where things can easily go wrong, leading to frustrations. Our starting point was to benefit from the motivational vector that stems from competing while correcting those common disadvantages. The core of the activity lies in that it stimulates an exceptional level of creativity and provides an early and manifold feedback in a repetitive fashion. In this paper we present interesting example tasks, discuss their classification, our experiences and recommendations and feedback from the participants.

Keywords: Robotics competition · Constructionism ·
After-school activities

1 Introduction

Our interest in the field of robotics originates in our passion for technology. Technology gives the power to reach places we could not reach, make useful things that would have been impossible, to help each other, to liberate ourselves, it helps to learn about ourselves, and it allows building a green, fair, peaceful and creative society with optimal use of resources, maximizing the happiness and productivity as contrasted to maximizing the profit and resources drain. We believe in technology that will make a better future for all the mankind,

© Springer Nature Switzerland AG 2020
M. Merdan et al. (Eds.): RiE 2019, AISC 1023, pp. 344–355, 2020.
https://doi.org/10.1007/978-3-030-26945-6_31

wildlife and the planet as well. We were lucky to be ignited and inspired for this as youngsters in social settings. In clubs, schools, competitions, and through sharing of ideas in the magazines, books, and other literature. We feel that sharing and social settings are crucial part for spreading the spirit of the belief in technology. With this attitude in mind, we have spent considerable efforts preparing or participating in organization of various events, where sharing and learning took place, individually, or together [1–3]. Summer schools and summer camps provided opportunities for the young people not only to be exposed to the technology, but included up to 24 h presence of savvy experts who had answers to almost all the questions. In a two week well-organized technology camp, the motivated children often advanced in their skills and knowledge more than during a whole year of regular participation in an after-school club, each one of the two serving as a catalyst for the other. A successful approach used in these schools relied on project work with a specific goal to be accomplished - typically building or programming a robot or another device to perform a chosen task. Even though the children cannot stay in the camp all year round, they can still work on projects. Hence the competitions. And hence the huge challenges they need to face: now they lack the access to tools, knowledge and skills, the coach has to organize their work despite their variable skills and interests. In some competitions, a team of 3–10 children is required to build a single robot in a several months long season. How much are all the members going to advance through hands-on learning, if the robot is obviously going to be designed by the one or two most talented builders in the team? Likewise for the programming part. Only through personal involvement and one's own mistakes, trials and errors, tinkering and experimentation the valuable learning takes place. Yet, in a team of 10, this could happen only in case of a group of ideal children. It is not the case in most of the groups. Other competitions suffer from the problem of recurrence. The category is the same every year, or slightly changes only occasionally. The successful teams come back in the following season with even stronger robots, being a very hard competitor for the newcomers. After they eventually leave, the level in the contest stagnates or even declines. A partial remedy for this is a requirement to submit all the technical documentation and publish it on-line. Thus the teams in the next season can build up on the previous experiences even from other teams. A huge problem for many robot competitions is the high participation cost, placing an extra burden on team coaches who often fail to find the sufficient resources. And despite the high fees, the tournaments and their organizers are still dependent on the sponsors and their unpredictable mood and behavior. And finally, the creativity and narrow focus: yes, the soccer robots are all different, but at the same time, they are all the same. We are seeing the same ideas, designs, and algorithms over and over. Industry 4.0 is going to depend on creative prototyping and versatility. We should reflect on that. In 2013, after the autumn FLL season has finished, we were asked by the coaches if we could suggest an activity for their robot clubs during the idle spring season. Collecting on our experiences from different types of contests, we founded the Robot League.

2   Robot League

Our answer to above mentioned problems is a completely new, off-line, distance competition called Robot League. The competition has a unique format allowing the teams to compete remotely from their home, school or club. They spend as much time as they need to construct and program a robot, the only limitation is to upload their solution till the deadline on the competition web server.

2.1   Organization

We publish a set of tasks every two weeks. Each round consists of two different tasks, so the teams can choose which one is more attractive to them. They can solve both, and the better solution counts in the scores. The team has to upload their solution to the special competition web, where it is then evaluated by the jury of at least three independent reviewers. Their average ranking is then assigned to the team.

Fig. 1. Physical exercise of the M-Team's robot.

Teams work at home or in the after-school robot clubs. When they are finished with their solution, they typically use mobile phone to take pictures and record videos that they upload to YouTube service. Some teams use video editing software to add interesting effects or narrative, explaining the principles of their solution. Some teams add comments to their programs. Since the solutions are invisible before the task deadlines, and the YouTube videos are marked as unlisted, it is guaranteed that the teams work independently. The more keen on watching their peers solutions they become as they are made available. The organizers, and the coaches also take their part of the benefit from the learning process.

We have observed that once a team exceeds a certain threshold of their motivation, then they are highly likely to attempt to solve all the tasks in the season. In this way, we also hope to contribute to their regular work habit. One of the main benefits of these activities is that they initiate an interaction between the adults and young learners, which tends to be neglected today. Yet, some of the teams are completely independent and work on their own, without accepting any advises from adults.

Fig. 2. Bulb replacement robot by the Legolas team.

The problem of the degree to which an adult can help the children solve the tasks is probably the largest we have encountered, but it is not specific to this contest. And there is no good solution to it except of wise coaches, which is something the organizers are not able to influence. Here we refer to an alternate approach used for instance in the Istrobot contest [3]: there is no age limitation, and the teams often consist of a father with a child, or similar. Such setup makes it possible for the young to learn even more in a project that gets completed to a level beyond his or her individual skills. Having this attitude and benefits in mind, it is even more questionable if it is beneficial to completely eliminate the help from elders.

Everything started in May 2013. During the first year we published a new task every week, however, the teams had time two weeks to solve it, thus they could already see the specification of another round while they were working on the previous one, and think in advance, let their ideas develop over time. Alternately, they could divide into dedicated groups, each solving one of the available tasks. The same principle remains till now, we just adjust the intervals of publishing new tasks to two weeks and we extend the solving period to four weeks.

Recently, we enjoy approximately 20 active teams and this is just enough to be able to evaluate all of them regularly. During the first years we tried to extend the period during the holiday time, but this was not successful attempt and most of the teams simply didn't work during the summer holidays.

During the years, it was necessary to develop a specific application to help maintaining the competition flow. Since the 2014 we use a specialized web application described in the Sect. 2.5.

The number of teams is summarized in Table 1. Not all the teams were working regularly, some of them just tried and didn't continue, some of them at least attempted to solve every single task. Last year, for example we have received and evaluated 151 unique solutions in total (on average 18–19 per round), however, not the quantity, but the endeavor, and the quality of the solutions were the most satisfying.

Table 1. Summary of participants since the origin of the competition, 2019 is estimate. Note: Teams 50 – Number of really active teams, i.e. teams solving at least half of tasks. 2019 – an estimation as the competition is not yet finished.

Year	2013	2014	2015	2016	2017	2018	2019
Number of tasks	10	10	10	10	8	8	8
Maximum score	30	30	30	30	24	24	24
Winners score [%]	75.3	94.5	95.9	89.7	99.0	99.5	*100*
Number of teams	6	18	26	24	20	19	*21*
Teams 50* [%]	33.3	38.9	61.5	62.5	75.0	78.9	*71.4*

2.2   About the Rules

In contrary to other robot contests, the Robot League focuses on constant and patient work on problem solving instead of single and energetic thrust culminating in a full stress. The contest is run in a friendly, open and trustful spirit with the main goal to have fun while learning something new.

The rules are kept as simple as possible with minimum limitations, see [4] or liga.robotika.sk for details.

The best teams in every round receive a diploma and a bag with spare plastic parts and sweets. A supportive NGO sponsor and the local distributor provide Lego Mindstorms robotics set to the overall winner with the highest score after all rounds. An average number of team members is 5–6, usually boys, with several pure girl teams (average team consists of 5 boys plus 1,75 girl). It is difficult to state exact numbers as some teams (especially school club) varied during the contest period. But the number of girls is higher than in most other contests, so we assume this type of design competition is also attractive for girls, but we didn't investigate this topic more deeply.

2.3 Tasks Classification

Our aim is to provide a wide variety of task types. One of our goals is to show examples of using robots in various curriculum-supporting scenarios, in subjects as Physics, Mathematics, Informatics, Art, or Biology, see [5–7] for more examples of inspiration. Tasks are being published in a random order as they are invented. Herein we classify them to different nondisjunct categories, see Table 2: *construction challenge* (C) is a task with a high demand on creativity in design and mechanical invention; *programming challenge* (P) requires non-trivial programming, sometimes use of data structures or files; *environmental interaction* (I) relies to somewhat larger extent on the interaction of the robot with its environment using sensors; *environmental manipulation* (M) requires the robot to successfully modify its environment; *Navigation* (N) needs strategies for moving the robot around its environment; *Measure* (E) tasks require measuring some physical properties using sensors; *Physics* (F) are tasks useful for discovering or demonstrating physics laws, or which need some physics insight; *Mathematical* (+) are tasks encouraging mathematical thinking; *Art* (A) are tasks focusing on artistic creativity; *Multi-robot* (2) are tasks with interaction of two robots; and *Static* (S) tasks have robots that do not travel in their environment.

Table 2. Classification of Robot League tasks.

Year/round	C	P	I	M	N	E	F	+	A	2	S	Year/round	C	P	I	M	N	E	F	+	A	2	S
2013	5	2	3	4	8						1	2017	5	2	2	6	7	1			2		5
2014	3	3	1	4	3	1		1			3	2018	6	3	3	3	6	3	1		1		7
2015	4	1	2	3	4		3		1		4	2019	9	4	1		5	2	1	1	1		9
2016	3	1	2	5	6	4	3	3			7	Total (96)	31	16	15	24	39	10	10	4	3	3	36

2.4 Example Tasks

In this section, we present few different tasks, each focused on a subset of competences. Formulation of the task is often very short, intentionally not specifying too much details. Solution is open-ended, leaving the final realization open just to the team creativity.

Pull-over. This is an example of task focused mainly on the mechanical construction and requires also some understanding of physics and mechanics. The task was to build a robot able to perform a pullover - a kind of gymnastic move performed on the high bar in which one pull its legs up and over the bar and its body rolls backward around the bar. To be honest, it was probably the most complicated tasks and no one was able to build the robot according to our vision. The closest one was M-Team, with two motors on legs, one for approaching the bar and last one for gripping the bar (see Fig. 1).

Bulb Exchange. The task specification is very short: build a robot that is able to replace a bulb in a table lamp. This requires creation of a mechanism for bulb grip, turning it an finding the right position. Many teams didn't think of complex solution, only one team designed also a bulb storage and their robot was able to store wrong bulb, take new one and replace it in the fixed lamp position (see Fig. 2). The gripper itself was able to adapt to different bulb diameters.

Spider Web. Task, inspired by Rodney Brooks bio-inspired robotics [8] asked to build a robot able to create simple spider net. Start of the fiber may be already fixed and the net has to be weaved using existing fixed objects (e.g. PET bottles, knitting needles etc.)

One of the best teams, RDS, created a plotter-like machine with the fiber bin and it was able to create not only the spider net, but almost any ornament (see Fig. 3).

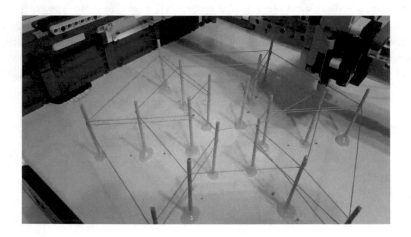

Fig. 3. Spider net by the RDS team.

Exchange Our Smiles. Some tasks are more focused on mechanical side, required just minimum programming, others are more challenging. An example of those is the task to find some randomly placed bricks in the front of the robot and to place them into the smiley shape. Robot should be able to work in many different starting configurations (see Fig. 4).

Fig. 4. Start and finish of the SiToMaFi team robot operation and the corresponding software in NXT-G programming language.

2.5 Supporting Software

In the first year, we have manually published the submitted solutions on-line, which even for 6 teams turned out to be a too demanding and time-consuming task. Later, students as part of the course Information Systems Development have specified, designed and implemented a web application as their team project. Currently, we are using the third complete rewrite, with new features added as the contest has been developing. The application allows the team leaders to register a new team at any time, view all past and current task specifications and submit and later edit a solution to a new task before the deadline. They provide a rich text description, upload pictures, programs or other attachments, and enter video links. Their solutions are nicely formatted and pictures are presented in a gallery, all integrated into the Robot League website. Judges can view the solutions, assign scores, and write their textual evaluation. Administrator then reviews and merges all the judges' comments, selects the best teams in each round, and makes the evaluation public. The resulting scores are computed automatically and shown in a table on the website. Organizers can enter their task descriptions with pictures and videos and maintain a pool of tasks to be used later in the contest, providing the starting and ending dates of the round. This system has simplified the organizers work and made it more convenient for the teams to adjust the presentation of their solution as they wish. The system currently supports Slovak, English and German languages, so that all tasks can be prepared in all three versions. It is open source and available at [9] (Fig. 5).

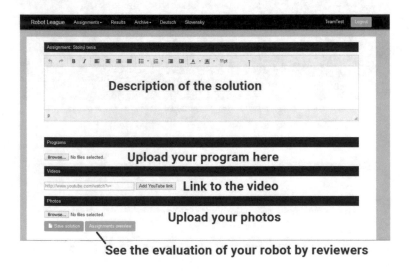

Fig. 5. The page for uploading a new team solution.

3   Discussion

Although we have some experiences and we believe this competition format is better than other in many ways, still some questions remains open.

Should the rules be limited in the same way as the FLL competition rules? After the requests from the participants, we have allowed to use an alternate software. Teams have then happily used NXC, Robot C, and MonoBrick software. It may also open further doors to creativity when we allow the use of additional components, not strictly limited to LEGO parts. Wider base here can later be limiting when teams start to prepare for regular FLL competition.

Children can easily forget some important limitations (this sensor was allowed, so why we can't use it now?), but FLL rules tend to be clear and the teams know they must check them thoroughly. Currently, we restrict these to be supportive objects only, for instance gymnastics rings, lighthouse, or an umbrella fabric. Some teams use 3D printing for making them. We keep the limitation to LEGO kits to address the issue of common platform and equal opportunities. However, we are considering a completely new parallel contest with Arduino platform and 3D printed designs. *Is there an "optimal" time interval for solving the tasks?* Are four weeks sufficient? Usually the clubs meetings are organized once a week, so they may need more time to discuss and solve the task. Currently, each round consists of two tasks. This leaves 4 tasks open for solutions at any time. A typical robot club has 6–15 members and thus the leader can split the children into groups of 2 or 3, each solving a different task. This makes the organization of work much more smooth, and manageable as contrasted for instance to running an FLL team with 10 children who are to build and program only a single robot in the span of four months. We believe leaving more time for a round would result in teams finding other activities, loosing focus. Lifting this time constraint could harm the efficiency and the skills of setting the priorities. It is a league after all. *Is the age limitation 9–16 years appropriate?* In FLL, we found that young teams from elementary schools have very hard time competing against 15–16 years old students of secondary schools. In fact, it is almost impossible for them to win, provided such a team with sufficient skills takes part. In our league, we try to design tasks, which can be solved by younger students too, if they try hard enough. A successful solution leads to full score – almost every round has some. Thus a focused younger team with full dedication can outperform older teams. Yet, they learn from each other by watching the solutions of other teams approaching the challenges from a different perspective. *Can the evaluation system be improved?* We also considered the idea of taking the 3 or 5 best tasks from all. The positive effect of giving teams a chance to skip a couple of tasks would result in too many teams having the full score. It could harm the motivation, and the habit of regular work. When communicating with teams, we try to give emphasis on learning, sharing, and having fun participating and that we are happy for all their solutions, even partial ones. Finally we provide a qualitative feedback from a team leader of one of the successful teams: *I have to say that our boys enjoyed the competition very much and not only this year. We have summarized our whole year last Tuesday, and Robot League is among the most interesting activities we do. We have a couple of new members this year, and they were immediately excited by the League and joined. Boys were looking at solutions of other teams after the deadline, and sometimes they could see a way they were considering, or were unable to complete, and it was super that they could see it from a different angle or in a different variation.*

4 Conclusions

During the seven years with the organization of creative constructionist robotics on-line competition, we have collected enough experiences and individual evidence for recommending it to others.

It was focused on young people in Slovakia aged 9–16, with 8–10 tasks each year. Its main aim is to support creative and design thinking of the pupils. We have experienced a growing interest and a positive feedback from the participants. In the 2019 we start to spread the competition in Austria too and we are open to a global international participation. Participation in the contest is free and easy, emphasizes learning, and creativity, it does not include stress and frustrations from failures at tournament day – which are inevitable in on-site single-day events. The amount of sharing ideas that takes place in this contest is among its strongest advantages. We efficiently utilize modern media – such as YouTube videos and a dedicated web application that allows the participants, referees, and organizers to maintain the contest automatically on their own without any other assistance from a system administrator.

Instead of increasing the size of this single competition, we would like to encourage leaders and organizers of robot competitions around the world to organize their own local versions. The software is open-source and can be found at Github repository [9]. We can provide assistance with its deployment. The prospective organizers can get inspired by the tasks solved in previous years of our competition. We would be honoured to provide consultations when starting new local contests.

Acknowledgements. Publication of this paper was supported from the project Smart mechatronic systems Nr. VEGA 1/0819/17, financed by the Slovak Scientific Grant Agency VEGA.

References

1. Petrovic, P., Balogh, R.: Educational robotics initiatives in Slovakia. In: Proceedings of Teaching with Robotics Workshop, SIMPAR 2008, Venice, Italy (2008)
2. Petrovic, P.: Ten years of creative robotics contests. In: ISSEP 2011, Bratislava (2011)
3. Petrovic, P., Balogh, R., Lucny, A.: Robotika.SK approach to educational robotics from elementary schools to universities. In: Proceedings of the 1st International Conference on Robotics in Education, RIE 2010, September 2010, Bratislava (2010). Jones, A.B., Smith, W.: Statistical Methods for Scientists. Wiley, New York (1984)
4. Petrovic, P., Balogh, R.: Summer League: Supporting FLL Competition, Constructionism in Action, Feburary 2016, Bangkok, Thailand (2016)
5. Bratzel, B.: Physics by Design with NXT Mindstorms. College House Enterprises (2009)
6. Ford, T., Perova, N., Church, W.: Physics with Robotics: An NXT and RCX Activity Guide. College House Enterprises (2009)
7. Cole, J., Terrell, A., Green, A.: Hands On Physics with LEGO MINDSTORMS NXT: 10 Lessons for the Classroom. CreateSpace Independent Publishing Platform (2013)

8. Brooks, R.A.: Flesh and Machines: How Robots will Change Us. Pantheon Books, New York (2002). Frank, D. (ed.)
9. Kunovska, E., et al.: Robot league web application. Open-source software (2016). https://github.com/TIS2016/Roboticka-Liga

Autonomous Driving Car Competition

João Pedro Alves[1,2,3], N. M. Fonseca Ferreira[1,3,4(✉)],
António Valente[2,3], Salviano Soares[2,5], and Vítor Filipe[2,3]

[1] Institute of Engineering of Coimbra/Polytechnic Institute of Coimbra,
Coimbra, Portugal
nunomig@isec.pt
[2] University of Trás-os-Montes and Alto Douro, Vila Real, Portugal
[3] INESC TEC-INESC Technology and Science, Porto, Portugal
[4] Knowledge Research Group on Intelligent Engineering
and Computing for Advanced Innovation and Development (GECAD)
of the ISEP, IPP, Porto, Portugal
[5] IEETA, Institute of Electronics and Informatics Engineering of Aveiro,
Aveiro, Portugal

Abstract. This paper presents the construction of an autonomous robot to participating in the autonomous driving competition of the National Festival of Robotics in Portugal, which relies on an open platform requiring basic knowledge of robotics, like mechanics, control, computer vision and energy management. The projet is an excellent way for teaching robotics concepts to engineering students, once the platform endows students with an intuitive learning for current technologies, development and testing of new algorithms in the area of mobile robotics and also in generating good team-building.

Keywords: Education · Project-based learning · Autonomous driving ·
Computer vision

1 Introduction

Robotics courses deal with kinematics, dynamics and control of manipulators [1], but it is fundamental to involve students in the construction and development of robotic platforms, and in project-oriented learning in order to forest creative and divergent thinking, and promote acquisition of self-learning and practical skills [2]. For this reason the use of a robotic teaching platform allows the students a fast understanding of the theoretical topics, and the use of open platform allows students to easily evolve acquired knowledge for project-based learning [3–11]. In this paper, we will focus autonomous driving with the aim of participating in the autonomous driving competition in the National Festival of Robotics in Portugal. The primary objective of the robot is the ability to drive autonomously. After studying the evolution of the main projects in the field of autonomous driving, a survey was conducted of the requisites necessary for the construction of a robot [12]. The robot was divided into modules in order to simplify its construction. The robot was equipped with a traction module (to enable control of the longitudinal displacement of the robot), a steering module (to enable steering control), a computer vision module (to detect the road's lane, crosswalk and traffic lights of the competition circuit) and a navigation application module (to enable the robot to drive autonomously around the circuit). Each module includes the hardware and software required for the performance of its function. With this in mind, in the future, different groups to test new solutions without having to know the full functioning of the robot. The odometry was used to calculate the position of the robot on the track. Based on the chassis of a modelling radio car and resorting to vision by the computer, the robot must be guided autonomously in a controlled environment. The development of the robot should contemplate future participation in the category Challenge (Challenge Class) of the autonomous driving test.

The National Robotics Festival is an event that has emerged through an initiative of the Portuguese Society of Robotics. This event aims to publicize the science and technology among young people who go from basic education to superior, as well as the general public, through competitions of robots, thus contributing to the development of research in the areas of robotics and automation. This festival has the characteristic of being held every year in a city has already covered several parts of the country since its inception in 2001. The events that take place throughout the festival involve the mastery of components with different degrees of difficulty, which is why the various tests distributed in two types of competitions, the senior exams, and the junior exams. The junior competitions are the search and rescue, dance and junior robotic football events. For its part, the tests that compose the senior competition are the tests of autonomous driving and the robotic football average robot league. The main objective of our project consists in the development of a robot with to participate in the autonomous driving test present at the National Festival of Robotics.

The paper is organized as follows. Section 2, describes the autonomous driving contest. Section 3, presents the mathematical model of the autonomous robot. Section 4 presents the design and implemented autonomous robot for the contest, and Sect. 5 present the adopted vision system and Sect. 6 an example of the lane lines recognition. Finally, Sect. 7 presents the main conclusions and future work.

2   The Autonomous Driving Contest

The autonomous driving contest is intended for fully autonomous, and this contest included two distinct categories by Rookie and Challenge Class, differing from each other on the difficulty level that they present. In the contest, the robots have to make a course along a closed lane, which bears similarities to the driving of a vehicle on a regular road. The circuit, shown in Fig. 1, has the approximate shape of an eight and is composed of two lanes delimited by two white solid lines and separated by a dashed line also white. In the middle of the circuit, there is a zebra crossing and a pair of traffic lights.

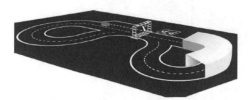

Fig. 1. The circuit

The traffic lights are intended to provide orders to the robot. The different orders are associated with the images, which the robot has to identify. However, only two of these signs, those that identify the order of stop and move on, are used in the Rookie category race. In the Challenge category, besides the signal panels, there is still a group of signals arranged at the edge of the track, which the robot has to identify. Figure 2 shows the signs used in the race in 2018.

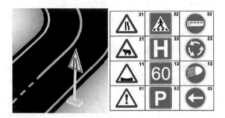

Fig. 2. Signage panels

There are also other challenges that are only present for the Challenge category, namely two obstacles that will be placed on the roadways, thus forcing the robot to identify them and to deviate from them if necessary. There is an area considered as a zone of works delimited by cones and a tape of signalling in all similar to those used in the identification of zones of work in real situations. There is also a tunnel, placed in one of the curves of the circuit, which suppresses the two continuous lines delimiting the track and darkening that area.

3   Mathematical Model

The Ackerman kinematic model describes how the robot moves in the road. In this model, the rear wheels are responsible for the movement and the front wheels are responsible for steering, Fig. 3.

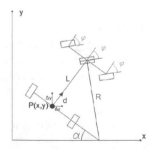

Fig. 3. Autonomous robot kinematic.

$$x(k+1) = x(k) + \Delta x \tag{1}$$

$$y(k+1) = y(k) + \Delta y \tag{2}$$

$$\alpha(k+1) = \alpha(k) + \Delta \alpha \tag{3}$$

$$\Delta x = \Delta d \, \sin \alpha \tag{4}$$

$$\Delta y = \Delta d \, \cos \alpha \tag{5}$$

$$\Delta \alpha = \frac{\Delta d}{R} \tag{6}$$

$$R = \frac{L}{\sin \varphi} \tag{7}$$

Δd is the displacement measured by the encoder, Δx is the displacement of the robot along the x-axis, Δy is the displacement of the robot along the y-axis, x(k + 1), y (k + 1) position of point P at the present moment, x(k), y(k) represent the position of point P at the previous moment, $\Delta \alpha$ represents the angular displacement relative to angle α. As can be seen in Fig. 3, which represents the navigation application, it is possible to see in the field of odometry the positioning of the robot with the origin of its movement. To obtain this information the navigation application uses the data of the linear displacement sent by the traction module and the inclination applied to the wheels coming from the steering module.

4 The Design Autonomous Robot

The autonomous robot named UTAD McQueen designed to consider the autonomy in order to be able to complete the contest. The decision making on track, the robot must be able to collect and analyze all the information about the track and traffic signals and also control its movements in order to perform the desired goal. Based on the analysis of the characteristics of the track, and also the solutions presented by the different teams over the last few years, and considering the rules of the contest, the energy autonomy can be guaranteed by two 12 V batteries each.

The robot use the right road lane and the dashed meddle line to guide along the track, and to identify this line, as well as the traffic lights that regulate the beginning of the race, the robot use computer vision. For navigation, the robot will resort to odometry to know where in the circuit also to activate or deactivate a specific task. For the communication the robot will use a raspberry PI3B as the primary processing tool and provide the necessary information to the PC and also the computer vision module. The adoption of modular architecture aims to simplify the development of the robot by dividing the problem into more straightforward and less complex problems. It is the architecture also allows making changes to one of the modules without changes the remainder of the modules. This modular architecture is composed of four modules: the traction module, the steering module, the computer vision module and the navigation application module.

- **Traction**
  This module covers all the elements that provide the movement of the robot, namely the traction motor, the controller and the module control application implemented in the PC, allows the (acceleration, speed, and displacement) is controlled in real time by simple commands [13].

- **The steering**
  This module covers all the elements responsible for by the electronic control, namely the servo motor, the micro controller and the control application of the module implemented in the PC.

- **Computer vision**
  This module is composed of one camera and the computer vision algorithms to process the images. The computer vision application running in the PC and in the PI3B is responsible for digital image acquisition, image processing and under-standing, including algorithms for lane segmentation, perspective geometry cor-rection, traffic lights detection and signal recognition. The processing of images provides information about the car's position in the track, on the traffic lights and the crosswalk supporting the robot's navigation [14].

- **The navigation application**
  This module is composed of an application implemented in ROS on the PC and in the PI3B, resulting from the merger of the control applications of each module. The navigation application interconnects and controls all other modules in a way that autonomous robot. In this application, the implemented task uses odometry to aid the navigation [15–18].

The platform used for the development of the robot comes from a chassis of a radio car modelling. This platform has a structure where the front wheels are attached to two steering bars, and an axle interconnects the rear wheels as show in Fig 4. In turn, the shaft is connected to a set of toothed wheels to which was coupled an electric motor. In order to allow the electronic control of the steering, a servo motor was used mechani-cally to the steering bars. In this way a platform that can be controlled electronically.

Fig. 4. The autonomous robot named *UTAD McQueen*,

5 The Adopted Vision System

Road perception is a major task in an autonomous driving robot, the robot's position on the track, the road lanes and the others objects are detected through image processing. As can be seen in Fig. 5, for simulation and real road, the spatial coordinates of an object identified in the images with the coordinates (x_{obj}, y_{obj}) are obtained by the following expressions:

$$x = x_{obj} - dx \tag{8}$$

$$y = y_{obj} \tag{9}$$

Fig. 5. The relation between the image position and the robot position

6 Example of Lane Lines Recognition Algorithm

For example, if a student choose to develop an algorithm based on CNN, it is necessary to obtain a set of training and test images. For this it is necessary to obtain images from several places of the track. Before starting CNN training, it is necessary to define some labels, which indicate what for example the dashed line, the right line, and other objects on the road, among other images. These images are segmented by clearly identifying each label in question. Then it is necessary to train the CNN, and this process is done only once. After obtaining the results, the network is stored and loaded into the memory on the PC, so CNN does the semantic segmentation of the obtained images acquire from the camera of the robot.

Once the segmentation is done the algorithm draws two lines one on the dashed and the other one on the line on the right side of the road and takes as reference the camera installed in the robot. When the robot start to move whenever there is a change in the configuration of the lane, it is detected by the camera installed, the algorithm calculates the angle from these two lines. For example, if the students want to trace lanes detection lines with the transform Hough algorithm or if they want to use a polynomial system, they can easily test and modify. After obtaining the value of the angle, these module sent this information to the steering module so that the direction of the robot can be controlled.

362   J. P. Alves et al.

7   Conclusions and Future Work

Summarizing, the proposed robotic platform allows the students, to improve and develop new ideas, also this type of robotic platform it is very good for serious research. The architecture of the robot, developed by modules allowed the students to test the operation of each module separately. The traction module was developed to allow displacement of the robot. The tests performed led to the conclusion that when feeding the traction module with both batteries the robot can reach an average velocity of 2 m/s. A steering module was developed to control the direction of the robot. In this module, a PID controller was used to allow the robot to follow the dashed line of the lane, based on the information collected by the computer vision module. The student can develop new algorithms without change the low level control, also these modules architecture and the ROS system allows the user friendly interconnection between the MATLAB® and the implemented hardware. Moreover, the reducing time of conception and also the facilitating learning of concepts acquired in other curricular units. Allows students to dedicate themselves to the task they have chosen to improve the project.

Acknowledgment. This work is financed by National Funds through the Portuguese funding agency, FCT - Fundação para a Ciência e a Tecnologia within project: UID/EEA/50014/2019.

References

1. Fonseca Ferreira, N.M., Freitas, E.D.C.: Computer applications for education on industrial robotic systems. Comput. Appl. Eng. Educ. **26**, 1–9 (2018). Special Issue Article. https://doi.org/10.1002/cae.21982
2. Zhao, Z.: The reform and practice of education mode in multi-disciplinary joint graduation design. Procedia Eng. **15**, 4168–4172 (2011). https://doi.org/10.1016/j.proeng.2011.08.782
3. Fonseca Ferreira, N.M., Moita, F., Santos, V.D.N., Ferreira, J., Candido, J., Santos, F.M., Silva, M.: Education with robots inspired in biological systems. In: Advances in Intelligent Systems and Computing, Robotics in Education Methods and Applications for Teaching and Learning (2019). ISBN 978-3-319-97085-1. https://doi.org/10.1007/978-3-319-97085-1_21
4. Couceiro, M.S., Figueiredo, C.M., Luz, J.M.A., Ferreira, N.M.F., Rocha, R.P.: A low-cost educational platform for swarm robotics. Int. J. Robots Educ. Art (2011)
5. Moura, T., Valente, A., Sousa, A., Filipe, V.: Traffic sign recognition for autonomous driving robot. In: IEEE International Conference Robot Systems and Competitions – ICARSC/Robotica 2014, pp. 303–308, Espinho 2014 (2014). https://doi.org/10.1109/icarsc.2014.6849803
6. Couceiro, M.S., Ferreira, N.M.F., Rocha, R.: Multi-robot exploration based on swarm optimization algorithms. In: ENOC 2011 - 7th European Nonlinear Dynamics Conference, Rome, Italy, 24–29 July 2011 (2011)
7. Couceiro, M.S., Ferreira, N.M.F., Machado, J.A.T.: Hybrid adaptive control of a dragonfly model. J. Commun. Nonlinear Sci. Numerical Simul. **17**(2), 893–903 (2012)
8. Vital, J.P.M., Rodrigues, N.N.M., Couceiro, M.S., Figueiredo, C.M., Fonseca Ferreira, N.M.: Fostering the NAO platform as an elderly care robot - first steps toward a low-cost off-the-shelf solution. In: 2nd International Conference on Serious Games and Applications for Health - SeGAH 2013, Vilamoura, Portugal, 2–3 May 2013 (2013)

9. Vital, J., Fonseca Ferreira, N.M., Valente, A.: NAO robot as a domestic robot. In: Proceedings of CLAWAR 2018, 21st International Conference on Climbing and Walking Robots and Support Technologies for Mobile Machines (2018). ISBN 978-1-9164490-0-8
10. Junior, F.C.F.M., Saraiva, A.A., Sousa, J.V.M., Fonseca Ferreira, N.M., Valente, A.: Manipulation of bioinspiration robot with gesture recognition through fractional calculus. In: IEEE LARS 2018 – 6th Brazilian Robotics Symposium (2018). https://doi.org/10.1109/lars/sbr/wre.2018.00050
11. Couceiro, M.S., Luz, J.M.A., Figueiredo, C.M., Fonseca Ferreira, N.M.: Modeling and control of biologically inspired flying robots. Robotica **30**(1), 107–121 (2012)
12. Balogh, R., Tapak, P.: Modelling the driver assistance systems using an arduino compatible robot. In: Advances in Intelligent Systems and Computing, Robotics in Education Methods and Applications for Teaching and Learning (2019). ISBN 978-3-319-97085-1. https://doi.org/10.1007/978-3-319-97085-1_36
13. Ivanov, V., Savitski, D., Shyrokau, B.: A survey of traction control and antilock braking systems of full electric vehicles with individually controlled electric motors. IEEE Trans. Veh. Technol. **64**(9), 3878–3896 (2015). https://doi.org/10.1109/tvt.2014.2361860
14. Costa, V., Cebola, P., Sousa, A., Reis, A.: Design of an embedded multi-camera vision system—a case study in mobile robotics. Robotics **7**, 12 (2018). https://doi.org/10.3390/robotics7010012
15. Costa, V., Rossetti, R., Sousa, A.: Simulator for teaching robotics, ROS and autonomous driving in a competitive mindset. Int. J. Technol. Hum. Interact. **13**, 19–32 (2017)
16. Costa, V., Rossetti, R.J.F., Sousa, A.: Autonomous driving simulator for educational purposes. In: 2016 11th Iberian Conference on Information Systems and Technologies (CISTI) (2016). https://doi.org/10.1109/cisti.2016.7521461
17. Costa, V., Rossetti, R., Sousa, A.: Simulator for teaching robotics, ROS and autonomous driving in a competitive mindset. Int. J. Technol. Hum. Interact. (IJTHI) **13**(4), 19–32 (2017). https://doi.org/10.4018/ijthi.2017100102
18. Costa, V., Resende, J., Sousa, P., Sousa, A., Lau, N., Reis, L.: Fostering efficient learning in the technical field of robotics by changing the autonomous driving competition of the portuguese robotics open. In: CERI 2017 Proceedings, pp. 7705–7711. ICERI 2017 (2017). https://doi.org/10.21125/iceri.2017.2049

Cross Topics in Educational Robotics

Cross-Age Mentoring to Educate High-School Students in Digital Design and Production

Yuval Walter[✉] and Igor Verner

Institute of Technology, 3200003 Haifa, Israel
walter.yuval@gmail.com, ttrigor@technion.ac.il

Abstract. This paper proposes a way to implement cross-age mentoring for laboratory instruction in technical drawing, computer aided design and production with 3D printer for secondary school students majoring in mechanical engineering. The study was conducted in two high schools, with twenty six 9th and 11th graders participated as mentees, and nine 12th graders served as mentors. The study used the mixed method to determine characteristics of the cross-age mentoring and evaluate learning outcomes of student mentors and mentees. As found, mentoring was effective when being conducted in small groups of mentees by mentors who took a preparatory instruction course and used 3D models that they produced as instructional aids. The mentees succeeded to design and print a larger number of models of higher quality, then been guided by the teacher only. The mentors deepened knowledge of the subject and developed instruction skills.

Keywords: Cross-age approach · Mentors · Mentees

1 Introduction

Studies of engineering design and digital production are mainly based on the concept that student is an active learner who constructs knowledge while experimenting in a technological environment and thus proceeding to theoretical learning and practical application of the acquired knowledge. The role of the teacher is to create an enabling environment to support meaningful learning of the subject and strengthen students' self-efficacy [5]. Teaching of engineering design and digital production disciplines in high schools is conducted in CAD\CAM labs, computer control and robotic labs and production workshops that include CNC machines and 3D printers [1]. Learning processes in these labs are carried out through the interaction of students with the technological systems, when the teacher tries to guide each student within the lesson time frame. In such setting the teacher and the student face the following challenges:

- The teacher has difficulties to give attention to each student throughout the lesson when he/she is the sole source of knowledge [11].
- The student has difficulties to learn through interaction with the technological system when he/she does not get immediate response to the questions arising during the interaction. Without real time guidance, the student can reach a lack of understanding, interest and motivation [11].

© Springer Nature Switzerland AG 2020
M. Merdan et al. (Eds.): RiE 2019, AISC 1023, pp. 367–375, 2020.
https://doi.org/10.1007/978-3-030-26945-6_33

- The students who learn CAD & CAM face difficulties in spatial reasoning, as many of them lack the required spatial skills and cannot mediate between the real objects and their computer drawings [3].

In practice, there are several ways to cope with these difficulties, based on active and passive assistance [10]. Within the active assistance framework, one way is to attach a tutor to assist the teacher during the lesson. However, the time devoted to each student is still limited. A second way is to integrate adolescent students as mentors into junior classes, so that each student mentor will assist a small group of students (mentees) during the lesson. In the passive assistance, the students learn the subjects with the help of dedicated working sheets that include exercises and structured tasks. The work sheets provide students with a framework for learning, but still cannot answer all the questions that the students could face.

Educators point out that a combination of an active and passive assistance, through cross-age mentoring and learning with work systems, provides a better solution. In particular, when teaching digital design and production, it would be helpful to engage mentors in guiding mentees in their learning practice with digital technology systems, individually or in small groups. An additional contribution of involving students as mentors can be to the development of their communication skills. Finally, mentors can serve as role models and convey to mentees interest in the subject, learning motivation, and proper habits of work in technology-intensive learning environments.

This study seeks to determine characteristics of a cross-age mentoring approach applied to educating students in digital design and production, in the framework of the mechanical engineering, which is an optional matriculation subject in secondary schools. The first author is involved as a teacher of the discipline and as a school coordinator of technology education. The topics studied in the cross-age instruction process were: descriptive geometry, computer-aided drawing, computer-aided design (CAD), 3D printing, preparation and manufacturing processes. The research goal was to examine ways to implement a cross-age mentoring approach in digital design and production studies in secondary schools.

2   Theoretical Background

According to the constructivism and the constructionism approaches, a rich techno-logical environment positively influences the learning and the outcomes of individual student and the group [8]. At the same time, such a rich environment also presents quite a few difficulties related to the fact that young students experience it for the first time. Those students do not know how to proceed in such an environment, so their learning abilities are not expressed. Students who arrive for the first time into a technology-rich learning environment, need guidance in order to mediate their interactions with the technological systems, as well as to get answers for their questions [11]. Underwood argues [12] that a technological environment for learning creates a kind of a game between the learner and the technological system, which increases interest in the subject and evokes emotions that drive self-development and motivation of the learner and the mentor.

In a cross-age mentoring, unlike peer mentoring, the mentors are older than the mentees (preferably two years). The mentors know the subject and undergo training in tutoring the subject and in interpersonal communication [6]. Learning through the cross-age mentoring approach takes place when a mentor guides a group of mentees in a digital design and production lab, thus implementing a social constructivist approach. In this approach, the students collaborate inside the group, with technological systems, and with the group mentor. Such collaboration enhances the learning experience and outcomes [7]. Collaboration in learning technological systems in digital design and production labs leads to the creation of mental models that are shared between the mentors and the mentees. It can result productive learning with satisfaction of the mentors and the mentees [2].

The cross-age mentoring program success requires several conditions [4, 6]:

- Good communication between the mentor and the group of mentees.
- Early training of the mentors in the material studied and how to communicate well and effectively within the group.
- Selection of a specific topic for guidance.
- Mentors discuss the reflections after each instruction session.

According to the cross-age mentoring approach, the mentor explains the content to the mentees and then observes how they perform the task. Such an observation contributes to their self-learning, and is supported by a theory of recursive feedback [9].

3   Study Frameworks

The study was conducted with high school students majoring in mechanical engineering and studying the programs of CAD/CAM or mechatronics. At the end of each program the students have to pass matriculation exams. The subjects learned include engineering mechanics, control and robotics, descriptive geometry, computer aided design and production, and strength of materials. The curricula integrate theoretical studies and extended experimentation in laboratories. The cross-age mentoring, investigated in our study, occurs in the laboratory and is influenced by the technological environment. In the study we considered two categories of participants: the group of mentors and the group of mentees.

The mentors were 12th grade students who studied the digital design and production subject, had good academic achievements and learning skills, and agreed to volunteer for mentoring younger students who started to learn the subject. In return for this work, the mentors earned community service credits. Before the practice the volunteers completed a questionnaire, in which they were asked about their previous experience in guidance and their willingness to mentor younger students. The answers indicated that none of them had prior experience in guidance, and that they were highly motivated to perform mentoring.

The study was conducted in two high schools offering long-standing technology education programs. The first school has an engineering technology center with laboratories equipped by 3D printers, conventional and CNC machines for experimentation in mechatronics and control, as well as for design and prototyping of robotic and digital technology systems. The laboratory facilities are presented in Fig. 1.

Fig. 1. Center's facilities (from left to right): Conventional and CNC machines; Design and 3D printing lab.

In the other high school there are technology education classrooms in which students practice with computer simulation, 3D printers, NXT & EV3 Lego kits (Fig. 2).

Fig. 2. Technology education classrooms for practice in: left - CAD/CAM; right - control and robotics.

The goal of our study was to examine ways to implement a cross-age mentoring approach in digital design and production studies in secondary schools. The research questions were:

1. What are the characteristics of the learning environment for digital design and production studies that support cross-age mentoring?
2. What are the indications of learning digital design and production among students involved as mentors and mentees in a cross-age mentoring approach?

Participants in the study from the two high schools included: nine 12th grade students as mentors and twenty six 9th and 11th grade as mentees. The 9th grade mentees learned the preparatory course of technology. They were moderate-achieving students. The 11th grade mentees studied mechatronics and were characterized as moderate to low achievers. The 12th grade mentors were high achieving students in the last year of the high school mechatronics program. The first author and another teacher of mechanical engineering trained the mentors in each of the schools.

To answer the research questions, we used the educational research tools specified in Table 1.

Table 1. Research questions and tools.

	Research tools	Research participants	
		1st high school	2nd high school
1.	Observations, video and audio records	9 grade mentees, 12 grade mentors	11 grade mentees and mentors
	Interviews	12 grade mentors	12 grade mentors
	Reflections	12 grade mentors	12 grade mentors
2.	Questionnaires	12 grade mentors	12 grade mentors
	Reflections	12 grade mentors	12 grade mentors
	Photographs, movies	9 grade mentees, 12 grade mentors	11 grade mentees and mentors
	Researcher's logbook, working sheets	11 grade mentees, 12 grade mentors	11 grade mentees and mentors
	Interviews	12 grade mentors	12 grade mentors

In the framework of this study, five laboratory assignments in technical drawing, computer aided design, and production with 3D printer were developed and adapted for learning through cross-age mentoring. The assignments are organized in the following learning sequence:

Assignment 1: Spatial vision and drawing projections of a 3D object. The assignment is intended to facilitate development of spatial vision and recognition of body projections. The assignment includes four tasks to be carried out by the mentees.

Assignment 2: Body isometry and projections in 3D space. The purpose is to develop skills of spatial vision. The assignment includes two tasks for mentees.

Assignment 3: Computer aided design. The purpose is to develop CAD skills and practice in sketching geometric objects. The assignment includes six tasks to be carried out by mentees.

Assignment 4: Advanced computer aided drawing. The purpose is to practice design of complex objects with given dimensions. The assignment includes six tasks for mentees' practice.

Assignment 5: Digital design and production using 3D printer. The purpose is to develop understanding and basic skills for designing and making products using MakerBot Print software and desktop MakerBot Replicator 2.

In order to enhance learning in the proposed setting, we followed the literature recommendation [13] and provided the mentors with special training, to deepen their subject matter knowledge, strengthen tutoring and communication skills. During the training, the mentors designed and produced 3D models that they used later to tutor the mentees. The study consisted of the following stages:

1. Pre-course interviews with student mentors to evaluate their subject matter knowledge, tutoring experience, and motivation to serve as mentors.
2. Mentors training to perform the five laboratory assignments.
3. Observations, video and audio recording of the cross-age mentoring processes.

4. Administering post-course questionnaires for mentors and mentees.
5. Data analysis.

Two such models are presented in Fig. 3.

Fig. 3. Models produced by mentors during the training.

The answers to the questionnaires were organized into tables, encoded, and analyzed. The founded characteristics were verified through triangulation with mentors' reflections and findings got from observations and records.

4   Findings

The findings of our study concern the characteristics of environments for learning digital design and production, and of the cross-age mentoring processes. In the post-course questionnaire mentees were asked to evaluate the characteristics of the learning environment, including work in small groups with CAD/CAM software, individual and group mentoring, integration of design with 3D printing, mentoring to support learning for understanding. The following characteristics of the environment were found essential for the cross-age mentoring process:

- Practice in digital design and production environments requires work with complex professional CAD/CAM software systems. The procedural knowledge needed for operating such systems can be acquired only through guided experimentation.
- Novice students can operate in this environment only under close mentoring, especially in the initial stages.
- Students have different abilities of learning in digital design and production environments, and therefore, while collaborating in a group, they need personal guidance which can be provided only when the learning group is small.
- As the environment is continuously upgraded, the mentor has to update knowledge and expertise through appropriate training by a professional teacher.

Evaluations of the learning process effectiveness by mentors and mentees were collected using the post-course questionnaires. Answers were received from 22 mentees and 9 mentors. One of the questions related to the level of mentees' attentiveness to explanations given by the mentors. The rationale for this question is that attentiveness is considered as one of the main indications of guided learning. The answers given by the mentees and the mentors to the question are presented in Table 2. In the data analysis we compared the evaluations of attentiveness given by the mentees and by their mentors. As found, for 8 out of 9 mentors, have good match evaluations. The correlation coefficient of that evaluations $r = 0.51$.

Table 2. Mentees' attentiveness to the mentors.

Attentiveness level	Mentees' answers (%)	Mentors' answers (%)
Very high	60	19
High	27	68
Moderate	13	0
Low	0	13

Findings related to learning digital design and production by student mentors and mentees are presented below:

- Student reports and reflections, supported by our observations, indicated that the mentoring assignment highly motivated mentors to train for tutoring the mentees.
- Mentors' reflection reports, post questionnaires and researcher observations indicated that mentors' focused observation of how the mentees performed the system operation assignments provided important feedback to evaluate and improve their content knowledge and tutoring skills.
- Post-course questionnaires, mentors' reports and analysis of mentees' performance of the assignments indicated that mentoring effectively facilitated development of mentees' self-directed learning skill. As mentees advanced in the course, their learning became more and more self-dependent and effective.
- The reports and reflections provided a detailed description of the mentees' learning throughout the course, which enabled authentic evaluation of their outcomes.

5   Discussion and Conclusions

In our study the two research questions related to the characteristics of the environment and to the indications of learning in the cross-age mentoring framework.

To answer the first research question, we observed that the learning practice involved designing models in a virtual domain and their implementation in the physical domain. The practice required students to understand the relationship between 3D images of the model and its projections, and correctly use the CAD software and 3D printing tools. Cross-age mentoring significantly helped the students in developing these skills. The student mentors used their physical models as teaching aids.

Mentoring helped the students to cope with the complexity of working with the CAD and 3D printer systems. Cross-age mentoring took place with 2–3 mentees per mentor. In this setting, mentees feel comfortable to interact with mentors and their peers and ask questions. This led them to achieve the academic goals.

To answer the second research question, we found indications of students' progress in learning digital design and production. Mentees performed the tasks faster and efficiently thus succeeded to design and print a larger number of models. Along the course, the nature of mentoring has changed so that the mentees have accepted full responsibility for performing their tasks. The role of mentor changed along the course from being helpers to being advisers. Mentors have deepened knowledge of the subject while preparing for the lessons and guiding the mentees. This finding is in line with the theory of recursive feedback [9].

The contribution of this study is both theoretical and practical: from the theoretical prospective there is a lack of empirical studies of cross-age mentoring in technology and engineering education in schools. Our empirical study adds to the body of knowledge regarding the processes of cross-age mentoring that actually take place in technology-education. From the practical prospective, results of the study can be used to train teachers of technology and engineering in using the cross-age mentoring approach in engineering design and robotics education.

References

1. Alimisis, D.: Educational robotics: open questions and new challenges. Themes Sci. Technol. Educ. **6**(1), 63–71 (2013)
2. Andres, P.: Technology-mediated collaboration, shared mental model and task performance. Organ. End User Comput. **24**(1), 64–81 (2012)
3. Chester, I.: Teaching for CAD expertise. Int. J. Technol. Des. Educ. **17**(1), 23–35 (2007)
4. De Smet, M., Van Keer, H., De Wever, B., Valcke, M.: Cross-age peer tutors in asynchronous discussion groups: exploring the impact of three types of tutor training on patterns in tutor support and on tutor characteristics. Instr. Sci. **37**(1), 87–105 (2009)
5. Harpaz, Y.: Meaningful learning conditions. Had-hachinuch, pp. 40–45 (2014)
6. Karcher, M., DuBois, D.: Cross-Age Peer Mentoring, pp. 233–257. Sage Publications, Thousand Oaks (2014)
7. Kundi, G.M., Nawaz, A.: From objectivism to social constructivism: the impacts of information and communication technologies (ICTs) on higher education. J. Sci. Technol. Educ. Res. **1**(2), 30–36 (2010)
8. Maxwell, J.: Re-situating constructionism. In: The International Handbook of Virtual Learning Environments, vol. 1, no. 11, pp. 279–298 (2006)
9. Okita, S.Y., Schwartz, D.L.: Learning by teaching human pupils and teachable agents: the importance of recursive feedback. J. Learn. Sci. **22**(3), 375–412 (2013)
10. O'Malley, K., Gupta, A.: Passive and active assistance for human performance of a simulated underactuated dynamic task. In: Proceeding of the 11th symposium on HAPTICS. Rice university, Houston (2003)

11. So, H.J., Kim, B.: Learning about problem based learning: Student teachers integrating technology, pedagogy and content knowledge. Australas. J. Educ. Technol. **25**(1), 101–116 (2009)
12. Steffens, K., Carneiro, R., Underwood, J.: Self-regulated learning in technology enhanced learning environments. In: TACONET Conference, Portugal, Lisbon, pp. 57–58 (2005)
13. Thrope, L., Wood, K.: Cross-age tutoring for young adolescents. Clgh.: J. Educ. Strat. Issues Ideas **73**(4), 239–242 (2000)

Robot Tutors: Welcome or Ethically Questionable?

Matthijs Smakman[1,2](✉) and Elly A. Konijn[1]

[1] Department of Communication Science, Media Psychology Program,
VU University Amsterdam, Amsterdam, The Netherlands
matthijs.smakman@hu.nl, elly.konijn@vu.nl
[2] Institute for ICT, HU University of Applied Sciences Utrecht,
Utrecht, The Netherlands

Abstract. Robot tutors provide new opportunities for education. However, they also introduce moral challenges. This study reports a systematic literature review ($N = 256$) aimed at identifying the moral considerations related to robots in education. While our findings suggest that robot tutors hold great potential for improving education, there are multiple values of both (special needs) children and teachers that are impacted (positively and negatively) by its introduction. Positive values related to robot tutors are: psychological welfare and happiness, efficiency, freedom from bias and usability. However, there are also concerns that robot tutors may negatively impact these same values. Other concerns relate to the values of friendship and attachment, human contact, deception and trust, privacy, security, safety and accountability. All these values relate to children and teachers. The moral values of other stakeholder groups, such as parents, are overlooked in the existing literature. The results suggest that, while there is a potential for applying robot tutors in a morally justified way, there are imported stakeholder groups that need to be consulted to also take their moral values into consideration by implementing tutor robots in an educational setting.

Keywords: Social robots · Moral values · Ethics · Robot tutors ·
Robot-assisted (language) learning · Child-robot interaction

1 Introduction

New technology provides important tools for modern education and can provide unique learning experiences to students, thereby improving their achievements. One such technology is the educational robot. The EduRobot Taxonomy classifies three types of educational robots, being: (1) Build Bots, (2) Use Bots and (3) Social Bots [1]. Build Bots are used for teaching students new subjects by letting them build and program robots, such as with LEGO Mindstorms. The second type (Use Bots), consist of robots that can be used immediately, that is, students don't need to build the robot. The third type (Social Bots), are for interacting with the robot as a social entity. The robot then appears to be perceived by children as a peer rather than a tool and – according to the children – the humanoid robots even seem to establish a kind of friendship-relation with them [2]. The Social Bots classification corresponds with the role of an educational robot as a *learning collaborator* described by Miller et al. [3]. This, often humanlike robot with social features, in the role of

© Springer Nature Switzerland AG 2020
M. Merdan et al. (Eds.): RiE 2019, AISC 1023, pp. 376–386, 2020.
https://doi.org/10.1007/978-3-030-26945-6_34

a learning collaborator is what this paper defines as a "robot tutor", which is a common understanding of the definition in robotic literature [4].

Although the robot tutor is said to provide great opportunities [4], it also introduces moral challenges. Potential risks related to applying robot tutors in an educational context are voiced through different channels, however, no systematic overview exists to date. Several studies on moral conceptions regarding this topic emphasise the need for moral considerations and guidelines [4–9]. In this paper, we present a systematic literature review aimed at identifying the opportunities and concerns for (moral) values regarding tutor robots.

In the following, we outlay our methodological approach to identify moral values, following the Value Sensitive Design methodology [10], which is often used to integrate moral values into technology. Then, we detail the selection procedure of the literature search and categorise the moral values based on the concerns and opportunities identified in applying robots in education.

1.1 Moral Conceptions Regarding Robot Tutors

Moral conceptions are "the basic notions of the right, the good, and moral worth" [11]. Moral conceptions define the relative (moral) values of activities and experiences, and they specify an appropriate ordering [11]. This paper uses a common definition of a value, being: "a value refers to what a person or group of people consider important in life" [10].

Thus far, there is no systematic literature review on the moral conceptions regarding tutor robots. The existing systematic reviews on robots in education, such as [4] and [12], do not address the moral conceptions. There are some systematic literature reviews on moral conceptions regarding general upcoming technologies which incidentally also mention robots in a classroom, such as [13]. However, a systematic literature review specifically addressing moral considerations regarding the implementation of robot tutors in an educational context is missing. Until now, researchers have used general reviews on technology and values as a basis to study moral conceptions regarding robot tutors, such as [9], in their study on the moral conceptions of teachers regarding tutor robots. A review by Sharkey [14] focused specifically on moral conceptions and robot tutors. However, the non-systematic nature of Sharkey's review makes it hard to evaluate. Given the nature of education and children being a vulnerable group, it is important to critically examine new technology intended to be used in education. Risks or pitfalls related to implementing robot tutors are still unknown and previous studies on moral conceptions regarding this topic stress the need for a systematic review on the academic literature regarding moral considerations that may provide a basis for desirable guidelines [4–9].

2 Methodology

Our methodological approach to identify the moral conceptions regarding tutor robots is based on the Value Sensitive Design methodology. Value Sensitive Design is a theoretically grounded methodology that accounts for values, from a multi stakeholder perspective, when designing and integrating new technology in a social context [10]. It provides a methodology to discover and conceptualise values related to that technology by identifying the concerns and opportunities at stake in the particular system from a multi stakeholder perspective [15]. The first step is to identify the stakeholders who will be affected by the technology. Second, for each stakeholder the concerns (disadvantages, downsides, drawbacks and risks) and opportunities caused by implementing a robot tutor are described. These opportunities and concerns are then linked to moral values, thereby identifying the moral values related to the implementation of robot tutors in education.

The first step of our systematic literature review was to identify relevant databases. A comprehensive search for relevant databases was conducted, resulting in databases from various academic fields, being: *IEEE Digital Library, SpringerLink, JSTOR, Science direct, ACM, NARCIS, EBSCO, Web of Science* and *Scopus*. Second, an initial search string was formed to identify synonyms for tutor robots.

To determine the initial search string, the keywords identifying robot tutors from a previous, initial review concerning robot tutors, were used [16]. This resulted in multiple search terms for tutor robots and various synonyms for concerns and opportunities. In several search rounds, we refined the search criteria such that most relevant references were selected, and irrelevant ones excluded. This resulted in our final search string as follows: ("robot tutor" OR "tutor robot" OR "robotic tutor" OR "teacher robot" OR "robot teacher" OR "robotic teacher" OR "education* robot") AND ("harm" OR "benefit" OR "positive effect" OR "negative effect").

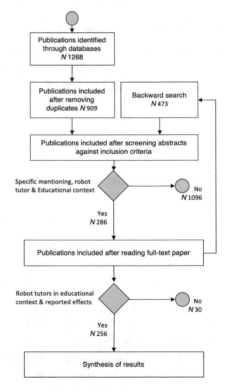

Fig. 1. Selection procedure

2.1   Selection Procedure

The first step in the selection procedure (shown in Fig. 1) was to exclude duplicates, resulting in 909 unique studies. Second, we checked if the abstracts did match our inclusion criteria, which were: (1) the context should be educational, and (2) the abstract should include a specific mentioning of a tutor robot. We also excluded publications that were not written in English. To identify the educational context, terms were included such as teacher, pupil, school, education, tutor, peer, assignment, learning, course, curriculum, kindergarten, and learning topics such as chess and language. Exclusion criteria for the educational context were: hospital, elderly, industry, robots learning from (human) teachers and reinforced learning. To identify various types of the robot tutor, inclusion terms were: learning collaborator, learning companion, learning peer, teaching assistant and physical agent. Exclusion criteria regarding the topic robot tutors were: as a programming project (e.g., Lego Mindstorms), as a learning focus, virtual agent, distance education, software robots, virtual reality, augmented reality, telerobot, therapy tool, constructivism, and robotic education. To focus on robot tutors, we excluded the constructionism literature because this focuses mostly on Build Bots and Use Bots. After this phase, we conducted a backward reference search which resulted in 473 possibly relevant studies. The abstracts of these studies were matched to the inclusion criteria, making the total of studies selected for full-text analysis 286. In the last step, 30 studies were excluded based on the missing educational context or missing full-text, resulting in a final list of 256 studies (available at the Open Science Framework [17]) which were included in the synthesis of the results.

2.2 Data Analysis

This review covers various scientific fields such as *Pedagogy, Education, Philosophy, Human-Computer Interaction, Robotics, Psychology*, and *Communication science*. Therefore, the 256 publications selected for full-text coding were diverse in their goal and methodology. The full-text data analysis was conducted in three cycles of coding, following Strauss and Corbin's process of (1) open coding, (2) axial coding, and (3) selective coding [18]. Applying these three cycles, we segmented the publications based on their main goal for comparison purposes and as such identified the categorisation of these studies. We identified five categories: (1) Conceptual studies, (2) Design studies, (3) Effect studies, (4) Exploratory cases, and (5) Perception studies. This categorisation does not only provide a framework for comparison purposes but also provides a systematic overview of the available studies till 2018 related to tutor robots.

For each individual category of these studies, through our full-text data analysis, we identified the concerns and opportunities discussed within each paper and linked their effects to a specific or multiple stakeholder group(s). The key stakeholders in education research are: the government, parents, staff, students, supervisory board, business, supplying schools, recipient schools, and other educational institutions [19].

3 Results and Discussion

The results of our systematic review are here presented in terms of the concerns and opportunities related to the introduction of a robot tutor from a stakeholder perspective as discussed in the papers under review. The identified concerns and opportunities, and the number of studies which reported on these concerns and opportunities per category (see Sect 2.2), are summarized in Table 1. Due to space limitations, the results are presented in this concise format to be further discussed in the next section. In general, we found that all concerns and opportunities discussed in the identified studies were related to children and teachers as stakeholders. Potential effects on other stakeholder groups were not discussed, therefore the remainder of this section will be structured around these two stakeholders, shown in Table 1 as Ch for Children and T for Teachers.

Table 1. Opportunities (O) and Concerns (C) per category for children (Ch) and teachers (T)

Opportunities and concerns			Categories (see Sect. 2.2)					
O/C	Ch/T	Description	Conceptual studies, N = 39	Exploratory cases, N = 87	Perception studies, N = 26	Design studies, N = 31	Effect studies, N = 73	Sum of H/B
O	Ch/T	Motivation and enjoyment	10	43	5	8	24	90
O	Ch	Reduced anxiety	1	9	2	1	1	14
O	Ch	Personalised learning	12	7	7	14	6	46
O	Ch/T	New opportunities for education, new social interactions, or beyond the classroom learning	11	21	13	6	9	60
O	T	Reduced workload	6	9	4	0	0	19
C	T	Cost of the robot	2	4	5	1	1	13
C	Ch	Privacy and security	2	0	3	3	0	8
C	Ch	Social implications, e.g. friendship, trust, respect, and deception	8	1	3	1	1	14
C	Ch/T	Discomfort, e.g. Uncanny Valley effect and stress	1	6	3	1	5	16
C	T	Technology is too complicated or low technology adaptation	1	4	3	0	0	8
C	Ch	Loss of motivation	4	6	1	1	3	15
C	Ch	Loss of human contact	2	2	5	0	0	9
C	T	Control and accountability issues	3	0	2	1	0	6
C	Ch/T	Disruption	0	2	2	1	3	8
C	T	Increase of workload	1	0	1	0	0	2
C	T	Technology is inadequate, ineffective or wrong expectations	6	18	7	4	6	41

The goal of this systematic literature review was to identify and categorise the concerns and opportunities linked to implementing robot tutors in an educational context as reported in the extant literature. Thereby, discovering the moral values affected by its introduction from a multi-stakeholder perspective. Following the steps of the Value Sensitive Design methodology, we evaluated and linked the effects of the concerns and opportunities onto moral values regarding new technology, design and robotics reported in earlier studies [9, 14, 20, 21]. Based on these studies [9, 14, 20, 21], we identified thirty-seven moral values. Of these possible values, fourteen were relevant to be related to the concerns and opportunities identified through our review. Some values were combined to form a clustered topic in merging them together, such as 'Psychological welfare & Happiness' and 'Friendship & Attachment' because these appear closely related. Thus, these fourteen values (summarized, clustered, and numbered in Table 2) are potentially undermined (i.e., negatively related) or positively related to the introduction of robot tutors and will each be discussed in the next section.

Table 2. Values related to the implementation of robot tutors

Positively related (+)	Negatively related (−)
(1) Psychological welfare & (2) Happiness	(1) Psychological welfare, (2) Happiness
(3) Efficiency	(3) Efficiency
(4) Freedom from bias	(4) Freedom from bias
(5) Usability	(5) Usability
	(6) Deception & (7) Trust
	(8) Friendship & (9) Attachment
	(10) Human contact
	(11) Privacy, (12) Security, (13) Safety and (14) Accountability

3.1   Values Attributed to the Introduction of Robot Tutors

Based on the opportunities reported, five positive values are created by the introduction of robot tutors: psychological welfare, happiness, efficiency, freedom from bias, and usability. However, all five values are also potentially negatively influenced by the robot tutor, making the total list of values that are potentially undermined by the robot tutor fourteen. In the following, we will describe the findings for each of these values in general without going into specific details due to space limitations. The detailed data-analysis, which can be retrieved from the Open Science Framework (https://osf.io/97uza), provides an overview of the 256 studies included in our systematic review.

Psychological Welfare & Happiness. Many studies report on opportunities and concerns that affect the values psychological welfare and happiness, for both children and teachers (e.g. [22–25]). These values refer to affective states such as mental health, comfort and peace. The robot's ability to comfort children, for example making children with ASD feel more at ease, directly relates to this value. The ability to create an enjoyable and fun educational context can also be linked to these values. However, concerns are reported in [26–28] that children sometimes fear robot tutors because of their appearance or sudden movements. Furthermore, the robot could lead to feelings of anxiety when children become too emotionally attached.

For teachers, the robot can take over dull or repetitive tasks which could lead to a potential increase in job satisfaction. Nevertheless, teachers are also reported to fear a loss of jobs by the introduction of robot tutors. However, the current state of technology is severely inadequate for a sophisticated level of natural and autonomous interaction with children.

Efficiency. Multiple studies report on opportunities and concerns that will affect the value of efficiency, referring to the relationship between the gains and means of resources and can affect both children and teachers (e.g. [29–31]). Some results suggest that robots can be a more effective tool compared to a computer-based tutoring system. However, since robot tutors are a novel technology and the empirical studies are often based on short interactions, the efficiency regarding specific learning topics needs further evaluation. It is further reported that the current robot tutors hardly meet the requirements posed by professionals [31]. Voice recognition and speech are just two of the technical components that need to be optimised. Furthermore, the robot's ability to efficiently, and appropriately, respond to social context is still lacking, which causes breakdowns in the interaction. Studies report that these shortcomings could lead to the robot being a costly and ineffective tool, causing a decline in efficiency in the learning process, for both child and teacher. However, its ability to support teachers in multiple activities, such as building e-portfolios and record data during assessments is seen as possibly enhancing this efficiency.

Freedom from Bias. Results of our review further showed that the introduction of robots may free possible unfair treatment of children due to biases, or one's perception thereof. A robot responds unbiased and in the same way to all children, without prejudice. Furthermore, the robot's capability to adapt to childrens' needs could lead to the removal of possible pre-existing social biases of teachers.

As designers and robot builders try to integrate human constructs into robot tutors, it's also possible for technical biases to occur. Studies report that programming bias-free self-learning systems, such as robots that can adapt to children's needs (i.e., personalisation), is one of the key challenges in Artificial Intelligence. Taking into account the nascency of the required technology for a robot tutor, designers should consider how biases could be excluded from educational robots to ensure each child gets a bias-free experience, and the robot does not potentially favour certain children over others.

Usability. In value sensitive design literature, the value of usability refers to making all relevant stakeholders successful users of technology and can be broken down into three challenges: (1) technological variety, (2) user diversity and (3) gaps in user knowledge [20]. Overlooking the results of the review, user diversity and gaps in user knowledge are reported in several studies (e.g. [32–34]) Results report that the robot tutor could be used by children of different age groups, skill levels, or children with disabilities, thereby positively impacting the value of usability. However, the potential gap between the knowledge of teachers to use robot tutors effectively should be attended to, to ensure that all teachers are capable of using the robots. Furthermore, in our opinion, interacting with robots early on in school could prepare children for a society in which robots could potentially play a big part, thereby making them able to access and use this technology in the future.

Deception and Trust. One of the design challenges for the robot tutor reported in the studies in our review, is to create trustworthy relationships with children. This would lead to a more stable and improved interaction between child and robot [35]. However, this could undermine the values of trust and avoiding deception. Studies (e.g. [9, 14, 35]) report on concerns about children that might potentially be deceived by the robot tutor; children could imagine that the robot really cares about them. As children are reported to be willing to share secrets with a robot, the value of trust could be undermined when the child finds out the teacher can access the data of the robot.

Friendship and Attachment. A concern which was mainly raised by the conceptual studies (Table 1) is that when children perceive a robot tutor as their friend. This might have a negative impact on the concept of friendship and attachment, according to several studies (e.g. [2, 36]). However, none of the effect studies report negative consequences regarding children's perception of friendship.

Human Contact. The value of human contact could be undermined, studies report, because the social bond children experience with a robot may lead to them preferring the companionship of a robot over that of their human peers. According to several studies, this could potentially lead to the loss of human contact [14, 35]. Although none of the studies report on the robot being designed to replace human contact, concerns are expressed about eventually reducing human-to-human contact in schools when teachers are replaced by robots.

Privacy, Security, Safety, and Accountability. Results of our review show that the introduction of robot tutors may impact the values of privacy, security and safety, and accountability (e.g. [14, 35, 37]). The physical presence of the robot and its ability to record data has an impact on these values. Audio and visual files of children, recorded by the sensors of the robot, could be unobtrusively stored and accessed by unauthorised individuals, which is a concern of teachers [9]. Who should be authorised to access these records, however, is an important open question. We believe this is especially important when such data contains private information of children, such as secrets, which the child told the robot in confidence. In line, who should be accountable for the impact of tutor robots and where the responsibility should lie, is reported as a concern [14, 35, 37], especially since the technology is reported to be costly [38].

4 Future Research

This paper shows the importance to address various sensitive moral considerations for children and teachers when designing and implementing robot tutors. Further qualitative and quantitative research is needed into how different stakeholders perceive and prioritise the moral values to allow schools to make calculated, well-informed decisions when implementing robot tutors, and to help the robotic industry to integrate moral values in their tutor robot design. As the current scientific literature on robot tutors does not include the values of all stakeholders affected by the introduction of robot tutors, future research should also focus on identifying their values and norms in an empirical manner. Specifically, the values of parents should be taken into account, in addition to the teachers and children, as they are the representatives of children and experience the effects of robot tutoring first hand.

References

1. Catlin, D., Kandlhofer, M., Holmquist, S., Csizmadia, A.P., Cabibihan, J.-J., Angel-Fernandez, J.: EduRobot taxonomy and papert's paradigm. In: Dagiene, V., Jasute, E. (eds.) Constructionism 2018 Constructionism, Computational Thinking and Educational Innovation: Conference Proceedings, Vilnius, Lithuania, p. 11 (2018)
2. Leite, I., Martinho, C., Paiva, A.: Social robots for long-term interaction: a survey. Int. J. Soc. Robot. **5**, 291–308 (2013). https://doi.org/10.1007/s12369-013-0178-y
3. Miller, D., Nourbakhsh, I., Siegwart, R.: Robots for education. In: Springer Handbook of Robotics, pp. 1283–1301 (2008). https://doi.org/10.1007/978-3-540-30301-5_56
4. Belpaeme, T., Kennedy, J., Ramachandran, A., Scassellati, B., Tanaka, F.: Social robots for education: a review. Sci. Robot. **3**, 10 (2018). https://doi.org/10.1126/scirobotics.aat5954
5. Baxter, P., Ashurst, E., Kennedy, J., Senft, E., Lemaignan, S., Belpaeme, T.: The wider supportive role of social robots in the classroom for teachers. In: 1st International Workshop on Educational Robotics at the International Conference on Social Robotics, Paris, France, p. 6 (2015)
6. Heerink, M., Vanderborght, B., Broekens, J., Albó-Canals, J.: New friends: social robots in therapy and education. Int. J. Soc. Robot. **8**, 443–444 (2016). https://doi.org/10.1007/s12369-016-0374-7
7. Lin, P., Abney, K., Bekey, G.: Robot ethics: mapping the issues for a mechanized world. Artif. Intell. **175**, 942–949 (2011). https://doi.org/10.1016/j.artint.2010.11.026
8. Serholt, S., Barendregt, W., Leite, I., Hastie, H., Jones, A., Paiva, A., Vasalou, A., Castellano, G.: Teachers' views on the use of empathic robotic tutors in the classroom. In: The 23rd IEEE International Symposium on Robot and Human Interactive Communication, pp. 955–960. IEEE, Edinburgh (2014). https://doi.org/10.1109/ROMAN.2014.6926376
9. Serholt, S., Barendregt, W., Vasalou, A., Alves-Oliveira, P., Jones, A., Petisca, S., Paiva, A.: The case of classroom robots: teachers' deliberations on the ethical tensions. AI Soc. **32**, 613–631 (2017). https://doi.org/10.1007/s00146-016-0667-2
10. Friedman, B., Kahn, P.H., Borning, A., Huldtgren, A.: Value sensitive design and information systems. In: Doorn, N., Schuurbiers, D., van de Poel, I., Gorman, M.E. (eds.) Early Engagement and New Technologies: Opening up the Laboratory, pp. 55–95. Springer, Dordrecht (2013). https://doi.org/10.1007/978-94-007-7844-3_4
11. Rawls, J.: The independence of moral theory. In: Proceedings and Addresses of the American Philosophical Association, p. 5. American Philosophical Association (1974). https://doi.org/10.2307/3129858
12. van den Berghe, R., Verhagen, J., Oudgenoeg-Paz, O., van der Ven, S., Leseman, P.: Social robots for language learning: a review. Rev. Educ. Res. (2018). https://doi.org/10.3102/0034654318821286
13. Heersmink, R., Timmermans, J., van den Hoven, J., Wakunuma, K.: ETICA Project: D.2.2 Normative Issues Report (2014)
14. Sharkey, A.J.C.: Should we welcome robot teachers? Ethics Inf. Technol. **18**, 283–297 (2016). https://doi.org/10.1007/s10676-016-9387-z
15. Spiekermann, S.: Ethical IT Innovation: A Value-Based System Design Approach. Auerbach Publications (2015). https://doi.org/10.1201/b19060
16. Smakman, M.: Moral concerns regarding tutor robots, a systematic review. In: ATEE Winter Conference, Technology and Innovative Learning, Utrecht, The Netherlands (2018). https://doi.org/10.13140/RG.2.2.33565.00482
17. The data-analysis data. https://osf.io/97uza/

18. Corbin, J.M., Strauss, A.: Grounded theory research: procedures, canons, and evaluative criteria. Qual. Sociol. **13**, 3–21 (1990). https://doi.org/10.1007/BF00988593
19. Konermann, J., Janssen, H., Vennegoor, G.: Naar een actief en open stakeholdersbeleid in het voortgezet onderwijs: stakeholders in kaart brengen, mobiliseren en managen, 's-Hertogenbosch (2010)
20. Friedman, B., Kahn, P.H., Jr.: Human values, ethics, and design. In: The Human-Computer Interaction Handbook, pp. 1177–1201. L. Erlbaum Associates Inc., Hillsdale (2003)
21. Van Den Hoven, J.: Responsible Innovation. Presented at the Third International Conference on Responsible Innovation, 22 May, The Hague, The Netherlands (2014)
22. Shih, C.-F., Chang, C.-W., Chen, G.-D.: Robot as a storytelling partner in the english classroom - preliminary discussion. In: Seventh IEEE International Conference on Advanced Learning Technologies (ICALT 2007), pp. 678–682. IEEE, Niigata (2007). https://doi.org/10.1109/ICALT.2007.219
23. Sumioka, H., Yoshikawa, Y., Wada, Y., Ishiguro, H.: Teachers' impressions on robots for therapeutic applications. In: Otake, M., Kurahashi, S., Ota, Y., Satoh, K., Bekki, D. (eds.) New Frontiers in Artificial Intelligence, pp. 462–469. Springer, Cham (2017). https://doi.org/10.1007/978-3-319-50953-2_33
24. Brown, L., Kerwin, R., Howard, A.M.: Applying behavioral strategies for student engagement using a robotic educational agent. In: 2013 IEEE International Conference on Systems, Man, and Cybernetics, pp. 4360–4365. IEEE, Manchester (2013). https://doi.org/10.1109/SMC.2013.744
25. Alemi, M., Meghdari, A., Haeri, N.S.: Young EFL learners' attitude towards RALL: an observational study focusing on motivation, anxiety, and interaction. In: Kheddar, A., Yoshida, E., Ge, S.S., Suzuki, K., Cabibihan, J.-J., Eyssel, F., He, H. (eds.) Social Robotics, pp. 252–261. Springer, Cham (2017). https://doi.org/10.1007/978-3-319-70022-9_25
26. Fridin, M.: Kindergarten social assistive robot: first meeting and ethical issues. Comput. Hum. Behav. **30**, 262–272 (2014). https://doi.org/10.1016/j.chb.2013.09.005
27. Mutlu, B., Forlizzi, J., Hodgins, J.: A storytelling robot: modeling and evaluation of human-like gaze behavior. In: 2006 6th IEEE-RAS International Conference on Humanoid Robots, pp. 518–523. IEEE, University of Genova, Genova, Italy (2006). https://doi.org/10.1109/ICHR.2006.321322
28. Ko, W.H., Ji, S.H., Lee, S.M., Nam, K.-T.: Design of a personalized r-learning system for children. In: 2010 IEEE/RSJ International Conference on Intelligent Robots and Systems, p. 6. IEEE, Taipei (2010). https://doi.org/10.1109/IROS.2010.5649668
29. Wei, C.-W., Hung, I.-C., Lee, L., Chen, N.-S.: A joyful classroom learning system with robot learning companion for children to learn mathematics multiplication. Turk. Online J. Educ. Technol. **10**, 11–23 (2011)
30. Han, J.: Robot-aided learning and r-learning services. In: Human-Robot Interaction. Intech Open Access Publisher (2010)
31. Huijnen, C.A.G.J., Lexis, M.A.S., Jansens, R., de Witte, L.P.: Mapping robots to therapy and educational objectives for children with autism spectrum disorder. J. Autism Dev. Disord. **46**, 2100–2114 (2016). https://doi.org/10.1007/s10803-016-2740-6
32. Reich-Stiebert, N., Eyssel, F.: Robots in the classroom: what teachers think about teaching and learning with education robots. In: Social Robotics, pp. 671–680. Springer, Cham (2016). https://doi.org/10.1007/978-3-319-47437-3_66
33. Ahmad, M.I., Mubin, O., Orlando, J.: Understanding behaviours and roles for social and adaptive robots in education: teacher's perspective. In: Proceedings of the Fourth International Conference on Human Agent Interaction, pp. 297–304. ACM Press, Biopolis (2016). https://doi.org/10.1145/2974804.2974829

34. Kennedy, J., Lemaignan, S., Belpaeme, T.: The cautious attitude of teachers towards social robots in schools. In: Robots 4 Learning Workshop at IEEE RO-MAN 2016, p. 6 (2016)
35. Pandey, A.K., Gelin, R.: Humanoid robots in education: a short review. In: Goswami, A., Vadakkepat, P. (eds.) Humanoid Robotics: A Reference, pp. 1–16. Springer, Dordrecht (2017). https://doi.org/10.1007/978-94-007-7194-9_113-1
36. Zawieska, K., Sprońska, A.: Anthropomorphic robots and human meaning makers in education. In: Alimisis, D., Moro, M., Menegatti, E. (eds.) Edurobotics 2016 2016. AISC, vol. 560, pp. 251–255. Springer, Cham (2017). https://doi.org/10.1007/978-3-319-55553-9_24
37. Salvini, P., Korsah, A., Nourbakhsh, I.: Yet Another Robot Application? [From the Guest Editors]. IEEE Robot. Autom. Mag. 23, 12–105 (2016). https://doi.org/10.1109/MRA.2016.2550958
38. Han, J., Jo, M., Park, S., Kim, S.: The educational use of home robots for children. In: ROMAN 2005. 2005 IEEE International Workshop on Robot and Human Interactive Communication, pp. 378–383. IEEE, Nashville (2005). https://doi.org/10.1109/ROMAN.2005.1513808

Teaching Robotics with a Simulator Environment Developed for the Autonomous Driving Competition

David Fernandes[1], Francisco Pinheiro[1], André Dias[1,2]([✉]), Alfredo Martins[1,2],
Jose Almeida[1,2], and Eduardo Silva[1,2]

[1] Instituto Superior de Engenharia do Porto, 4200-072 Porto, Portugal
[2] INESC TEC, Institute for Systems and Computer Engineering,
Technology and Science, Porto, Portugal
adias@lsa.isep.ipp.pt

Abstract. Teaching robotics based on challenge of our daily lives is always more motivating for students and teachers. Several competitions of self-driving have emerged recently, challenging students and researchers to develop solutions addressing the autonomous driving systems. The Portuguese Festival Nacional de Robótica (FNR) Autonomous Driving Competition is one of those examples. Even though the competition is an exciting challenger, it requires the development of real robots, which implies several limitations that may discourage the students and compromise a fluid teaching process. The simulation can contribute to overcome this limitation and can assume an important role as a tool, providing an effortless and costless solution, allowing students and researchers to keep their focus on the main issues. This paper presents a simulation environment for FNR, providing an overall framework able to support the exploration of robotics topics like perception, navigation, data fusion and deep learning based on the autonomous driving competition.

Keywords: Teaching robotics topics · Autonomous driving · Simulation environment

1 Introduction

In last years, robotics has expanded its intervention areas, which has led to the existing of different robotic research topics (perception, navigation, motion control, artificial intelligence, etc.). At the same time, this increase also requires that the students must explore more robotic topics in less time. This not mean that all students must be expert in all topics, but should have enough background to work in different robotic topics. How should the teachers cover all areas? From our point of view, they should expose the students to heterogeneous challenges that require the exploration by the students of transversal robotic knowledge. This is not possible, for several reasons, like budget, equipment and time. Robotic

© Springer Nature Switzerland AG 2020
M. Merdan et al. (Eds.): RiE 2019, AISC 1023, pp. 387–399, 2020.
https://doi.org/10.1007/978-3-030-26945-6_35

simulation environments can play an important role in answer to this issues, they are able to provide a tool not only for testing systems but also to teach different robotics topics by being able to emulate real-world processes. Recently, autonomous driving systems have become a major research and teaching field for the scientific community engaged with autonomous mobile robotics.

Many robotic competitions such as, Audi Autonomous Driving Cup [1], Autonomous Driving Challenge [2], Open Zeka MARC [3] have surged as an important playground to learn, develop and explore robotic topics, and test new strategies for self-driving vehicles.

The Autonomous Driving Competition (ADC) in the Festival Nacional de Robótica (FNR) is a field-test example, where fully autonomous robots have to navigate in a traffic road shape track and deal with different difficulty levels. The challenge requires the capability to detect and react based on indicating signals, adapt the robot's trajectory upon obstacle and working zone detection, and park the robot in a predefined area [4]. Considering the challenger of teaching robotics and at the same time have a solution to address the ADC this paper presents a simulation environment developed for FNR, providing an overall framework able to support the exploration of robotics topics like perception, navigation, data fusion and deep learning based on the autonomous driving competition. The use of simulation in robotics development has a widespread and long history. It is particularly useful for algorithm testing and validation prior to real implementation. Multiple simulation tools have been used from generic simulation software (such as Simulink or Modelica) to dedicated robotic simulators such as Player/Stage, Webots [5], VREP [6], Gazebo [7–9], MORSE [10–12] or USAR-Sim [13]. A paradigmatic example is DRC-DARPA Robotics Challenge [14] and VRC-Virtual Robotics Challenge [15]. Several simulation-based robotic competitions have been created, emphasizing the development of robot control software and artificial intelligence. A few examples are: the RoboCup simulation leagues (from 2D and 3D robotic soccer to rescue league simulation [16]) and the IEEE/NIST virtual manufacturing automation simulation competition focusing on industrial robotics applications [17]. Robotic challenges based on simulation can differ in application scope, physics fidelity or realistic scenario [15,18,19].

Researchers face numerous challenges when designing their solutions using real robots. Some are associated with mechanical and/or hardware aspects, commencing from cost and maintenance of the platform itself, time spent to set up experiments or the difficulty to replicate and measure experiments with statistical relevance. Such constraints easily drive away attention from the main objective: designing the concept and test systems such as navigation, motion control, perception, etc. the core of a robust autonomous driving system. Additionally, hardware development may overwhelm students driving them into frustrated results [20]. Simulation tools offer to researchers the possibility to avoid hardware constraints at the initial stage of the development, allowing not only the validation of concepts and solutions but also a staged evaluation, ensuring a greater probability of success on real robots implementation [21,22].

Fig. 1. Autonomous Driving Competition at Festival Nacional de Robótica (FNR) 2017 in Coimbra, Portugal.

A typical 3D robotics simulator such as Gazebo, USARSim or Modular Open Robots Simulation Engine (MORSE), permits complex environment definition with detailed 3D models from objects and terrain. It also permits to define the robot's shape and structure, as well as the positioning and configuration of multiple sensors. In addition, modern robotic simulators offer advanced mathematical models capable to realistically represent dynamic systems or the possibility to use components datasets at runtime - "hardware-in-the-loop" [20]. Physics engines are used such as Bullet Physics [23] or ODE [24] to add physical characteristics to the simulation such as mass or friction (Fig. 1).

There are two approaches to implement realistic 3D simulator; one is to develop a simulator from scratch using physics engines and graphics libraries; the other is to adapt and use 3D game engines already providing a complete software framework. One can refer Gazebo open source robotic simulator, as an example of a fully developed simulator, initially planned to integrate Player/Stage robotic software framework and more recently experienced vast improvements including tight integration with Robot Operating System (ROS) middleware [25]. Other examples are V-REP simulator and Webots simulator and robotics development software. Game-engines are introduced to streamline the development of computer game software. They incorporate environment 3D simulation, integrates physics engines (with different levels of precision) and easy methods to import 3D models from games characters, environments, and scenarios, objects and elements. Scripting capabilities and language interfaces are also available to improve some features, scaling the software development or simplifying the development process [10].

This paper is outlined as follows: in Sect. 2 the Autonomous Driving Competition (ADC) is presented. The autonomous driving simulator is described in Sect. 3. Section 3, the autonomous driving simulator is present with respect to the simulated scenario, the logic brick that allows us to configure the different ADC challenges and the ability to introduce Gaussian noise on top of each robot sensor. Section 4, some case studies are present, with respect to the ADC challenges that the students can address with the simulator. Finally, some concluding remarks and future work are presented.

2   Autonomous Driving Competition

The Autonomous Driving Competition (ADC) is one of research challenge competition from the FNR. The competition is composed of several challenges grouped into three categories: driving, parking, and vertical traffic sign identification challenges [4]. The driving category has four rounds with increasing difficulty degrees. In each round the robots must complete two laps within the shortest time. In the first round is performed with robot at pure speed. In the following round, the robot shall complete the lap and simultaneously react upon signals identification, such as stopping or changing lane. In the third round, a tunnel is placed in the track and obstacles are placed at unknown locations. In the final round, a road working area is added. Regarding the parking contest; upon parking signal is switched on, the robot must drive to parking area: parallel parking zone or parking bay with difficulty level increased by adding obstacles. In the end, a special challenge is addressed to participants, where six from twelve possible vertical signs are positioned along the track for proper identification.

The competition addresses relevant educational robotic topics, like localization, motion control, path planning, mapping, deep learning and computer vision.

3   Autonomous Driving Simulator

3.1   Simulator Engine: MORSE

The 3D simulator chosen for the ADC-FNR is the Modular Open Robots Simulation Engine (MORSE), an open-source robotics simulator based on the Blender Game Engine [10]. Several other 3D simulators engines were considered, such as Gazebo [26] or VREP, but MORSE gathers important characteristics like modularity, flexibility, and reusability [12,19]. Aspects supporting the choice on MORSE are the possibility to model components, environments, and robots in Blender; the use of Blender's 3D Game Engine; and the clean installation process when compared with other simulator engines. The capacity of supporting multi-robots simulations, large simulation scenarios, and better use of computational resources, are additional aspects supporting the choice of MORSE [19,27].

MORSE robotics simulator lies on a component-based architecture to simulate sensors, actuators, and other components modeled in Blender and coded in Python scripts libraries. The possibility of defining several levels of abstraction is another aspect distinguishing MORSE's flexibility. Based on the *software-in-the-loop* philosophy, MORSE permits the algorithm's evaluation embedded in the software of the robot and yet, it is independent of any robot system architecture or middleware [19].

The MORSE simulator engine is built on top of Blender, making use of its 3D modeling and rendering features, as well as the built-in Blender's Game Engine. This OpenGL-based game engine renders meshes in real-time, supports shaders, advanced lighting, and multi-texturing. The Blender Game Engine also offers a flexible graphic interface called Logic Bricks which permit to script actions (in Python) to elements present in the scene or even define properties to each object (Logic Properties) [19].

To achieve a realistic physics behavior of the robot and all the other elements involved in the simulation scenario, Blender also integrates the Bullet Physics Engine [23] to simulate physics, providing the possibility of replicating real-world effects, such as gravity, forces, friction, or static properties as mass, rigid body collisions and many others.

3.2 Autonomous Driving Competition Simulation Scenario

The Autonomous Driving Competition scenario, detailed in Figs. 2 and 3 was entirely designed in Blender fulfilling the FNR specifications. The scenario includes the design of the track, parking areas, obstacles, tunnel, working zone and indicating signals. The visual aspect takes an essential role in simulation, particularly on robotic vision simulation, since the virtual world captured by cameras can be realistic enough to test algorithms [19], and successfully apply them in reality. To produce a more realistic simulation, the scenario includes textures, light effects, shadows, and other objects around the track.

Fig. 2. Autonomous driving simulation environment.

3.3 Game Engine - Logic Bricks

The FNR competition detailed in Sect. 2, is composed of several challenges with different configurations. Using Logic Bricks functionalities, it is possible to configure, in the same Blender file, all the specifications that characterize each round of ADC-FNR and interactively adapt the scenario for a specific challenge. The

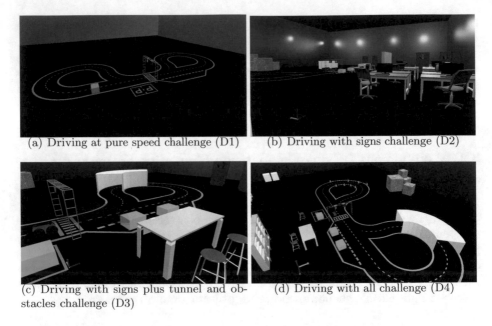

(a) Driving at pure speed challenge (D1)   (b) Driving with signs challenge (D2)

(c) Driving with signs plus tunnel and ob-   (d) Driving with all challenge (D4)
stacles challenge (D3)

Fig. 3. Autonomous driving simulation environment for each challenger.

Logic Bricks graphical interface it's the tool were objects behaviors or configuration properties are defined.

The interface is divided by Sensors, Controllers, Actuators, and can be also supported by python scripts. The tunnel, obstacle and working zone can be moved in the scene using predefined keyboard keys. The six different indicating signals are also controlled by keyboard keys, that make it appears on the two screens simultaneously. The spotlights have a different animation controller. Using Logic Properties, property sensors and actuators with the animation system, the controller developed permits change the spotlights energy on five predefined values.

3.4 Physics

The mesh objects in the simulation environment can have their physics properties, like mass and friction, adjusted. This configuration is done in Blender in the *Properties >> Physics* panel. There are other configurations used to bring more realism to the simulation, such as the usage of Static objects. These objects will not move, even if the robot hits them, as an example the monitor stand, and yet those objects are considered obstacles.

Other objects populating the simulation scene such as the green obstacle boxes or traffic cones signalizing the working-area, are configured as obstacles, subjected to Forces and Collisions.

The mobile robots, including the frame mesh, sensors, and actuators have defined their physics properties also set like mass, friction, and collision bounds.

These parameters will be used by Bullet Physics Engine to compute realistic interaction between the robot and the simulation scenario.

3.5 Simulation Builder

MORSE simulator lies on a modular structure, based on a library of components that can be combined or assembled together. These components consist of two elements: a Blender file with the mesh object, and a Python file which defines the class object with all its methods characterizing the behavior of that component.

Each component can register services that are made public outside the simulator engine through middlewares like ROS topics.

There are three types of components in MORSE: Robots, Sensors, and Actuators. All component inherits from an abstract class *Morse*. The Python script for MORSE simulation builder - *Builder* script - will instantiate the class object associated with the simulated robot, as well as the actuators and sensors. Components can support two types of transformations: translation and rotation. The *Builder* script can modify or add properties to sensors and actuators, with repercussion in the game engine. In real-world both sensors and actuators are affected by noise. For instance, Accelerometers and Gyroscopes are affected by time-varying bias and wide-band noise. In order to approach the simulation to the real-world constraints, *Modifiers* can alter the quality of the data to achieve a more realistic behavior. This is a desirable feature to test high-level algorithms.

```
my_odometry.alter('Noise', pos_std=1, rot_std=1, _2D=
"True")
```

To enable the usage of Accelerometer, it necessary to rewrite the script *accelerometer.py* located in MORSE's distribution package to allow publishing time-stamped messages with the data stream:

```
from geometry_msgs.msg import TwistStamped
(...)
twist = TwistStamped()
```

4 Teaching Robotics with the Autonomous Driving Simulator

For teaching robotics, the Autonomous Driving Simulator provides the required framework structure in order to ensure that the students can explore robotic research topics. The students can explore topics like computer vision algorithms to identify features from the simulation scenario like lanes, objects, and landmarks and use them as observations on a filter supporting other subsystems like localization, path planning, etc.

In the scope of the FNR-Autonomous Driving Simulator, a basic framework supported on ROS is proposed, depicted in Fig. 5, with nodes publishing and subscribing to the most relevant data-streams accessing sensors and actuators.

Fig. 4. A and B - Lane detection; C and D - Signals identification

These same nodes can be easily ported to a real robot. The architecture present in Fig. 5 was designed to adapt the simulation environment to the type of challenge that a research group intends to evaluate. In order to validate the simulation environment and infer the potentialities in the FNR - Autonomous Driving Competition, a case study will be present by testing and validating multiple software components, such as image processing (see Fig. 4), motion control and deep learning.

4.1 Image Processing

With respect to image processing, an important block in the autonomous driving system, tasks like *lane − recognition* and *signals identification* were developed and validate. Taking advantage of mono-camera emulators available in MORSE, it was possible to generate a geometric camera and capture RGB images. The camera parameterization was made in order to simulate a real camera sensor. The images are being converted to HSV and submitted to a simple segmentation process giving attention to time-processing constraints. The segmentation is done in 3 steps; noise reduction using a 2D Gaussian blur filter (transit-signs recognition); then image binarization. Alternatively to the traditional binarization from gray-scale images, the binary image is obtained directly from HSV color space.

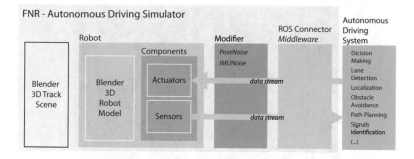

Fig. 5. Autonomous driving proposed architecture.

This process, more robust to *luminance* variations, is achieved by defining upper and lower boundaries per channel; the third step consists in removing small areas from the binary image using morphological operations [28] *Erode* and *Dilate*. After image segmentation, the homography transformation is used to generate a top-view image perspective, which simplifies the extraction of some features, namely lane-detection. Two images with different perspective of a plane can be related with an homography matrix (Fig. 6).

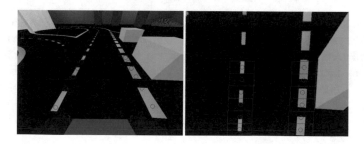

Fig. 6. Homography transformation with sliding window lane-detection. Blue boxes represents the lane detection.

The top-view image resulting from homography transformation, is used in the lane-detection. The process is achieved through two main steps: first it is evaluated the segmented/transformed image to produce an histogram from the closest region of interest. Histogram's peaks are thereafter detected corresponding to lane's base points. A sliding windows is defined and horizontally aligned with the peaks found. The lane detection algorithm proved to be robust enough, allowing the vehicle in the simulator to complete the track without getting away from its lane - pure speed challenge (Fig. 7). In order to evaluate stability of the proposed solution, Gaussian noise was added to the camera sensor.

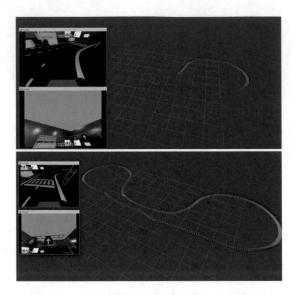

Fig. 7. Vehicle performing pure speed challenge

4.2   Motion Control - Lane Following

The lane-following low-level control implemented, robot is forced to converge and follow a line generated by the lane detection algorithm previously describe. There is no localization process to estimate robot's pose, the control is only relative to the tracking-lane trajectory. The tracking-lane camera, is geometrically placed in the center/front of the robot avoiding any translation. A longitudinal waypoint position should converge to the camera image center, ensuring the robot position within the lane. From the lane-detection algorithm the desired tracking-lane is computed (see Fig. 7).

4.3   Deep Learning - Crosswalk Identification with YOLOv2

Another case study that was performed, was to evaluate the simulator as a tool for the students to explore Neural Networks as a robotic topic. In this case study, we implement the crosswalk detection with the YoloV2 Neural Network. For that, we apply the available Yolo Network and train the network with the Autonomous driving simulator images.

The system was capable to process the crosswalk identification, as depicted in Fig. 8, with a frame rate of 20 fps, using a computer system Intel Core i7-7700HQ - CPU 2,8 GHz (x8) with a GeForce GTX 1050TI (768 CUDA cores). Base in the results, we are able to integrate in the Neural Network other objects like, arrows, vertical signals and traffic cones.

Fig. 8. Neural Network (YOLOv2) implementation for crosswalk identification.

5   Conclusion

A simulation environment based on the open source MORSE robotic simulator was developed for the FNR Autonomous Driving Competition. The simulation environment proves to be an interesting teaching tool to address different robotic topics, from perception, path planning, motion control, and even deep learning. The MORSE modularity allows students to create their own robots, sensors, and actuators, expanding the possibilities of breakthrough innovations. The FNR-Autonomous Driving Simulator robustness supports the integration of neural networks and deep learning. This feature provides a relevant added value in the development of the innovative and intelligent system. The FNR-Autonomous Driving Simulator proposed gathers all necessary elements to allow students to develop solutions addressing each Autonomous Driving Competition challenge. The presented results also confirm the simulator as a valuable tool to be used in this competition and demonstrate the versatility of the MORSE/Blender framework for robotics development. In the future, we intend to evaluate with the students and the teachers what was the impact of the simulator in the robotic learning process. Identify what were the main difficulties that they faced and what were the benefits they identified in the simulator. This will allows to improve the simulator and provide to the robotic community.

References

1. Audi autonomous driving cup. https://www.audi-autonomous-driving-cup.com/. Accessed 10 Jan 2019
2. Autonomous driving challenge. http://www.autonomousdrivingchallenge.com/. Accessed 10 Jan 2019
3. Open zeka marc. https://openzeka.com/marc/. Accessed 10 Jan 2019
4. Sociedade Portuguesa de Robótica. Rules for autonomous driving. Technical report (2016)

5. Michel, O.: Webots^{TM}: professional mobile robot simulation. J. Adv. Robot. Syst. **1**(1), 39–42 (2004)
6. Freese, M., Singh, S., Ozaki, F., Matsuhira, N.: Virtual robot experimentation platform V-REP: a versatile 3D robot simulator. In: Ando, N., Balakirsky, S., Hemker, T., Reggiani, M., von Stryk, O. (eds.) Simulation, Modeling, and Programming for Autonomous Robots. Lecture Notes in Computer Science, vol. 6472, pp. 51–62. Springer, Heidelberg (2010)
7. Folgado, E., Rincón, M., Álvarez, J.R., Mira, J.: A multi-robot surveillance system simulated in gazebo. In: Mira, J., Álvarez, J.R. (eds.) Nature Inspired Problem-Solving Methods in Knowledge Engineering. Lecture Notes in Computer Science, vol. 4528, pp. 202–211. Springer, Heidelberg (2007)
8. Koenig, N., Howard, A.: Design and use paradigms for gazebo, an open-source multi-robot simulator. In: Proceedings of the 2004 IEEE/RSJ International Conference on Intelligent Robots and Systems (IROS 2004), vol. 3, pp. 2149–2154, September 2004
9. Yao, W., Dai, W., Xiao, J., Lu, H., Zheng, Z.: A simulation system based on ROS and Gazebo for RoboCup middle size league. In: 2015 IEEE International Conference on Robotics and Biomimetics (ROBIO), pp. 54–59, December 2015
10. Echeverria, G., Lassabe, N., Degroote, A., Lemaignan, S.: Modular open robots simulation engine: MORSE. In: 2011 IEEE International Conference on Robotics and Automation (ICRA), pp. 46–51, May 2011
11. Echeverria, G., Lemaignan, S., Degroote, A., Lacroix, S., Karg, M., Koch, P., Lesire, C., Stinckwich, S.: Simulating complex robotic scenarios with MORSE. In: Noda, I., Ando, N., Brugali, D., Kuffner, J.J. (eds.) Simulation, Modeling, and Programming for Autonomous Robots. Lecture Notes in Computer Science, vol. 7628, pp. 197–208. Springer, Heidelberg (2012)
12. Dias, A., Almeida, J., Dias, N., Silva, E., Lima, P.: Simulation environment for multi-robot cooperative 3D target perception. In: SIMPAR 2014 – 4th International Conference on Simulation, Modeling, and Programming for Autonomous Robots. Lecture Notes in Computer Science. Springer, Cham (2014)
13. Carpin, S., Lewis, M., Wang, J., Balakisky, S., Scrapper, C.: USARSim: a robot simulator for research and education. In: 2007 IEEE International Conference on Robotics and Automation, pp. 1400–1405 (2007)
14. Hsu J.M., Peters, S.C.: Extending open dynamics engine for the DARPA virtual robotics challenge. In: Brugali, D., Broenink, J.F., Kroeger, T., MacDonald, B.A. (eds.) Simulation, Modeling, and Programming for Autonomous Robots, pp. 37–48. Springer, Cham(2014)
15. Agüero, C.E., Koenig, N., Chen, I., Boyer, H., Peters, S., Hsu, J., Gerkey, B., Paepcke, S., Rivero, J.L., Manzo, J., Krotkov, E., Pratt, G.: Inside the virtual robotics challenge: simulating real-time robotic disaster response. IEEE Trans. Autom. Sci. Eng. **12**(2), 494–506 (2015)
16. Budden, D.M., Wang, P., Obst, O., Prokopenko, M.: Robocup simulation leagues: enabling replicable and robust investigation of complex robotic systems. IEEE Robot. Autom. Mag. **22**(3), 140–146 (2015)
17. Balakirsky, S., Madhavan, R., Scrapper, C.: NIST/IEEE virtual manufacturing automation competition: from earliest beginnings to future directions, pp. 214–219 (2008)
18. Defense Advanced Research Projects Agency (DARPA). https://www.subtchallenge.com/. Accessed 16 Jan 2019

19. Echeverria, G., Lassabe, N., Degroote, A., Lemaignan, S.: Modular open robots simulation engine: MORSE. In: Proceedings - IEEE International Conference on Robotics and Automation, pp. 46–51 (2011)
20. Costa, V., Rossetti, R., Sousa, A.: Simulator for teaching robotics, ROS and autonomous driving in a competitive mindset. Int. J. Technol. Hum. Interact. **13**(4), 19–32 (2017)
21. Castillo-Pizarro, P., Arredondo, T.V., Torres-Torriti, M.: Introductory survey to open-source mobile robot simulation software. In: Proceedings - 2010 Latin American Robotics Symposium and Intelligent Robotics Meeting, LARS 2010, July 2014, pp. 150–155 (2010)
22. Osório, F., Wolf, D., Castelo Branco, K., Pessin, G.: Mobile robots design and implementation: from virtual simulation to real robots. Technical report, Introductory Survey to Open-Source Mobile Robot Simulation Software, Bordeaux (2010)
23. Bullet real-time physics simulation. https://pybullet.org/wordpress/. Accessed 17 Jan 2019
24. Open dynamics engine. https://www.ode.org/. Accessed 17 Jan 2019
25. Quigley, M., Conley, K., Gerkey, B., Faust, J., Foote, T., Leibs, J., Wheeler, R., Ng, A.Y.: ROS: an open-source robot operating system. In: ICRA Workshop on Open Source Software, vol. 3, no. 3.2 (2009)
26. Costa, V., Rossetti, R.J.F., Sousa, A.: Autonomous driving simulator for educational purposes. In: 2016 11th Iberian Conference on Information Systems and Technologies (CISTI), pp. 1–5, June 2016
27. Noori, F.M., Portugal, D., Rocha, R.P., Couceiro, M.S.: On 3D simulators for multi-robot systems in ROS: MORSE or Gazebo? In: SSRR 2017 - 15th IEEE International Symposium on Safety, Security and Rescue Robotics, Conference, pp. 19–24 (2017)
28. Chudasama, D., Patel, T., Joshi, S., Prajapati, G.I.: Image segmentation using morphological operations. Int. J. Comput. Appl. **117**(18), 16–19 (2015)

User-Driven Design of Robot Costume for Child-Robot Interactions Among Children with Cognitive Impairment

Luthffi Idzhar Ismail[1,3]([✉]), Fazah Akhtar Hanapiah[2], and Francis wyffels[1]

[1] Ghent University – imec, Technologiepark-Zwijnaarde 126, 9052 Ghent, Belgium
luthffiidzharbin.ismail@ugent.be
[2] Faculty of Medicine, Universiti Teknologi MARA, 47000 Sungai Buloh, Malaysia
[3] Faculty of Engineering, Universiti Putra Malaysia, UPM,
43400 Serdang, Selangor, Malaysia

Abstract. The involvement of arts and psychology elements in robotics research for children with cognitive impairment is still limited. However, the combination of robots, arts, psychology and education in the development of robots could significantly contribute to the improvement of social interaction skills among children with cognitive impairment. In this article, we would like to share our work on building and innovating the costume of LUCA's robot, which incorporating the positive psychological perspectives and arts values for children with cognitive impairment. Our goals are (1) to educate arts students in secondary arts school on the importance of social robot appearance for children with cognitive impairment, and (2) to select the best costume for future child-robot interaction study with children with cognitive impairments.

Keywords: Robot · Arts · Psychology · Education ·
Child-robot interaction

1 Introduction

Robots, educations, arts and psychology need to be integrated in social robot research since it could bring a huge impact to the affected children especially children with cognitive impairment (CWCI) [6]. Most of the physical robots are lacking of arts values despite of their advance technological capabilities and intelligence. The gaps between arts and robotics are huge. However, the elements of arts could add more value to the appearance of social robots [8] in term of creating more interesting, attractive and appealing robots. The appearance of some robot could be a positive agent and function as a "catalyst" towards improving social interaction skills in child-robot interaction [9]. In social robotics research, psychology perspectives are very important since it could affect the output of child-robot interaction in a positive or negative substances [4]. Thus, providing good psychological perspectives [11] in child robot interaction are crucial and

M. Merdan et al. (Eds.): RiE 2019, AISC 1023, pp. 400–405, 2020.
https://doi.org/10.1007/978-3-030-26945-6_36

vital in ensuring the study shall benefits the target group such as children with cognitive impairment.

In this study, we worked together with the secondary art students to build the robot's costume for future child-robot interaction between LUCA robot and CWCI. In fact, one of the art students, was also diagnosed to have mild autism. However, she had followed some intervention programs and has improved her mild impairments a lot. This user-driven design of robot costume is also laterally performed in order to wind up our aspiration in educating the secondary art students in Ghent, Belgium on the significance of arts and psychology aspects in child-robot interaction. The objectives of this project are (1) to educate art students in Secondary Arts School on the importance of arts elements in social robotics appearance for children with cognitive impairment, and (2) to select the best two costumes for future child-robot interaction study with children with cognitive impairments. Thus, this paper discusses the importance of youngster's education and creating more awareness towards integration of robots, education, arts and psychology.

2   Incorporating Arts and Psychology in Social Robots

As mentioned in earlier section, we were trying to be innovative in this study by integrating component of arts, psychology, robotics and education. Thus, we modified an open source robot based on the OPSORO platform [10] since it is very affordable and could achieve the objectives of our study. The robot, namely as LUCA is a special robot since it could do some basic facial expressions such as happy, sad, angry and etc. In this study, we asked some advises from certified clinical occupational therapists on what are the important aspects in psychological appearance [2] of the robot when interacting with children diagnosed with cognitive impairment. Visual and appearance aspects of the robot are among the important things when designing the costume of the robot for child-robot interaction study. Thus, the color of robot's costume for this study was selected based on finding from Grandgeorge et al. [3]. Students were given several color's option to be used in their design such as blue, red, yellow and brown. They were free to choose these colors to match with their design of LUCA's costume.

3   Design of Robot's Costume

3.1   Costume Design Criteria

In this process, it is interesting to see how secondary arts students brainstorm their ideas and innovative thought together based on inputs given to them by postgraduate student, teacher and professor. As mentioned earlier, the input that was given to the students is summarize as below:

1. **Psychological perspectives**: The selection of colour for robot's costume must have some psychological effects [3] towards children with cognitive impairment.

2. **Design simplicity**: The design should be simple but interesting [7].
3. **Design safety**: No dangerous materials shall be used in fabricating the costume such as metal and hazardous items [5].
4. **Robot's fitting**: The fitting of the costume shall not distract the robot's capabilities and functions [1] (i.e. facial expressions, no design obstacle for robot's head to turns left and right).

3.2   Costume Design Process

Various aspects and perspectives have been considered in integrating arts, robots, education and psychology. In this study, the work of creating the robot's costume were carried out by young, bright and talented secondary arts students. There are several important processes in building the costume for LUCA robot (1) Measure the robot's dimension (2) fabrics selection and cutting (3) Sewing (4) Fitting (5) Final trimming.

Fig. 1. Figure shows the student tried their best to test their robot's costume on the actual robot.

Fig. 2. All costumes prototypes made by art students in Ghent, Belgium

Measure: During this process, students came to the robotics laboratory and manually took the dimension of lower body and upper body of the robot.
Fabrics selection and cutting: Most of the students decided to used soft fabrics as it followed the requirement of safety [5].
Sewing: In this stage, art students were trained by their teacher to sew the fabrics to the shape of the robot.
Fitting: Once the costume is ready, the students tried to fit their design to the actual robot as shown in Fig. 1.
Final trimming: In this final stage, most of the students need to do the final trimming to fit the eye of the robot.

4 Selection of Robot Costume for Child-Robot Interaction

In previous section, we discussed the design criteria and processes of creating the prototype of LUCA's costume. The selection of costumes was done based on their presentation and justification of their innovation. Most of the students had very idealistic reasons and good motivations in creating the costume such as selection of colors, arts values and most importantly the theme of their innovative design. Overall, 11 designs were made by 11 students in this study as illustrated in Fig. 2. Most of the students were able to relate their design concept from arts value, psychological perspectives, design safety and the future task of the robot. Additionally, most of the students also able to relate some of the potential psychological effects in their presentation story which makes their costume very useful for future child-robot interaction study.

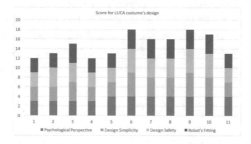

Fig. 3. Bar chart representation of all costume design score

Fig. 4. (a) LUCA costume RED, (b) LUCA costume blue

In achieving the second objective of the study, we only select the best two designs for future (actual) experiment of child-robot interaction between LUCA and children with cognitive impairment. The final two LUCA's costumes design was selected based during their final design presentation. We evaluated their costumes design based on specified design criteria as mentioned in previous section which are (1) Selection of colour for robot's costume must have some psychological effect towards children with cognitive impairment, (2) The design should be simple but interesting, (3) No dangerous materials shall be used in fabricating the costume such as metal and hazardous items, (4) The fitting of the costume shall not disturb the robot's capabilities and functions [1] (i.e. facial expression, head to turn left and right).

In the selection process, marks were given to all costumes during their costume presentation based on mentioned criteria. 5 full marks were allocated for each criteria, which yield to the total score of 20 marks for 4 criteria. The scale of the marks depend on the achievement of each criteria. For example (1) Very Bad (2) Bad (3) Average (4) Good (5) Very Good. The overall score of design costume can be seen as illustrated in Fig. 3. Costume design number 6 and 9 obtained the highest marks. Their costumes design are shown in Fig. 4a and Fig. 4b. These were the best among others and had qualified them to obtained the highest score as compared to others. Their design was successfully incorporated arts to the robot's costume while considering positive psychological perspective towards social interaction in future child-robot interaction. It is amazing that the art students could relate arts, with psychology and robotics in their costume's design. Moreover, this make them feel more excited and be proud of their design since their design will be used in future child-robot interaction study. This is very meaningful for young and talented secondary arts students and they feel honored if their design could be useful for children with cognitive impairment.

5   Conclusion and Future Work

In this article, we presented a study which incorporated robots, arts, psychology and education. We had worked together with art students to build interesting robot costume with some psychological perspectives. By working together the results of the research finding can be greatly impact on the children with cognitive impairment and beneficial towards society. Research that integrate robots, arts, psychology and education is still very limited. Thus, we tried to contribute to the research community the integration of robot, arts, psychology and education by sharing our research work and finding. Our goal (1) to educate art students in secondary art students on the importance of arts elements in social robotics appearance for children with cognitive impairment has been significantly achieved and proved that nothing is impossible if we work together and guide the student educationally. Moreover, the second objective (2) to select the best costume for future child-robot interaction study with children with cognitive impairments also successfully achieved since all the costume are good and impressive. Most importantly, their design did not distract the functionality of the robot and safe to use for future child-robot interaction. To conclude, it is our hope that more future work that incorporate arts, psychology, robotics and education shall be carried out in the future work so that more people will get the benefits while learning things across multiple field of knowledge.

Funding Information. Luthffi Idzhar Ismail received a Postgraduate Education Fund from Majlis Amanah Rakyat, MARA (MARA REF: 330407445608). This work was partially funded by the Niche Research Grant Scheme (NRGS):600-RMI/NRGS 5/3 (11/2013).

Conflict of Interest. The authors declare that they have no conflict of interest.

References

1. Dautenhahn, K.: Roles and functions of robots in human society: implications from research in autism therapy. Robotica **21**(4), 443–452 (2003)
2. Goetz, J., Kiesler, S., Powers, A.: Matching robot appearance and behavior to tasks to improve human-robot cooperation. In: Proceedings of the 12th IEEE International Workshop on Robot and Human Interactive Communication, pp 55–60. IEEE Press Piscataway (2003)
3. Grandgeorge, M., Masataka, N.: Atypical color preference in children with autism spectrum disorder. Front. Psychol. **7**, 1976 (2016)
4. Kennedy, J., Baxter, P., Belpaeme, T.: The robot who tried too hard: social behaviour of a robot tutor can negatively affect child learning. In: Proceedings of the Tenth Annual ACM/IEEE International Conference on Human-robot Interaction, pp. 67–74. ACM (2015)
5. Lasota, P.A., Fong, T., Shah, J.A., et al.: A survey of methods for safe human-robot interaction. Found. Trends® Robot. **5**(4), 261–349 (2017)
6. Pachidis, T., Vrochidou, E., Kaburlasos, V., Kostova, S., Bonković, M., Papić, V.: Social robotics in education: state-of-the-art and directions. In: International Conference on Robotics in Alpe-Adria Danube Region, pp. 689–700. Springer (2018)
7. Robins, B., Dautenhahn, K., Dubowski, J.: Does appearance matter in the interaction of children with autism with a humanoid robot? Inter. Stud. **7**(3), 479–512 (2006)
8. Scassellati, B., Admoni, H., Matarić, M.: Robots for use in autism research. Ann. Rev. Biomed. Eng. **14**, 275–294 (2012)
9. Tapus, A., et al.: Children with autism social engagement in interaction with Nao, an imitative robot: a series of single case experiments. Inter. Stud. **13**(3), 315–347 (2012)
10. Vandevelde, C., Wyffels, F., Vanderborght, B., Saldien, J.: An open-source hardware platform to encourage innovation. IEEE Robot. Autom. Mag. **1070**(9932/17), 2 (2017)
11. Young, J.E., Hawkins, R., Sharlin, E., Igarashi, T.: Toward acceptable domestic robots: applying insights from social psychology. Int. J. Social Robot. **1**(1), 95 (2009)

Setup of a Temporary Makerspace
for Children at University:
MAKER DAYS for Kids 2018

Maria Grandl$^{(\boxtimes)}$ ⓘ, Martin Ebner ⓘ, and Andreas Strasser

Graz University of Technology, Münzgrabenstraße 36/1, 8010 Graz, Austria
maria.grandl@tugraz.at

Abstract. The maker movement has become a driving force for the *new industrial revolution*, whereby all learners should have the opportunity to engage. Makerspaces exist in different forms with different names and a variety of specializations. The MAKER DAYS for kids are a temporary open makerspace setting for children and teenagers with the goal to democratize STEAM education and social innovation and to empower young learners, especially girls, to shape their world. This publication presents the setup and results of a temporary makerspace at Graz University of Technology with more than 100 participants in four days in summer 2018 and discusses the role of new technologies as a trigger of making in education. Moreover, the MAKER DAYS implemented an innovative evaluation concept to document the participants' activities in open and unstructured learning environments.

Keywords: Maker movement · Maker education · Maker space · STEAM education · Computer science education · Design education

1 Introduction

"I am calling on people across the country to join us in sparking creativity and encouraging invention in their communities" [1]. Ever since Barack Obama proclaimed the 18th of June as *National Day of Making* in the USA and hosted the first Maker Faire in the White House, the maker movement took roots in various learning contexts with children, adolescents or adults as the main group of learners. Making incorporates many different areas, focusing on the process of creating, designing, tinkering, exploring and experimenting rather than consuming information and technologies. Makerspaces exist in different forms with different names and a variety of specializations [2]. Sheridan et al. identified three aspects of the maker movement, combining arts, engineering and entrepreneurship [3]. They describe "*making* as a set of activities, *makerspaces* as communities of practise, and *makers* as identities of participation" [4]. Successful makerspaces are more than a just a collection of tools and activities. They are based on a supportive environment and community engagement, that empowers people and democratizes science and (social) innovation [2, 5]. The maker movement gets people inventing again, with more opportunities for young people to experience

M. Merdan et al. (Eds.): RiE 2019, AISC 1023, pp. 406–418, 2020.
https://doi.org/10.1007/978-3-030-26945-6_37

self-efficacy, gain confidence, develop creativity and interest for STEAM (science, technology, engineering, arts and math) related subjects and topics.

Graz University of Technology offers various summer courses for children and teenagers to raise interest and to enhance girls' participation in STEAM. Reaching from the *AI summer lab* over the *robotic club* to girls-only courses, such as the *girls coding week*, the courses are suitable for beginners as well as (more) advanced students.[1] In August 2018, the course program was extended by the MAKER DAYS for kids, a four-day long temporary maker space, created for children and teenagers between the ages of 10 and 14.

The contribution of this work is to comment on the motivation and goals behind the MAKER DAYS, to describe and reflect on the implemented design and evaluation concept and to present and discuss the results, based on personal observations and collected data. Created by Sandra Schön and supported by Martin Ebner, the MAKER DAYS took place in Bad Reichenhall (Germany) in 2015 [6]. Based on the award-winning concept of Sandra Schön, the authors redid the MAKER DAYS at Graz University of Technology, that attracted more than 100 participants and acted as experimental playground for new workshop concepts and new educational technologies.

2 Theoretical Foundation

In recent years, a lot of research has been carried out on how makerspaces function as learning environments in formal and informal educational contexts. Maker activities promote learning in a less teacher driven and multidisciplinary way, where cognitive, technical and social-emotional skills are addressed. The Horizon 2020 project *DOIT - Entrepreneurial skills for young social innovators in an open digital world* (2017–2020), co-financed by the European Union, builds upon the idea that social innovations within makerspace settings allow authentic and tangible learning experiences, that empower young people "to turn creative ideas into entrepreneurial action" [7].

Working mainly in the field of didactics in computer science and technology enhanced learning, the authors think of the maker movement as a driving force for the "new industrial revolution", whereby all learners have the opportunity to engage in their personal way [8]. It is obvious: New digital technologies are no longer in the hands of computer scientists. They can support young learners' imagination and enhance what they can create, build and learn [9]. This is where Seymour Papert comes in. In his famous book *Mindstorms*, Papert describes how computers can affect the way humans think and learn (in a positive sense) [10]. Influenced by the important works of Jean Piaget and John Dewey, Papert established the theory of constructionism as "learning by doing", where learners use tools, that he called *objects-to-think-with*, to build their own knowledge structures. When learners become actively engaged in exploring, designing or creating things, they think of *the way they think* or *the way others* (mentors, peers or computers) *think*. Tools that come with an "intersection of

---

[1] Overview of summer courses offered at the Graz University of Technology: www.tugraz.at/go/sommerkurse.

cultural presence, embedded knowledge, and the possibility for personal identification" help learners to explore important concepts (especially in STEAM) and to reflect on their own learning process and problem-solving skills [10]. Making mistakes is, of course, part of this process.

Maker education is closely related to the learning theory of constructionism. Following the advice of Halverson et al. [4] the authors hosted a temporary makerspace for kids in summer 2018: *"If we believe that making activities and maker identities are crucial for empowerment, then it is, in part, our job to set up situations whereby all learners have the opportunity to engage"*.

3 Austrian Proof-of-Concept: MAKER DAYS for Kids 2018

Inspired by the concept and results of the first MAKER DAYS in Bad Reichenhall (Germany) in 2015, the authors applied for funding to bring the MAKER DAYS to Graz University of Technology with the overall goal to help children and teenagers become more fluent and expressive with new innovative technologies as well as traditional tools such as the soldering iron and the sewing machine. As soon as the authors had received a positive answer about the funding, they started with the organization of the event in March 2018. They set up a web page on the existing weblog[2] for basic computer science education, where they announced the date of the event and provided an online form for registration. The following data was collected: name and age of participants, name, e-mail-address and telephone number of a parent, desired date(s) of participation (It was also possible to participate over all 4 days.), consent to transfer and store electronic data in terms of the data protection law. Since the promotion of the TU Graz summer courses did not start before June, there were only two registrations for the MAKER DAYS at the end of May. The courses were announced in the local newspaper, the internal newsletter, on different websites, in social media, in schools and at different local events. Consequently, the number of registrations increased, and the authors had to limit the number of daily participants to 50.

Announced as a creative (digital) workshop with an open approach the MAKER DAYS aim

- to offer low-threshold activities (no specific previous knowledge is required),
- to involve young learners in the design and preparation process,
- to increase participation of children and teenagers as guides and peer tutors,
- to support design thinking and early social entrepreneurial education,
- to promote sustainability,
- to trigger a self-directed development of digital competencies.
- to support girls' participation in maker activities,
- to implement a gender-sensitive setting and
- to select accessible tools and software.

---

[2] The authors' weblog addressing basic computer science education: https://learninglab.tugraz.at/informatischegrundbildung/.

To meet these goals, a team of 20 people, including researchers, teacher candidate students and employees of the service department Educational Technology was responsible for registration of participants, guided tours for new participants, taking photos and making videos (documentation of the participants' activities), taking photos of the participants' products, organisation and coordination, doing workshops (over all four days) and doing workshops (on a single day). In addition, six children and teenagers (participants of other summer courses at TU Graz) were asked to act as peer tutors. They were introduced to the concept and activities the week before the event.

3.1 Room Design and Rules

Conveniently situated on the campus "Neue Technik" with a nearby tram and bus station, the participants of the MAKER DAYS were received on the ground floor of a modern building, where all available lecture rooms were reserved for the duration of the event and the preparatory work (see Fig. 1). With three bright and colourful lecture rooms, a huge foyer for registration and photo documentation, a storage room and short walking distance between toilettes, kitchen, and the maker space, the building was the perfect choice for this event. The two biggest lecture rooms with a size of 165 m^2 and 95 m^2 were transformed in two temporary makerspaces with separated workshop areas. Both lecture rooms were equipped with colourful chairs. This contributed to a more child-oriented atmosphere. The tables could be easily rearranged and one of the lecture rooms was even equipped with colourful tables on wheels and a couch. To create a cosier setting, (borrowed) beanbag chairs, carpets and pillows of the authors' office were used. Pompoms and self-made pennant chains were used for decoration.

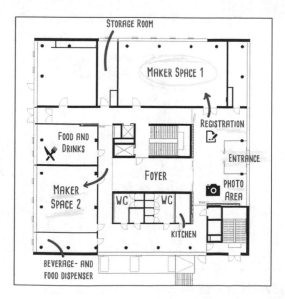

Fig. 1. During the summer holidays, the ground floor of the building Inffeldgasse 10 at Graz University of Technology served as event location for the MAKER DAYS for kids.

First, the authors had to ensure that there were enough electric sockets or portable multi sockets for all devices. The position of the electric sockets in the room influenced the arrangement of the different workshop areas. Magnetic blackboards were used for the announcement of the workshops. The authors put tags on each table or spot to indicate the different workshop areas (see Fig. 2). For the construction of the Lego® City, a space of about 10 m² was defined, clearly marked on the floor with tape. Additional tables and chairs (Creative Area 1, 2, 3, 4) were provided for additional workshops by (peer) tutors.

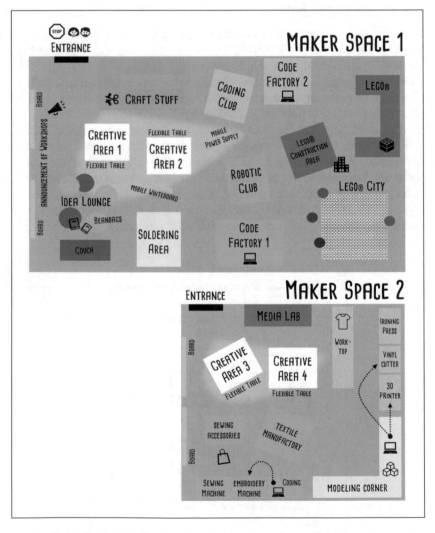

Fig. 2. Two lecture rooms at Graz University of Technology turned into a temporary maker space with separate workshop areas, comfortable places to think and relax and a great variety of craft stuff, tools, and sewing accessories, that were freely available to all participants.

By rule, adults were not allowed to enter the makerspace, except for the tutors, the media team and the team that was responsible for registration, organisation and coordination. The MAKER DAYS are an event for young learners, helping them to find their own path within a given topic. The authors' aim was to create a sheltered environment that evokes trust in one's own competences. The participants should not be intimidated or get diverted by a (grand) parent. They should be the only ones that demand the (peer) tutor's time and effort. Each team member got a special shirt that they had to wear for the duration of the event. This way, the participants instantly knew who they could ask for help and support.

On each day, after registration, the new participants were asked to wait for a (peer) tutor. The (peer) tutor's task was to guide a group of newcomers through the maker space, explaining the tools and activities as well as the guiding principles of the MAKER DAYS for kids. *The 9 principles of the MAKER DAYS* arc an adaption of the maker movement manifesto by Hatch [11], done by Schön et al. [12]. For the MAKER DAYS 2018, the principles were printed on a poster and put up in the makerspace (see Fig. 3).

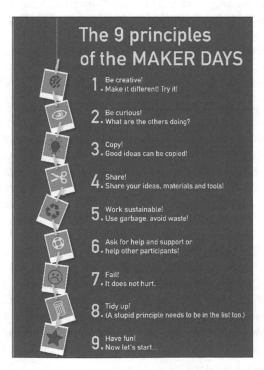

Fig. 3. During the MAKER DAYS, participants had to follow these rules that should nurture a positive maker mindset.

3.2   Workshops and Tools

According to Davee et al. [2], makerspaces can be classified as either "dedicated" (all the devices, materials and tools are pooled in a single space), "distributed" (e.g. a computer lab and a room for handicrafts at a school) or "mobile" (e.g. selected materials and tools in a box). The MAKER DAYS for kids implemented a *dedicated* makerspace of 260 m². Registration was open from 9 am and the makerspace closed at 4 pm, with a lunch break from noon to 1 pm. One of the main questions was, how to structure the learning experience within the makerspace. The authors decided to follow the *mini-workshop* approach of the MAKER DAYS in 2015, where the (peer) tutors announced short workshop sessions addressing a specific tool, activity or learning goal. For the announcement of these workshops, the (peer) tutors were asked to fill out a printed workshop card for every single workshop. At the MAKER DAYS in Graz, these cards were placed on the blackboard in makerspace 1. One thing is for sure: Not every participant read these workshop cards. There were other triggers that encouraged participants to start *making*.

Looking at maker education theory and practise in the Netherlands, Douma et al. [13] observed that "the trigger to start making was mostly caused by some form of intrinsic motivation" and the "provided materials and (digital) fabrication tools [...] triggered the students to explore, discover and iterate or to "make artefacts while using digital oriented technology". Sheridan et al. [3] identified (evocative) products, processes and materials as children's "ways in" to making. Table 1 provides an overview of the materials, tools and devices used during the MAKER DAYS 2018.

Lego® City. The purpose of the Lego® City was to combine the results of different workshops with the needs of a livable and future-oriented city. The authors defined a *daily theme* to draw the participants' attention to relevant features of sustainable urban development planning. On the first day, tutors and participants discussed the following questions: *What makes a city? Which buildings and facilities does a city need?* The participants started to build houses of different types and started to model trees and other objects with the 3D modelling software Tinkercad™. On the second day, the focus was on *mobility*: *What do you understand by mobility? Which vehicles are there? Which buildings and facilities still need to be built? Which facilities/vehicles are not yet available?* The participants built roads for the Ozobots, that acted as self-driving cars and created traffic lights made out of cardboard, LEDs, jumper wires and controlled by a BBC micro:bit. Moreover, a train station and bridges were built. Referring to the discussions of the first two days, the last two days focused on the development of a *child-friendly* and *eco-friendly* city, where the participants created playgrounds, a zoo and parks. The participants' contributions to the daily discussions were documented with the help of posters. The tutors put these posters up on the wall next to the Lego® construction area.

Educational Robotics. To raise students' interest in robotics and coding, the authors decided on two different types of programmable boards. The Calliope mini and the BBC micro:bit come with a microprocessor and further electronical and mechanical components [14]. From a technical point of view, the Calliope mini and the BBC micro:bit do not differ significantly. In contrast to the BBC micro:bit, the Calliope mini is equipped with a built-in loudspeaker, a microphone and a RGB-LED.

Table 1. Overview of devices, materials and tools used during the MAKER DAYS for kids

Name of workshop area	Provided materials and tools	Examples of (digital) products/results
Craft Stuff *tables to do arts and crafts*	*Freely available to all participants:* paper of different colour and shapes, cardboard, acrylic paint, wooden picks, wiggly eyes, etc. brushes, rulers, rubbers, pencils, scissors, box cutters, hot glue gun, glue, etc. *available on request:* LEDs, crocodile clips, jumper wires, 3D-Doodler, etc.	Drawings, buildings of paper, wooden picks or drinking straws
Creative Area 1, 2, 3, 4 *mobile tables for additional workshops*	Depending on the workshop (MakeyMakey, Cardboard and lenses)	Depending on the workshop (controller for a video game, VR cardboard goggles)
Soldering Area	Soldering iron and accessories, batteries, small vibration motors, LEDs, luster terminals, cable scrap, etc.	Vibrobots, electric circuits, more advanced projects, e.g. electric roulette
Coding Club	30 pc. BBC micro:bit, 20 pc. Calliope mini, crocodile clips, aluminium foil, craft stuff	Milky monster, electric piano, oracle, reaction game
Robotic Club	Ozobot, Thymio, felt pen, sheets of white paper, electrical tape	Road system on a piece of paper with intersecting roads and colour codes to control the Ozobot, automatic boom gate made from Lego® and controlled by the Thymio robot
Code Factory 1, 2	4 + 6 laptops, connected to the internet	Programs for BBC micro:bit, Calliope mini and Thymio, Games developed with Scratch
Lego® Construction Area	Lego® bricks in different colours and shape	Football stadium with two BBC micro:bit used as score boards, house facades with blinking LEDs, houses, playgrounds, museum, train station
Textile Manufacture	Programmable embroidery machine, 2 laptops, sewing machine, sewing accessories, fabrics	Embroidered t-shirts, cloth bags and fabrics, homemade bags
Modelling Corner	5 laptops, 3D printers, vinyl cutter, ironing press	3D printed keychains, trees, objects for the Lego® city, t-shirts and bags with vinyl patterns

(*continued*)

Table 1. (*continued*)

Name of workshop area	Provided materials and tools	Examples of (digital) products/results
Media Lab	VR glasses, Amazon Alexa, iPad for playing Osmo Coding and Osmo Tangram, Virtuali-Tee, Dash™ educational robot	Experimenting with new, innovative technologies
Dark room *situated in an adjacent building*	Camera, different sources of light	Light paintings (photos)

At the *Coding Club*, children often made use of the input/output pins that can emit and receive micro currents (to check whether an electric circuit is closed or not, for example). To support activities in the field of physical computing, crocodile clips, jumper wires, LEDs and servo motors were provided. For the programming part, the (peer) tutors introduced the corresponding visual development environment and provided *cheat sheets* with useful code snippets that implemented and explained basic programming concepts, such as conditions, loops or variables.

Taking the didactical settings of a *mini-workshop* within the MAKER DAYS into account, the authors opted for educational robots that require only a short explanation. The Ozobot is a small mobile robot with a colour sensor that can be controlled with a specific set of colour codes. Once the Ozobot recognizes a black line, the robot follows this line and stops as soon as he reaches the end of the line. Besides the Ozobot, the tutors also offered workshops to explain the functions of Thymio, a mobile robot equipped with different sensors, including five distance sensors and two grounds sensors. As well as that, the Thymio can be combined with Lego® bricks and Lego Technic® constructions. In the first step, learners can experiment with the six pre-programmed behavior patterns to learn how the Thymio is programmed.

3.3   Evaluation of Activities in an Open Teaching and Learning Setting

Thinking of maker spaces as learning environment that raise enthusiasm for learning, collaboration, innovation and STEAM, more research on what children and teenagers learn from their participation in makerspaces is required [9, 15]. In comparison to a formal school setting, it is more difficult to document and assess learning processes in an open learning setting, where students work on their own (digital) products, supported by peers or tutors. Figure 4 illustrates the implemented evaluation concept of the MAKER DAYS 2018. Every participant got a name badge with an *ID*, that was used for the documentation of the *mini-workshops* as well as the participants' (digital) results and reflections. For each result, the participants had to fill out a *result card*. A *result* might be a specific, physical or digital, product, for example, a machine-sewed pillow, a 3D-printable model or a program for the Thymio robot. Together with their tutors, the participants had to decide, what they consider a result or final product. In some cases, the tutors asked the participants to add some more features. A result could also be a takeaway from a sequence of actions, for example, how to control the behavior of the Ozobot with color codes or how to talk with a chatbot. A result could also describe an intended goal, that could not be achieved for certain reasons. Each (physical) result was tagged with a *result ID* and documented with a photo or video. The photos and videos of the results were uploaded on the event's web page[3] and tagged with the corresponding IDs. This way, everybody gets an individualized but anonymous online portfolio. As well as that, all participants were asked to fill out two questionnaires, including questions regarding their interests and previous knowledge/experience and a computational thinking task: The first questionnaire was handed out after they had participated in the guided tour and the second one before the final presentation of the results.

4 Results

With 110 participants, 206 daily visits, 126 workshops (14% done by peers) and 468 (digital) results, the MAKER DAYS for kids at Graz University of Technology attracted children and teenagers from all parts of Styria over 4 days. On average, the students attended the MAKER DAYS for a duration of 1.9 days and produced 4.3 results.

On the first two days, the number of female participants was relatively low (see Table 2). Macdonald suggests the "50:50 rule" for mixed groups, but this was not possible due to the higher number of male registrations. Not surprisingly, the *Textile Manufacture* and the possibility to do craft and arts attracted more than twice as much girls than boys (see Table 3). As many girls as boys took part in the *Coding Club*. The *Soldering Area* was strongly male-dominated, with nearly six times more boy than girls. To support girls' participation in maker activities, Schoen et al. [6, 16] highlight the importance of female (peer) tutors and role models. Therefore, the authors ensured, that more than half of all (peer) tutors are female. This may have contributed to a balanced number of boy and girls in the *Coding Club*.

For the participants, it was important not to end up with empty hands. Often, the focus was more on the final product or result than on the creative process itself. Even some of the tutors found it hard to give students enough room to think, make, improve, reflect and iterate. Learning in school is mostly teacher-driven with disciplinary boundaries and pre-defined goals. Maker education changes this vision on learning. At the beginning, most of the participants were afraid of making mistakes or failing. Some workshops followed a step-by-step approach (e.g. How to create a vibrobot/milky monster/3D printed key chain?), meaning that the (peer) tutors explained every step from the beginning (e.g. which materials to use) to the final product. Participants got impatient, when the program or result did not work as expected. In many cases, they wanted the (peer) tutor to fix the problem instead of trying to manage difficulties on their own or asking the right questions. It seemed that they missed some form of intrinsic motivation or did not find their own path within more open-ended assignments.

---

[3] Webpage of the event: https://learninglab.tugraz.at/informatischegrundbildung/makerdays/.

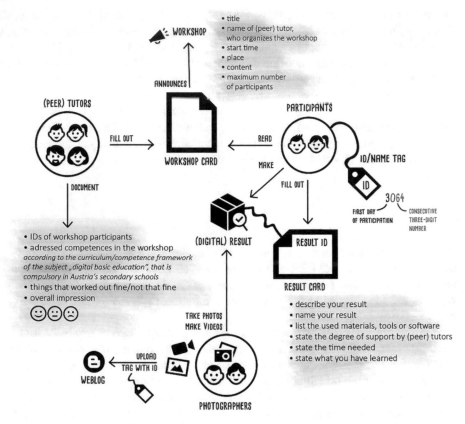

Fig. 4. An innovative evaluation concept for the documentation of the participants' activities within an open makerspace setting.

Table 2. Number and age of participants at the MAKER DAYS for kids 2018

	Monday		Tuesday		Thursday		Friday	
Number of participants	55		52		47		52	
Number of girls/boys	13	42	12	40	22	25	23	29
Number of girls/boys in %	24%	76%	23%	77%	47%	53%	44%	56%
Average age of participants	11.15		11.21		11.47		11.56	

Table 3. Number of results produced by male and female participants in different workshops

	Number of results	By boys	By girls
Coding Club	32	16	16
Soldering Area	41	35	6
Robotic Club	44	28	13
Lego® Construction Area	55	31	10
Textile Manufacture	58	15	43
Craft stuff	64	16	48
Modelling corner	133	92	41

5   Conclusion

Makerspaces have the potential to fuel deep and sustainable learning. New technologies can act as trigger of making in education, but a simple collection of materials, tools and devices does not define a makerspace and does not empower young people to shape their world. Learning and teaching are different in an open makerspace setting. The assignments and learning experiences need to be authentic and need to ask for creative solutions. The authors are currently analysing the data gathered from questionnaires, workshop cards, result cards, photos and videos of the MAKER DAYS 2018 to improve the concept of a temporary open makerspace setting that increases girls' participation, fosters social innovation and collaboration, and supports the development of skills and competences in STEAM. The MAKER DAYS will take place again in summer 2019 at Graz University of Technology – with the current goals of reaching more girls and providing tasks with a more open-ended approach. Results have already shown, that girls have mainly participated in those workshops that provided a scope for imagination and creativity.

The authors' long-term goal is to pave the way for maker education in schools, democratizing STEAM education and innovation development. Halverson et al. put it straight: *"Bringing the maker movement into the education conversation has the potential to transform how we understand "what counts" as learning, as a learner, and as a learning environment. An expanded sense of what counts may legitimate a broader range of identities, practices, and environments—a bold step toward equity in education"* [4]. Let us *make* this happen!

Acknowledgement. The event "MAKER DAYS for kids", that took place from the 13th to the 18th of August 2018 at the Graz University of Technology has received funding from organizations, including *Land Steiermark, WKO Steiermark, Federation of Styrian Industries, SFG* and the *Software and Data Council Styria*. We would like to thank the *TU Graz FabLab*, the *Institute of Software Technology*, the non-profit society *BIMS e.V.*, the *Centre for Social Innovation* in Vienna, and *Salzburg Research* for their great support.

References

1. White House: Presidential Proclamation - National Day of Making (2014). https://ob amawhitehouse.archives.gov/the-press-office/2014/06/17/presidential-proclamation-national-day-making-2014. Accessed 7 Jan 2019
2. Davee, S., Regalla, L., Chang, S.: Makerspaces. Highlights of Select Literature (2015). https://makered.org/wp-content/uploads/2015/08/Makerspace-Lit-Review-5B.pdf. Accessed 7 Jan 2019
3. Sheridan, K., Halverson, E.R., Litts, B., Brahms, L., Jacobs-Priebe, L., Owens, T.: Learning in the making: a comparative case study of three makerspaces. Harvard Educ. Rev. (2014). https://doi.org/10.17763/haer.84.4.brr34733723j648u
4. Halverson, E.R., Sheridan, K.: The maker movement in education. Harvard Educ. Rev. (2014). https://doi.org/10.17763/haer.84.4.34j1g68140382063

5. Kurti, R.S., Kurti, D.L., Fleming, L.: The philosophy of educational makerspaces: part 1 of making an educational makerspace (2014). http://www.teacherlibrarian.com/wp-content/uploads/2014/07/Kurti-article.pdf. Accessed 10 Jan 2019
6. Schön, S., Ebner, M., Reip, I.: Kreative digitale Arbeit mit Kindern in einer viertägigen offenen Werkstatt. Konzept und Erfahrungen im Projekt "Maker Days for Kids". Medienimpulse (1) (2016)
7. Schön, S., Voigt, C., Jagrikova, R.: Social innovations within makerspace settings for early entrepreneurial education - the DOIT project. In: Proceedings of EdMedia+ Innovate Learning, Amsterdam, Netherlands, pp. 1716–1725. Association for the Advancement of Computing in Education (AACE) (2018)
8. Anderson, C.: Makers. The New Industrial Revolution, 1st edn. Crown Business, New York (2012)
9. Lindstrom, D., Thompson, A.D., Schmidt-Crawford, D.A.: The maker movement: democratizing STEM education and empowering learners to shape their world. J. Digit. Learn. Teach. Educ. (2017). https://doi.org/10.1080/21532974.2017.1316153
10. Papert, S.: Mindstorms. Children, Computers, and Powerful Ideas. Basic Books, New York (1980)
11. Hatch, M.: The Maker Movement Manifesto. Rules for Innovation in the New World of Crafters, Hackers, and Tinkerers. McGraw-Hill Education, New York (2014)
12. Schön, S., Ebner, M., Kumar, S.: The maker movement. Implications of new digital gadgets, fabrication tools and spaces for creative learning and teaching. eLearning Pap. **39**, 14–25 (2014)
13. Douma, I., van der Poel, J., Scheltenaar, K., Bekker, T.: Maker Education Theory and Practice in the Netherlands. White paper for Platform Maker Education Netherlands, Amsterdam (2016). https://www.researchgate.net/publication/329572539_Maker_Education_Theory_and_Practice_in_the_Netherlands_White_paper_for_Platform_Maker_Education_Netherlands_Maker_Education-Theory_and_Practice_in_the_Netherlands. Accessed 7 Jan 2019
14. Grandl, M., Ebner, M.: Kissed by the muse: promoting computer science education for all with the calliope board. In: Proceedings of EdMedia: World Conference on Educational Media and Technology, Amsterdam (2018)
15. González González, C., Aller Arias, L.G.: Maker movement in education: maker mindset and makerspaces. In: Jornadas de HCI, IV
16. Gappmaier, L.: MakerDays for Kids – Analyse und Konzepterstellung. Diploma thesis, Graz University of Technology (2018). https://itug.eu

The Uncanny Valley of the Virtual (Animal) Robot

Alexandra Sierra Rativa[✉], Marie Postma[✉],
and Menno van Zaanen[✉]

Department of Cognitive Science and Artificial Intelligence, Tilburg University,
Tilburg, The Netherlands
{asierrar, marie.postma, mvzaanen}@uvt.nl

Abstract. In this paper we explore whether the uncanny valley effect, which is found for human-like appearances, can also be found for animal-like virtual characters such as virtual robots and other types of virtual animals. In contrast to studies that investigate human-like appearance, there is much less information about the effects concerning how a virtual character's animal-likeness influences their users' perception. In total, 162 participants evaluated six different virtual panda designs in an online questionnaire. Participants were asked to rate different panda faces in terms of their familiarity, commonality, naturalness, attractiveness, interestingness, and animateness. The results show that a robot animal is perceived as less familiar, common, attractive, and natural. The robot animal is interesting and animate to users, but no big differences with the other images are found. We propose future applications for the human-(animal) robot interaction as tutorial agents in videogames, virtual reality, simulation robot labs using real-time facial animation.

Keywords: Uncanny valley · Virtual animals · Human-computer interaction · Virtual panda · Animal-likeness · Familiarity · Commonality · Naturalness · Attractiveness · Interestingness · Animateness · Virtual robot · Zoomorphic robots

1 Introduction

There is a growing body of literature that recognizes the importance of human-robot interaction in educational settings. A recent study on the diversity of robots in education conducted by Belpaeme, Kennedy, Ramachandran, Scassellati, and Tanaka [3] provides an overview of which robots are used most in pedagogical studies. They found that the NAO robot (with human-like appearance) was the most used often (in 48% of the studies). Other types of robots that were used (in a range between 6–4% of the studies), were Keepon (with animal appearance), Wakamaru (human appearance), Robovie (human appearance), Dragonbot (animal appearance), iCat (animal appearance), and Bandit (android appearance). Interestingly, although there was a predominance of robots with human characteristics, robots with an animal appearance were also found. Additionally, data from several studies with HRI reported more affective effects (66%) than cognitive effects (34%).

© Springer Nature Switzerland AG 2020
M. Merdan et al. (Eds.): RiE 2019, AISC 1023, pp. 419–427, 2020.
https://doi.org/10.1007/978-3-030-26945-6_38

Studies on human-(animal) robot interaction show positive emotional effects on users. For instance, the seal robot "Paro" has demonstrated to help people to reduce stress, anxiety, depression, dementia, and other behavioral and psychological symptoms [33]. Previous research with seal and dog companion robots found that loneliness of the users can be reduced as robot animals have an effect comparable to a live animal [2, 27]. Interestingly, some physical robot animals also have a virtual model prototype. For instance, (1) the bear robot called "Keio U Robot-phone" has a digital version called "Keio U Robot-phone animation" [15]; (2) the dinosaur robot called "Pleo" has a digital version called "Pleo animation" [9, 28, 29]; (3) the dog robot called "Sony Aibo" has a digital version called "Simulated Aibo" [5, 7, 13]; the cat robot called "Philips iCat" has a digital version called "Philips iCat animation" [11, 14, 17, 25]; and the rabbit robot called "NTT Cyber Solution Lab robot" has a digital version called "NTT Cyber Solution Lab animation" [32].

One of the reasons that animal robots may result in a friendly user interface is due to their funny and attractive behavior as a possible robot pet [10]. Other reasons may be that virtual robots or agents can offer abilities similar to a physical robot with lower prices, fewer problems with installation and additional technical requirements, personalization (also in curricula), and can be used by a large number of students into a classroom concurrently. However, previous research which compared physical robots with virtual models have found that virtual robots lead to less engagement, empathy, and smaller effects in educational outcomes [3]. This can be associated with problems with the robot's appearance or behavior in virtual environment. For this reason, we investigate whether the uncanny valley can give answers to these problems.

The Uncanny Valley in Virtual Animals
The effects of the uncanny valley theory related to virtual humans (which are characters with anthropomorphic features) are currently well-studied. Virtual humans are found in the form of virtual android robots, bots, chatbots, teaching avatars, and related agents [6, 18]. However, one may wonder whether the uncanny valley theory can also be applied to non-human characters like "stuffed animals" with human-like features. Unfortunately, it is unclear why the research in the uncanny valley area has been mostly focused on virtual humans only, when virtual non-human characters are also regularly used in the industry, but not yet widely analyzed in the academic field. In particular, there is a lack of research regarding the uncanny valley theory on whether virtual animals and their resemblance can spark a feeling of familiarity. Specifically, we are interested in the questions that are related to the existence of the uncanny valley theory in the area of virtual animals. Additionally, we would like to know more about the possible effect of the uncanny valley theory on the design of the appearance of the virtual animal and its relationship with the perception of users. This study sets out to investigate whether the uncanny valley theory can be applied to virtual animals. The main research question addressed in this paper is: Can we identify the effect of the uncanny valley theory in the familiarity, commonality, naturalness, attractiveness, interestingness, and animateness perception towards a virtual animal?

2 Methods

Several researchers have investigated the uncanny valley effect and its relation to their familiarity [19, 30, 31, 34]. According to Schwind, Leicht et al. [31], one may expect the uncanny valley effect to occur when the agent is familiar (similarly to human agents). The reason for this is that animals are very familiar to humans. As such, we selected a panda due to it being known worldwide. We expected the panda to inspire a sense of familiarity in participants of the study. The panda is considered a charismatic species [8, 12]. Moreover, people perceive pandas as attractive, cute, and charming. This means we can analyze the panda not only on its familiarity aspect but also on other features such as attractiveness, commonality, naturalness, interestingness, and animateness in a virtual setting. For this study, an image of a real panda and five different virtual pandas were used as stimuli.

2.1 Survey Procedure

Participants accessed the online survey using a hyperlink in Qualtrics. The survey took less than 5 min to complete. Participants were asked for their age, gender, current location, and how frequently they played videogames. After these demographic questions, all participants saw six different images of pandas and for each image they answered questions using semantic differential scales for each of the properties: familiarity, commonality, naturalness, attractiveness, interestingness, and animateness. The participants indicated their answers on a 9-point scale where the extremes were labeled (e.g. 1 = Very strange and 5 = Very familiar).

2.2 Participants

We recruited participants via Facebook and directly from the student population of Tilburg University. A total of 162 participants came from age groups of 10 to 20 (15.43%), 21 to 30 (74.07%), 31 to 40 (8.03%), and 41 to 50 (2.47%) years old.

2.3 Stimulus

As this study focuses on virtual animals, we propose a scale similar to the human-likeness scale: the animal-likeness scale. We placed six different faces of a panda (animal) on the animal-likeness scale, similar to that developed by Mori (Fig. 1) [21, 22, 35]. A real panda face [24] is located on the high side of animal-likeness scale. Subsequently, a face used with a photorealistic design [23] is placed lower. Next, models of a zombie panda, robot panda, stuffed panda, and mechanical panda are placed on the scale, gradually being less animal-like. The mechanical panda is considered to have the lowest number of animal-like features in this range.

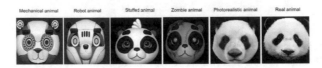

Fig. 1. Images of pandas adapted from Mori's uncanny valley graph in animal-likeness scale.

3 Results

The main question in this study explores whether there is an uncanny valley effect in the perception of familiarity, commonality, naturalness, attractiveness, interestingness, and animateness perception towards a virtual animal (Fig. 2). We found that the uncanny valley effect is present for measures of familiarity, commonality, naturalness, and attractiveness perception. However, no uncanny valley effect was found for perception of interestingness, and animateness. The robot animal is found in the uncanny valley for almost all the variables along with the zombie, except for perception of interestingness, and animateness.

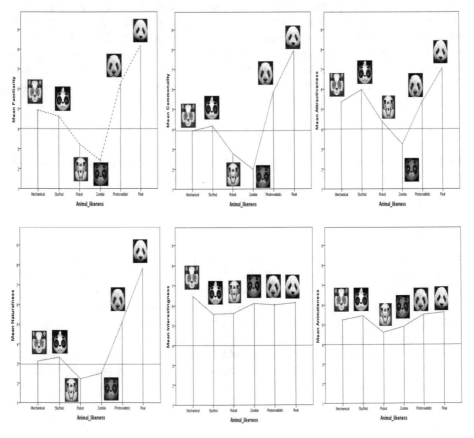

Fig. 2. Mean perception of the six stimuli of the virtual panda organized on the x-axis on different properties (y-axis).

4 Discussion

Unexpectedly, the robot panda (virtual robot) is found in the uncanny valley for the properties familiarity, commonality, attractiveness, and naturalness. Contrary to our expectations, these results show that robot animals are perceived less familiar, common, attractive, natural, animate, and interesting when compared to photorealistic and real animals. The robot animal is interesting and animate to users, but no big differences with the other images are found. This result may be explained by the fact that although many researchers are investigating how to increase the empathy and emotional connection of social interaction with robots for long term interaction, this relationship is not resolved yet [36]. The robots still lack the social interaction skills of humans because machines cannot operate perfectly natural under the different real-world conditions. Additionally, these robots require more natural or biological appearances due to the uncanny valley effect. This finding, while preliminary, suggests that is better to use a virtual robot animal with natural appearance than an artificial appearance.

5 Future Work

5.1 Applications for the Human-(Animal) Robot Interaction as Tutorial Agents

Previous research developed by Li, Kizilcec, Bailenson, and Ju [16] found that students who watched video lectures with an animated robot (virtual NAO robot) had a better knowledge recall compared with a recording with a real robot (NAO robot). They concluded that video instructions can be affected by the appearance of the tutor (human or virtual appearance). For this reason, in future investigations, we are interested in using dynamic images (e.g., animations) compared to static images of the virtual animals to analyze whether the movement of virtual robot has similar uncanny valley effects for the user. Moreover, we are curious to understand how this research can be applied practically on robotics in education. For instance, we can use different versions of virtual animals (in particular robot animals) as virtual tutors in video instruction, where they may lead to different effects on affective and cognitive outcomes (see Fig. 3) depending of their visual appearance. In these future investigations, the animal robot can have other types of effect to that we found in this study of uncanny valley of virtual animals.

Fig. 3. Example of a future application of the images into animal-likeness scale as virtual tutors.

5.2   Virtual Animal Robots in Serious Games and Virtual Reality

The results of the uncanny valley appearance of the robot animal extend our knowledge of the empathic reactions of users towards robots [22]. Previous studies have explored the relationships between empathy and robots, especially in uncanny valley studies [20, 26, 37]. These investigations are limited to physical prototypes of social robots, however, less is known of digital prototypes. Serious games and virtual reality simulations can be an attractive alternative as it provides interactive experiences that are similar to the real world but now with robots animals. Further research should focus on determining whether users can have empathic reactions towards robots in virtual environments. This may have a number of important implications for future practice in new technology environments for teaching with robots in the classroom.

5.3   Simulation Robot Labs: Virtual Animals

Simulation labs are fundamental tools in robotics education. In these simulation labs, students perform laboratory exercises, including mathematical models, electrical, mechanical, hydraulic systems, robotic models, physics, automated systems, and others. The android and mechanical robots are currently the major area of interest within the field of simulation lab [1]. Up to now, too little attention has been paid to investigate the benefits of employing robot animals. As such, further investigation and experimentation into simulation robot labs with animal models is strongly recommended.

5.4   Interactive Robotic Avatars with Real-Time Facial Animation

One of the latest advances in technology is the "modeling for real-time facial animation" [4]. This software allows the natural use of avatars. Facial recognition software tracks the eyes, mouth and face posture of a human and maps this onto the interface (an image of humans, robots, animals, or objects). This software can be used by students and teachers, where they can create robotic avatars to make class presentations or interactive activities more engaging with more motivating effects (see Fig. 4). These interactive robotic avatars can have a great potential in human-robot interaction. However, users should experience these avatars as human or animal like.

Fig. 4. Example of a future application robotic panda as a real-time facial interactive animation.

6 Conclusion

In this investigation, the aim was to explore the presence of the uncanny valley effect associated to virtual animals, especially in robot animals. The results of this study indicate that robot animals fall into uncanny valley on the properties of perception of familiarity, commonality, naturalness, and attractiveness. One unanticipated finding was that for the robot animal no uncanny valley effect is found for the properties of perception of interestingness, and animateness. The generalizability of these results is subject to certain limitations. For instance, in our experiment, we analyzed the perception of a panda as a virtual animal, but we are not sure whether similar uncanny valley effects are found when other types of animals are used. Keeping the knowledge of the uncanny valley effects into account, further research should explore the effects of virtual animal robot interaction when used as tutorial agents, in video games, simulation labs, and used for real-time facial animation.

Acknowledgements. We thanks our research to Esteban Plazas for help in the design and develop of the panda images.

References

1. Balamuralithara, B., Woods, P.: Virtual laboratories in engineering education: the simulation lab and remote lab. Comput. Appl. Eng. Educ. **17**(1), 108–118 (2009)
2. Banks, M.R., Willoughby, L.M., Banks, W.A.: Animal-assisted therapy and loneliness in nursing homes: use of robotic versus living dogs. J. Am. Med. Dir. Assoc. **9**(3), 173–177 (2008)
3. Belpaeme, T., Kennedy, J., Ramachandran, A., Scassellati, B., Tanaka, F.: Social robots for education: a review. Sci. Robot. **3**(21) (2018)
4. Bouaziz, S., Wang, Y., Pauly, M.: Online modeling for realtime facial animation. ACM Trans. Graph. (ToG) **32**(4), 40 (2013)
5. Carpin, S., Lewis, M., Wang, J., Balakirsky, S., Scrapper, C.: USARSim: a robot simulator for research and education. Paper Presented at the IEEE International Conference on Robotics and Automation, April 2007
6. Ciechanowski, L., Przegalinska, A., Magnuski, M., Gloor, P.: In the shades of the uncanny valley: an experimental study of human–chatbot interaction. Future Gener. Comput. Syst. **92**, 539–548 (2018)
7. Coghlan, S., Waycott, J., Neves, B.B., Vetere, F.: Using robot pets instead of companion animals for older people: a case of 'reinventing the wheel'? Paper presented at the Proceedings of the 30th Australian Conference on Computer-Human Interaction, December 2018
8. Ducarme, F., Luque, G.M., Courchamp, F.: What are "charismatic species" for conservation biologists. BioSci. Master Rev. **10**, 1–8 (2013)
9. Fernaeus, Y., Håkansson, M., Jacobsson, M., Ljungblad, S.: How do you play with a robotic toy animal? A long-term study of pleo. Paper presented at the Proceedings of the 9th International Conference on Interaction Design and Children, June 2010
10. Goris, K., Saldien, J., Lefeber, D.: Probo, a test bed for human robot interaction. In: 2009 4th ACM/IEEE International Conference on Human-Robot Interaction (HRI). IEEE (2009)

11. Heerink, M., Kröse, B., Wielinga, B., Evers, V.: Measuring the influence of social abilities on acceptance of an interface robot and a screen agent by elderly users. Paper presented at the Proceedings of the 23rd British HCI Group Annual Conference on People and Computers: Celebrating People and Technology, September 2009

12. Kandel, K., Huettmann, F., Suwal, M.K., Regmi, G.R., Nijman, V., Nekaris, K.A.I., Lama, S.T., Thapa, A., Sharma, H.P., Subedi, T.R.: Rapid multi-nation distribution assessment of a charismatic conservation species using open access ensemble model GIS predictions: red panda (Ailurus fulgens) in the Hindu-Kush Himalaya region. Biol. Cons. **181**, 150–161 (2015)

13. Kertész, C., Turunen, M.: Exploratory analysis of Sony AIBO users. AI Soc. 1–14 (2018)

14. Leite, I., Pereira, A., Martinho, C., Paiva, A.: Are emotional robots more fun to play with? Robot and human interactive communication. Paper presented at the 17th IEEE International Symposium on RO-MAN, August 2008

15. Li, J., Chignell, M.: Communication of emotion in social robots through simple head and arm movements. Int. J. Soc. Robot. **3**(2), 125–142 (2011)

16. Li, J., Kizilcec, R., Bailenson, J.N., Ju, W.: Social robots and virtual agents as lecturers for video instruction. Comput. Hum. Behav. **55**, 1222–1230 (2016)

17. Looije, R., Neerincx, M.A., Cnossen, F.: Persuasive robotic assistant for health self-management of older adults: design and evaluation of social behaviors. Int. J. Hum. Comput. Stud. **68**(6), 386–397 (2010)

18. Lugrin, J.L., Ertl, M., Krop, P., Klüpfel, R., Stierstorfer, S., Weisz, B., Ruck, M., Schmitt, J., Schmidt, N., Latoschik, M.E.: Any "Body" there? Avatar visibility effects in a virtual reality game. Paper Presented at the 2018 IEEE Conference on Virtual Reality and 3D User Interfaces (VR), March 2018

19. MacDorman, K.F.: Subjective ratings of robot video clips for human likeness, familiarity, and eeriness: an exploration of the uncanny valley. Paper presented at the ICCS/CogSci-2006 Long Symposium: Toward Social Mechanisms of Android Science, July 2006

20. Misselhorn, C.: Empathy with inanimate objects and the uncanny valley. Mind. Mach. **19**(3), 345 (2009)

21. Mori, M.: The uncanny valley. Energy **7**(4), 33–35 (1970)

22. Mori, M.: The uncanny valley [from the field]. IEEE Robot. Autom. Mag. **19**(2), 98–100 (2012)

23. MotionCow: Photo realistic panda bear: rigged & animation ready. http://motioncow.com/index.php/product/panda-bear/. Accessed 01 Mar 2018

24. Oso panda. http://saveanimalsworld.blogspot.nl/2013/09/oso-panda.html. Accessed 01 Mar 2018

25. Pereira, A., Martinho, C., Leite, I., Paiva, A.: iCat, the chess player: the influence of embodiment in the enjoyment of a game. Paper presented at the Proceedings of the 7th International Joint Conference on Autonomous Agents and Multiagent Systems, May 2008

26. Riek, L.D., Rabinowitch, T.C., Chakrabarti, B., Robinson, P.: How anthropomorphism affects empathy toward robots. Paper presented at the Proceedings of the 4th ACM/IEEE International Conference on Human Robot Interaction, La Jolla, CA, USA (2009)

27. Robinson, H., MacDonald, B., Kerse, N., Broadbent, E.: The psychosocial effects of a companion robot: a randomized controlled trial. J. Am. Med. Dir. Assoc. **14**(4), 661–667 (2013)

28. Rosenthal-von der Pütten, A.M., Krämer, N.C., Hoffmann, L., Sobieraj, S., Eimler, S.C.: An experimental study on emotional reactions towards a robot. Int. J. Soc. Robot. **5**(1), 17–34 (2013)

29. Ryokai, K., Lee, M.J., Breitbart, J.M.: Children's storytelling and programming with robotic characters. Paper presented at the Proceedings of the Seventh ACM Conference on Creativity and Cognition, October 2009
30. Schwind, V.: Implications of the uncanny valley of avatars and virtual characters for human-computer interaction. Ph.D., University of Stuttgart, Stuttgart, Germany 2018
31. Schwind, V., Leicht, K., Jäger, S., Wolf, K., Henze, N.: Is there an uncanny valley of virtual animals? A quantitative and qualitative investigation. Int. J. Hum. Comput. Stud. **111**, 49–61 (2018)
32. Shinozawa, K., Naya, F., Yamato, J., Kogure, K.: Differences in effect of robot and screen agent recommendations on human decision-making. Int. J. Hum. Comput. Stud. **62**(2), 267–279 (2005)
33. Takayanagi, K., Kirita, T., Shibata, T.: Comparison of verbal and emotional responses of elderly people with mild/moderate dementia and those with severe dementia in responses to seal robot, PARO. Front. Aging Neurosci. **6**, 257 (2014)
34. Tinwell, A., Grimshaw, M., Nabi, D.A., Williams, A.: Facial expression of emotion and perception of the uncanny valley in virtual characters. Comput. Hum. Behav. **27**(2), 741–749 (2011)
35. Tobe, F.: Podcast: our relationship with the uncanny valley. https://www.therobotreport.com/our-relationship-with-the-uncanny-valley/. Accessed 01 Mar 2018
36. Yang, G.Z., Bellingham, J., Dupont, P.E., Fischer, P., Floridi, L., Full, R., Jacobstein, N., Kumar, V., McNutt, M., Merrifield, R., Nelson, B.J., Scassellati, B., Taddeo, M., Taylor, R.H., Veloso, M., Wang, Z.L., Wood, R.: The grand challenges of science robotics. Sci. Robot. 3(4) (2018)
37. Złotowski, J.A., Sumioka, H., Nishio, S., Glas, D.F., Bartneck, C., Ishiguro, H.: Persistence of the uncanny valley. In: Geminoid Studies: Science and Technologies for Human Like Teleoperated Androids, pp. 163–187 (2018)

Cyber-Physical System Security: Position Spoofing in a Class Project on Autonomous Vehicles

Gregory C. Lewin[✉]

Worcester Polytechnic Institute, Worcester, MA, USA
glewin@wpi.edu

Abstract. We present an experiment to gauge student awareness of security threats to cyber-physical systems. Students in a third-year engineering course were tasked with designing, building, and testing small, robotic vehicles that could perform basic goal seeking. Unbeknownst to the students, the indoor positioning system for the project was deliberately configured to report incorrect positional information, similar to the effect of GPS spoofing. When asked to conjecture reasons for the spurious behaviour of their robots, none of the students considered the possibility that the feedback system was sending spoofed data, despite being given case studies in cyber-physical system security earlier in the term. The results suggest that students need more direct education in threats and design considerations for security of cyber-physical systems.

Keywords: Cyber-physical system security · GPS spoofing · Indoor localization

1 Introduction

The physical aspect of cyber-physical systems introduces new concerns for security of engineered systems. While cyber-physical systems have many advantages in terms of functionality and efficiency, the physical components and connectivity of disparate systems create vulnerabilities related to security and reliability [3]. Unlike purely cyber-systems, the consequences of security breaches include the possibility of physical damage and potential for injury. In addition, the physical aspects of the systems introduce new avenues of attack, for example through spoofing of sensors. It is important that the next generation of engineers are not only aware of existing threats, but are able to anticipate and counteract new types of threats in the future. Further, it is important that security concerns are addressed throughout the life cycle of a system and that engineers understand security at both the component and system levels [3].

Autonomous navigation is a prime example of a developing technology being subjected to threats that were either unforeseen or ignored. There are a number of examples where a navigation system was compromised or reportedly compromised. In a high-profile example, a researcher was able to exploit security holes

© Springer Nature Switzerland AG 2020
M. Merdan et al. (Eds.): RiE 2019, AISC 1023, pp. 428–438, 2020.
https://doi.org/10.1007/978-3-030-26945-6_39

in software to take control of basic control functions of a Jeep [5]. Shoukry *et al.* demonstrated that the anti-lock braking system of a vehicle could be spoofed without access to the software, though physical access to the system was needed [8]. Others have shown that signals from the global positioning system (GPS), a cornerstone technology for autonomous vehicles of all sorts, can be spoofed to cause vehicles to miscalculate their position – an attack that requires no direct access to the system [9].

Here we report on a class project where a local positioning system – similar in functionality to a GPS – was programmed to feed erroneous information to robotic vehicles for the purpose of assessing students' awareness of possible threats. The experiment was inspired by a case study of a cyber-attack on a small-scale manufacturing system, where instructors at a large university assigned students the task of designing, milling, and testing a tensile strength specimen [10]. Unknown to the students, the computer controlling the mill had software on it that altered the tool paths of the mill. Only one group (out of eight) correctly noted that the tool path was incorrect, and none of them concluded that the reason was malicious software.

Often, class activities centered on GPS are confined to simulations due to budgetary constraints. GPS signals are generally not available inside a classroom and the accuracy of low-cost GPS receivers are on the order of a meter, which requires vehicles to traverse large distances for effective activities, which is impractical for small vehicles built under tight budget constraints. Accordingly, this experiment required the design of an indoor localization system that would give sufficient feedback to enable activities where students could learn about techniques in localization, sensor fusion, and autonomous navigation with a physical system. A number of researchers have developed methods for localization indoors, with various degrees of accuracy and cost. Indeed, the University of Virginia already has several high-accuracy VICON camera systems in a handful of research labs, but interfacing with them requires complex hardware and software that are beyond the scope of the course. Others have used IR beacons, RSSI, and other techniques with varying levels of complexity and accuracy.

For this project, we implemented a simple, low-cost, indoor localization system that uses off-the-shelf components to monitor the position of small, robotic vehicles and relay the coordinates back to them in a format that replicates the one-way data flow of GPS. With the system, not only can the students gain experience with localization and sensor fusion techniques, but the instructor can demonstrate GPS-spoofing and other attacks for the purpose of increasing awareness of potential threats to autonomous systems. In the case here, the signals were altered without the students' knowledge to assess their awareness of potential threats.

1.1   Course and Project Objectives

The author teaches several courses in the Technology Leaders Program (TLP) at the University of Virginia. The TLP is an interdisciplinary program that focuses on design of complex and connected systems. In Cyber-physical Systems, the third course in the TLP sequence, students explore communication and connectivity, autonomy, and sensor fusion, among other topics. In addition, students discuss important concepts in cyber-physical security, including how the physical aspect of cyber-physical systems affects both the consequences and avenues of attack. Early in the term, student teams are each assigned a paper that discusses a specific security breach and they make a summary presentation to the rest of the class. One of the examples is a paper on GPS spoofing [4]. As will be noted in the results, simply having the students report on a specific attack was not enough to make them suspect that their instructor would use a similar attack against them.

For their final project in the course, the students design, build, and program small, autonomous vehicles. In earlier editions of the course, the vehicles would use line following to navigate a network of "roads", travelling from a warehouse to a house to fulfill deliveries placed on an online webpage. Other variations include implementation of platooning and physical implementation of shortest-path algorithms. Students were introduced to the (extended) Kalman filter as a sensor fusion technique, using simulations of a differential drive vehicle with positional feedback to demonstrate the probabilistic nature of sensor fusion, as well as show how certain states (in this case orientation angle) can be estimated even though they are not directly observed. Practical constraints prevented implementation on a physical system until the most recent edition of the course.

2   Local Positioning System

A key aspect of the local positioning system is that it is easy to use from the students' perspective. With a GPS, the complex calculations needed to determine location from satellite signals are performed by the receiver itself and generally hidden from the user. Most GPS receivers communicate through standard protocols, for example streaming NMEA strings across a UART connection. While processing GPS-like signals through trigonometric equations could be a useful exercise in certain contexts, it would likely distract from the more important objectives of the class and add unnecessary complications to the project.

While there are many potential ways to emulate a GPS, the current system uses a camera to detect AprilTags [6] that are mounted on the vehicles to determine their positions. Coordinates are then relayed to the vehicles via radio on a channel specific to each vehicle to eliminate on-board processing of unnecessary signals. The functionality is very similar to an authentic GPS system: the microcontroller merely communicates with a peripheral device (in this case on an SPI bus) to get coordinates formatted in a simple text string. By using unique AprilTags and radio channels, several vehicles can use the system simultaneously, which could support exercises involving fleets of vehicles in future editions (e.g., optimal task allocation).

2.1   AprilTag

Developed at the University of Michigan, AprilTag is a visual fiducial system, useful for a wide variety of tasks including augmented reality, robotics, and camera calibration [6]. AprilTag detection algorithms compute the 3D position, orientation, and identity of tags relative to the camera, and the system is designed to be easily included in other applications, including embedded devices. Figure 1 shows a tag from the 16H5 family mounted on a student-designed vehicle. AprilTag is most often used in an "inside-looking-out" system, where a camera on board a robot detects tags at known locations and uses that information to localize itself. Here, the concept is used in an "outside-looking-in" configuration, where an external camera detects tags and reports the location back to the robot.

2.2   System Description

The complete system consists of an off-the-shelf camera that is mounted over a small arena in which the autonomous vehicles can drive freely. Each vehicle carries a unique AprilTag mounted on top of the vehicle, an example of which is shown in Fig. 1. When the camera detects a tag, the coordinates are relayed to the vehicle via a radio. Several low-cost camera systems were considered, including the PiCam for the Raspberry Pi and off-the-shelf webcams. For this first version, the OpenMV M7 [1] was chosen because it is very easy to use and its performance in tests was no worse than the other options considered.

Fig. 1. One of the student-designed robotic vehicles with an AprilTag mounted on top.

The OpenMV camera runs MicroPython, a python variant optimized for microcontrollers, and the IDE comes with a number of example codes, including codes for recognizing AprilTags. Only minor modifications were needed to optimize performance for the lighting conditions in the lab space and to output tag detections to a second microcontroller via a UART. This second microcontroller, an ATmega32U4 (on a SparkFun Pro micro breakout board), was used to communicate with an RFM69HCW radio operating in the ISM band centered on 915 MHz.[1] Each vehicle was equipped with its own radio, and the node ID of each radio was mapped to the vehicle's tag ID so that it would only process coordinates sent to that vehicle.

The camera was mounted approximately 3.5 m above the floor of an experiential design lab at the university, which resulted in an arena of approximately 2.4 m by 3.2 m. The camera operates in VGA mode with a resolution of 640×480 pixels, giving an average resolution of 5 mm/pixel. No correction was made for lens distortion or errors caused by mounting the AprilTags on top of the vehicles, several cm off the ground. Such considerations complicate the observation/correction portion of the EKF by requiring the calculation of Jacobians of the observation function with respect to the states and noise. While a higher fidelity system would account for these effects, the effects were considered small and don't affect the general trajectory of the vehicles. A class dedicated solely to autonomous vehicles would benefit from a more complete analysis, but such detailed analysis was beyond the objectives of this edition of the course.

With no correction for lens or other distortions, the transformation from physical to pixel coordinates is given by a linear observation matrix, H:

$$\begin{bmatrix} u \\ v \end{bmatrix} = H \begin{bmatrix} x \\ y \\ \theta \end{bmatrix} = \begin{bmatrix} 0 & 200 & 0 \\ 200 & 0 & 0 \end{bmatrix} \begin{bmatrix} x \\ y \\ \theta \end{bmatrix}$$

Figure 2 shows the correspondence between the physical (x, y) and pixel (u, v) coordinates, and the camera system reports the *pixel coordinates* of each tag, which must be interpreted by code on the vehicle. Although the system can easily calculate orientation of the tags, it only reports location.

The overall latency of the system was tested to be under 100 ms, which was considered passable, though ideally the latency would be significantly reduced. Driving at 20–30 cm/s results in lag errors of 2–3 cm, equivalent to several pixels. The lags become less pronounced as the vehicles approach their targets, however, since the controllers, which are standard PID algorithms, automatically slow the vehicles as they near their destinations.

2.3   Vehicle Design

The design criteria for the vehicles are provided to the students at the start of the project. Students are provided with components to build a working vehicle, including:

---

[1] The equivalent band in Europe is centered on 868 MHz.

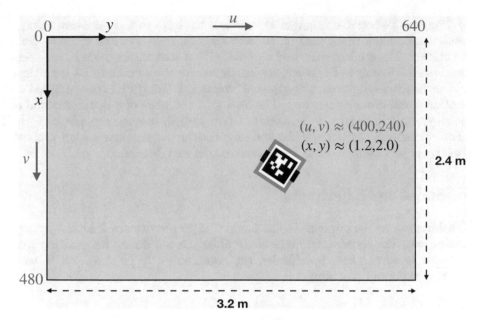

Fig. 2. Diagram of the arena, showing the correspondence of coordinate systems and a hypothetical robot position.

- A SAMD21 breakout board from SparkFun (utilizing the Atmel SAMD21 ARM processor)
- (2) high-powered, 50:1 micro metal gear motors from Pololu
- (2) wheel/encoder kits from Pololu
- A TB6612FNG motor driver breakout board
- An RFM69HCW 915 MHz radio
- Breadboards and passive electronic components
- Acrylic and hardware for assembling the chassis

Students design and cut out their own chassis from acrylic sheets using a laser cutter. The dimensions of the vehicles are limited to 5.75 in. (14.6 cm) in any dimension.

As part of the course, students receive instruction and do workshops and tutorials in motor control, PID control, sensor fusion and the extended Kalman filter, radio communication, CAD, event-driven programming, and other topics needed for completing the project. The general flow of the course is for students to explore a concept in a lab exercise and then demonstrate that they can implement the material into their own system. For example, in one lab the students characterize the performance of a motor by measuring the motor speed while lifting known weights, which introduces them to concepts in motor control (power and direction) and the process of reading encoders (interrupts and state machines). They follow up by constructing the lower section of their vehicles (motors, controller, and battery) and then implement a dead reckoning algorithm. Similar exercises cover the EKF, radio communication, etc.

The final deliverable is a vehicle that will navigate to a destination using a sensor fusion algorithm based on encoder data and observations received from the camera. The students also build a base station that allows them to send destination coordinates to the robot, as well as receive state information from their vehicle. Students implement "go-to-goal" behaviour using PID control based on positional and directional errors. The final exam consists of a demonstration of their system followed by a discussion of how well their system performed, tho technical aspects of their vehicle, and how they managed their design process. Each team sets aside an hour for demonstration and discussion.

3   Spoofing Experiment

Unbeknownst to the students, code was added to the camera feedback system that allowed the instructor to selectively alter the coordinate information that was sent to each robot. Specifically, the coordinates, (u', v'), relayed to each robot were altered according to:

$$u' = u$$
$$v' = v + \varepsilon u$$

where ε is a user (instructor) controlled parameter and the primes refer to the spoofed coordinates. The spoof has the effect of transforming the coordinate system as shown in Fig. 3: a rectangle in the actual coordinate system becomes a parallelogram in the spoofed coordinate system. A spoof initiated when the robot is at the arrow will cause it to deviate from its expected destination (dashed circle) to a new one (solid circle). With one exception, no spoofing was done while the students were developing and testing their systems so that it would not cause issues during the design phases.[2] During their final demonstrations, however, the instructor was able to engage and disengage the spoofing arbitrarily, using his own computer on the pretext that he was observing the system outputs. The instructor would enter values of ε ranging from 0 (spoofing off) to 0.2. When altered while in transit, the vehicles would typically undergo a noticeable change in direction, followed by some wavering as the vehicle adjusted to the new coordinate system. If the spoof was engaged when the destination had a large value in the u coordinate (y coordinate in physical space), the vehicle would stop a noticeable distance from the expected destination. A couple of times the spoof was engaged immediately after a vehicle had reached its destination, after which the vehicle would lurch a short distance to reach the spoofed coordinate.

---

[2] One student nearly discovered the spoof before the final demonstrations. The instructor was testing the system and the student noticed that the coordinates reported by the camera were clearly incorrect. The instructor quickly went to his computer on the pretext of "seeing for himself" and disabled the spoof, after which the student discounted the phenomenon as a glitch.

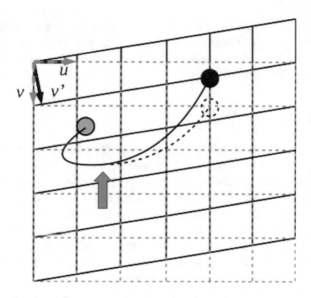

Fig. 3. Grid transformation due to a typical spoof and the hypothetical path alteration due to a spoof while in transit (arrow).

3.1 Student Reactions

Though no objective measurements were collected, the behaviour of the vehicles was noticeable enough to cause concern among the students. Unexpected deviations or restarts typically elicited comments from the students such as, "That was odd." Occasionally, students would look at the output from the vehicles on their base station, but none of them took the time to thoroughly compare the actual coordinates of the vehicle to the reported coordinates, despite repeated engaging and disengaging of the spoof.

After the demonstrations, students were asked about the performance of their vehicles. Typical responses indicated that they were generally pleased with the performance, despite the fact that the behaviour was often not as expected (indeed, without the spoof, the robots performed very well). Students frequently defended the performance by noting that their vehicle "worked perfectly this morning when we tested it" or similar. Others simply discounted the behaviour as "just a glitch." When pressed to come up with possible explanations for the phenomena, only after much leading did one student even consider the possibility that the GPS was sending erroneous information, and she quickly followed that up with a laugh and moved on to other possibilities. Some proposed explanations included temporary loss of camera signal, failures of the encoder hardware, coding errors, or errors due to camera distortion. While all of these issues could cause errors, it is straightforward to show that none of them would fully explain the observed behaviour.

After continued discussion, the students were asked about the cyber-physical security examples they presented earlier in the term. Only when they brought up the paper on GPS spoofing [9] did they begin to suspect what had happened. After revealing the true cause of the behaviour, students generated ideas for possible countermeasures, though implementation of countermeasures was not done due to time constraints.

4   Lessons Learned

It is clear that awareness of cyber-physical security threats needs to be strengthened in educational curricula. Despite covering security concerns as part of the course objectives, the possibility of an actual attack on their systems never occurred to the students. Several indirect references to the possibility were made throughout the term, most notably presentations of security breaches, but also in a MATLAB demonstration of resilience against the anti-lock braking spoof mentioned earlier. As part of the security case studies, students discussed the need to consider security early in the design process, though they were never directly told that an attack on their systems was possible.

Some students posited that the likelihood of a GPS being spoofed is so low that it's no surprise that no one considered it a possibility in this exercise. It made more sense for them to focus on the parts of the system that they believed were more likely to cause the aberrations. In particular, as students learning complex theory for the first time, it is natural that they would focus first on their own contributions, in both software and hardware. For example, having struggled with the discontinuity in θ as the vehicle makes a 360° rotation, it is reasonable that when the vehicle makes a sudden deviation in direction, the students first consider whether or not their turning algorithm is correct.

4.1   Future Work

A number of improvements could be made to both the localization system and the project. The arena size of the current system is only a few square meters, which limits the range the vehicles can travel. If a camera were angled to cover more floor space, it is likely that a higher resolution camera would be necessary. In addition, a more complicated homography would be needed to translate between pixel and physical coordinates. Depending on the objectives of the class, this homography could be left as an exercise for the students or computed by the instructor. In either case, a more authentic implementation of the non-linear correction steps in the EKF would be needed. Alternatively, multiple cameras could be used, though this would require more processing and better system calibration to ensure that transitioning from different camera regions was seamless.

Even without changing the camera geometry, proper calibration of the camera would improve the accuracy of the system, which had notable errors in this first edition. If course objectives warrant it, camera calibration by the students can be a useful exercise, as well as implementation of a non-linear observation function.

The time lag in the system could also be reduced. When driving full speed, the vehicle could travel several centimeters before receiving coordinate information from the camera. As the vehicles approached their destinations, however, they naturally slowed down, so the time lags had less effect. Still, the system could be improved with faster processing or the inclusion of timestamp information in the radio communications. The latter would increase the complexity of the code developed by the students. Such techniques are important in real systems, however, so they would be a good addition to the exercise.

Finally, implementation of countermeasures could add significant value to the experience. Simple countermeasures include adding and monitoring a bias term to the wheel speeds or monitoring the innovation or covariance matrix in the correction step of the EKF. More reliable (and complex) techniques include recursive state estimators that detect and discount measurements that are outside the noise profile of a sensor [2, 7]. It is unlikely that the current system has the precision needed to reliably implement such measures, but with improvements they could be possible. Another practical approach would be to include an IMU on the system and/or provide orientation feedback from the camera, which would allow the systems to fuse information from additional sensors, enabling the identification of those sensors that have been compromised.

Finally, it is important to develop more case studies and give students material on how to detect and counteract threats in a broader sense. This experiment was one of many ways that cyber-physical systems can be attacked, and it is important that students see a variety of threats and develop a systematic approach to security.

Acknowledgments. This work was supported by a grant from Virginia's 4-VA fund.

References

1. Openmv. https://openmv.io. Accessed 01 Oct 2018
2. Bezzo, N., Weimer, J., Pajic, M., Sokolsky, O., Pappas, G., Lee, I.: Attack resilient state estimation for autonomous robotic systems. In: IEEE International Conference on Intelligent Robots and Systems, pp. 3692–3698, October 2014
3. National Research Council: Interim Report on 21st Century Cyber-Physical Systems Education. The National Academies Press, Washington, DC (2015)
4. Gift, S.: A simple demonstration of one-way light speed anisotropy using Global Positioning System (GPS) technology. Phys. Essays **25**, 387–389 (2012)
5. Greenberg, A.: Hackers remotely kill a jeep on the highway. http://www.wired.com/2015/07/hackers-remotely-kill-jeep-highway/. Accessed 30 Sept 2017
6. Olson, E.: AprilTag: a robust and flexible visual fiducial system. In: Proceedings of the IEEE International Conference on Robotics and Automation (ICRA), pp. 3400–3407. IEEE, May 2011
7. Pajic, M., Weimer, J., Bezzo, N., Sokolsky, O., Pappas, G.J., Lee, I.: Design and implementation of attack-resilient cyberphysical systems: with a focus on attack-resilient state estimators. IEEE Control Syst. Mag. **37**(2), 66–81 (2017)

8. Shoukry, Y., Martin, P., Tabuada, P., Srivastava, M.: Non-invasive spoofing attacks for anti-lock braking systems. In: Bertoni, G., Coron, J.-S. (eds.) Cryptographic Hardware and Embedded Systems - CHES 2013, pp. 55–72. Springer, Heidelberg (2013)

9. Warner, J.S., Johnston, R.G.: A simple demonstration that the global positioning system is vulnerable to spoofing. J. Secur. Admin. **25**(2), 19–27 (2002)

10. Wells, L.J., Camello, J.A., Williams, C.B., White, J.: Cyber-physical security challenges in manufacturing systems. Manuf. Lett. **2**(2), 74–77 (2014)

Author Index

© Springer Nature Switzerland AG 2020
M. Merdan et al. (Eds.): RiE 2019, AISC 1023, pp. 439–441, 2020.
https://doi.org/10.1007/978-3-030-26945-6